PHILOSOPHERS AT WAR

PHILOSOPHERS AT WAR

THE QUARREL BETWEEN NEWTON AND LEIBNIZ

A. RUPERT HALL

Imperial College of Science and Technology

CAMBRIDGE UNIVERSITY PRESS

CAMBRIDGE

LONDON   NEW YORK   NEW ROCHELLE

MELBOURNE   SYDNEY

PUBLISHED BY THE PRESS SYNDICATE OF THE UNIVERSITY OF CAMBRIDGE
The Pitt Building, Trumpington Street, Cambridge, United Kingdom

CAMBRIDGE UNIVERSITY PRESS
The Edinburgh Building, Cambridge CB2 2RU, UK
40 West 20th Street, New York NY 10011–4211, USA
477 Williamstown Road, Port Melbourne, VIC 3207, Australia
Ruiz de Alarcón 13, 28014 Madrid, Spain
Dock House, The Waterfront, Cape Town 8001, South Africa

http://www.cambridge.org

First published 1980
First paperback edition 2002

A catalogue record for this book is available from the British Library

Library of Congress Cataloguing in Publication data
Hall, Alfred Rupert, 1920–
Philosphers at war.
Includes index.
1. Calculus – History.
2. Newton, Isaac, Sir, 1642–1727.
3. Leibniz, Gottfried Wilhelm, Freiherr von,
1646–1716.   I. Title.
QA303.H16   515′.09   79-15724

ISBN 0 521 22732 1 hardback
ISBN 0 521 52489 X paperback

TO
D T W

Après tout ces Anglais ont leur mérite

N. Rémond to Leibniz
2 September 1714

Contents

PREFACE

IN TELLING THIS STORY of the bitter quarrel between two of the greatest men in the history of thought, the most notorious of all priority disputes, I have not attempted to enter into the ,technical details of the evolution of the differential and integral calculus and have tried rather to trespass as little as may be into the province of the professional historian of mathematics. My interest has been in the course of the quarrel, rather than in the technical nature of its subject, in mathematicians rather than in mathematics.

So far as I am aware, there is no earlier history of the calculus dispute of any size, though it is discussed in general histories of mathematics and in biographies of the participants, nor has there been any reissue of the *Commercium Epistolicum* since that edited by J-B. Biot and F. Lefort in 1856 (Paris: Mallet-Bachelier); a Spanish version of its documents was published by J. Babini in 1972 (*Gotifredo Guillermo Leibniz, Isaac Newton. El cálcula infinitesimal. Origen. Polemica*, Buenos Aires) and an Italian one by G. Cantelli in 1958 (*La disputa Leibniz-Newton sull'analysi*, Turin and Florence: P. Boringhieri). Older works such as F. Cajori's *History of the Conceptions of Limits and Fluxions in Great Britain from Newton to Woodhouse* (Chicago and London: Open Court Publishing Co., 1919) and J. M. Child's *Early Mathematical Manuscripts of Leibniz* (London: Open Court, 1920) are very out of date. As a guide to the sources and to modern studies, the reader may consult the works listed in the Notes, pursuing particularly the many additional references given in the books of J. E. Hofmann and D. T. Whiteside (especially *The Mathematical Papers of Isaac Newton*, Vol. VIII, to appear shortly), including other writings by these scholars. A number of contributions are noted in the Short Titles preceding the Notes section at the end of this book. The reader should also refer to P. and R. Wallis, *Newton and Newtoniana 1672–1975, a bibliography* (Folkestone, Eng.: Dawson, 1977), where modern publications on the origins of the calculus are listed in the section devoted to "Fluxions."

In quoting from letters and drafts originally composed in English I have often lightly modified the orthography, because it is pointless to trouble the reader with the idiosyncrasies, invariably inconsistent among themselves, that are to be found in the printed sources I have used. Most translations from other languages are my own.

I am necessarily greatly indebted to the historians of mathematics, as will appear later. Among living colleagues I am particularly indebted to Laura Tilling, who patiently explored the calculus dispute in our joint work on *The Correspondence of Isaac Newton*, to Adolf Prag, who read and improved the whole manuscript, and to Tom Whiteside for counsel and a generous sharing of information, as in so many previous years, notably by allowing me to read part of the manuscript of the eighth volume of his *Mathematical Papers of Isaac Newton*. I thank my wife for forbearance, encouragement, and constructive criticism, and both Marjorie Downs and Celia Richards for producing the typescript.

At the end of the book I have reprinted from the *Philosophical Transactions* of 1715 Newton's anonymous "Account of the Book entituled *Commercium Epistolicum*," the only narrative of the whole affair to come from the pen of either contestant.

A. R. H.

CHRONOLOGICAL OUTLINE

1661   Newton goes up to Cambridge; Leibniz enters the University of Leipzig.

1664–6   Newton's *anni mirabiles*: Cambridge and Lincolnshire.

1666   Leibniz composes *On the combinatory Art*; Newton (October) his second tract on fluxions.

1669   Newton's *On Analysis* imparted to Barrow and Collins.

1670   Leibniz opens correspondence with the Royal Society.

1671   Newton writes *On the Method of Series and Fluxions*. His telescope sent to London.

1672   Leibniz moves from Mainz to Paris, forms acquaintance with Huygens. Newton's optical letters published.

1673   Leibniz's first visit to London (January-March); returning to Paris, his mathematical skill develops rapidly.

1675   (October) Leibniz conceives the first idea of his differential calculus.

1676   (June) Newton writes his *First Letter* to Leibniz; followed (October) by his *Second Letter*. (October) Leibniz briefly returns to London on his way to Hanover.

1677   (June) Leibniz receives the *Second Letter* and in reply outlines the differential method to Newton.

1682   Leibniz publishes his arithmetical quadrature of the circle.

1684   Leibniz publishes a brief outline of the differential calculus.

1687   Newton in the *Principia* describes calculation by "moments," mentioning the *Second Letter* and Leibniz's reply to it.

1691   Newton writes "On the Quadrature of Curves."

1695–9 Newton's calculus letters published, with amplifications, in Wallis's *Works*.

1697   Johann Bernoulli issues the brachistochrone challenge problem, solved anonymously by Newton.

1699   Fatio de Duillier asserts the priority and superiority of fluxions.

1704   Newton publishes (with *Opticks*) "On the Quadrature of Curves" and "On the Enumeration of Lines of the Third Order."

1705   Leibniz reviews *Opticks* and the mathematical essays.

1710   Leibniz in *Theodicée* attacks Newtonian attraction theory; John Keill accuses Leibniz of plagiarism.

1711   Leibniz unsuccessfully demands redress from the Royal Society. Jones publishes Newton's mathematical essays.

1712   Newton prepares the *Commercium Epistolicum*, issued in February 1713.

1713   Second edition of *Principia* published. (Summer) Leibniz prepares and circulates the *Charta Volans*, endorsing Bernoulli's opinion that Newton had plagiarized Leibniz's calculus. Bernoulli publishes mathematical criticisms of the *Principia*. Keill takes up the systematic defense of Newton.

1715   Conti and others fail to reconcile Newton and Leibniz.

1

INTRODUCTION

I BECAME interested in the theme of this book while editing Newton's correspondence during the years of his controversy with Leibniz and Leibniz's supporters. Although the outline of its story has often been told, the great richness of materials bearing upon it that has appeared during recent years made a more detailed study seem worthwhile, and more than one scholar has asked that it should be made. Moreover, a historian of today can approach the calculus dispute with a more detached perspective than his Victorian predecessors could do. He will not be shocked to discover that even Leibniz and Newton could display human faults. Again, the historian who (like myself) has no intention of investigating in technical detail the origins, development, and applications of calculus methods in mathematics can safely rely on modern work devoted to precisely these questions. Although he will not overlook his debt to the pioneers, notably C. I. Gerhardt, he must be particularly grateful for the interpretations and especially the documentation provided by J. E. Hofmann, H. W. Turnbull, and D. T. Whiteside, not to mention other equally reliable scholars who have examined the lesser mathematicians contemporary with Leibniz, James Gregory, and Newton.[1] What happened, mathematically speaking, in the 1660s and 1670s is no longer in doubt — as it certainly was to many a century ago and perhaps to some more recently still — and therefore consideration of the quarrel between the two great rivals need be clouded by no hesitation as to the actual historical facts upon which the quarrel turned. It was certainly Isaac Newton who first devised a new infinitesimal calculus and elaborated it into a widely extensible algorithm, whose potentialities he fully understood; of equal certainty, the differential and integral calculus, the fount of great developments flowing continuously from 1684 to the present day, was created independently by Gottfried Wilhelm Leibniz. Whatever we may feel of the relations between these two men, we cannot but admire their analogous creative achievements with as much impartiality as our emotions will admit.

Although quarrels and rivalries between painters, poets, and musicians have at best been taken to promote artistic skills or at worst been treated as subjects for comedy, the altercations of the learned have in the past, at any rate, been regarded as so reprehensible that they should be dismissed in silence. It is not easy to see why this should have been, because a philologist or a positivist is no less human than an artist and certainly no less prone to embrace absurd hypotheses with enthusiasm. Learning and science do not necessarily improve a man's judgment or his character. At all events, it is clear that grave doctors have very, very frequently taken the easy path from disagreement to dispute. Newton's own colleague (and opponent) in the Royal Society, John Woodward, engaged in an unpremeditated public sword fight with another distinguished physician, Richard Mead; if the unusual incidents of Woodward's life stamp him as a stormy petrel,[2] consider the battle-scarred career of Richard Bentley, another friend of Newton's, Master of Trinity, one of the most learned and one of the most hated men of Newton's era. Or recall – all within the same living experience – the "Battle of the Books," wherein Sir William Temple and his vitriolic protégé, Jonathan Swift, defended the ancients against the pretensions of modern learning and science, a battle that, of course, led Swift (like many lesser scribblers) to satirize the Royal Society, of which Newton was president, in his Tale of a Tub and Gulliver's Travels. And though Newton took part in no public conflict with any one but Leibniz, the German philosopher fought (both directly and indirectly) in many scholarly skirmishes, and Newton's own life was not free from struggles behind the scenes. And if all this suggests, not unjustly, the intellectual violence of the age in which the Newton-Leibniz dispute was set, that picture is surely confirmed by the concomitant political and religious excesses even of temperate England, where dissenters were sentenced to the pillory and nonjurors to the Tower.

If folly, self-love, and aggression are by no means incompatible with the highest intellectual powers (and few historians nowadays, perhaps, would claim a total exemption from such vices of pedantry on behalf of Isaac Newton and G. W. Leibniz), one should not forget either that, despite polite contrary conventions, success in the scholarly or academic world depended far more on a militant combativeness then than it does now. For most it began with the ability to put down opponents in university disputations and was confirmed by excellence of wit. The dull scholar, however learned, was not likely to get beyond a rural rectory. Patronage helped a few shy men to live productive scholarly lives in retirement, and some may see the secret of Newton's early success in Barrow's patronage, though it would be a misreading, in my view, to deny to the young man the toughness, energy, and determination so evident in the mature Newton; for most scholars, however, the lighting of their own brilliance required the dowsing of another's flame. In so tiny an intellectual world, where the highest rewards were so scarce (and often disposed of by those who appreciated an epigram better than a monograph), competition was inevitably unrelenting, and the more so for those, like Newton and Leibniz, endowed with no high social advantage in the first place. To put it crudely, an achievement in scholarship, science, mathematics, or medicine was a marketable commodity, a highly personal property: The recognition it conferred might be a first step toward attainment of a bishopric or an office of state. And the rules of the marketplace were both capricious and very different from those that now prevail. From the late nineteenth century, peer evaluation has been the rule of science and learning in the civilized world; and laymen have largely accepted the judgment of the internal experts. In the lifetimes of Newton and Leibniz what counted most was not the opinion of one's peers but the direct impression made upon princes and ministers, prelates and magnates, who exercised enormous personal powers of appointment.

Hence the competition, subtly weighted by all sorts of other considerations of family connection and personal character, was sharp between book and book, brain and brain, constituting (as Leibniz himself was to remark) almost a gladiatorial spectacle for the entertainment of the sophisticated. Philologists did not quite fight duels with Latin hexameters, as their successors were to try their skills on the Rosetta stone, but mathematicians fairly regularly battled over challenge problems, from the rivalry of Cardan and Tartaglia in the early sixteenth century through the celebrated cycloid puzzles of Pascal to the series of such duellos associated with the early development of Leibniz's calculus in the pages of the *Acta Eruditorum* of Leipzig. Peter Ramus's *Testamentum* (1576) had codified competition in the arrangements for the new chair of mathematics that he founded at the Collège Royale in Paris: Professors were to be the winners of public competitions, and if any mathematician should challenge and defeat an incumbent professor, he was to be allowed to take his post. A similar spirit of naked, personal ascendancy permeated the Cambridge Mathematical Tripos until after the middle of the nineteenth century.

I do not mean to suggest that the quarrel between Newton and Leibniz was not shocking to contemporaries. Many clearly found it so. Charges of intellectual theft and personal dishonesty exchanged between two eminent graybeards, both quite close to the new Hanoverian crown of Britain, one the greatest of living philosophers (and hardly less a mathematician), the other the greatest living mathematician (whose philosophical views certainly commanded attention), could only reduce the dignity of learning. But I think the sheer egoism of the dispute, to us, perhaps, its most unpleasant characteristic, was less striking to contemporaries. In terms of the Augustan age, when matters came to a certain pass, it was right for a man to assert his intellectual property rights in a manner that would now be thought disgracefully self-assertive or self-regarding. Far from the development of scholarship, art, or science being a social phenomenon – of which an Einstein or a Picasso may be almost a passive vehicle – as some would have it now, any originality, any creative success, was judged to be the cause of uniquely personal merit, deserving personal reward, and therefore it was natural, rather than egotistical, to defend both merit and reward against rival claimants.

If the structure of society did not favor a sociological theory of success, its rather primitive psychological conceptions nevertheless strengthened an individualistic attitude toward achievement. For example, when ancients were compared with moderns, no one attempted to measure the literacy of Aristotle's Athens against that of Queen Anne's London; it was, rather, philosopher weighed against philosopher, physician against physician. If one man succeeded better than another it was because he had better natural endowments and a stronger character – though education (*formation*, as the French still significantly say) was not neglected as a factor, being, of course, the education of the individual child by an individual parent, not education as a social process. Our tendency to speak of a certain discovery or solution of a learned difficulty as "inevitable" was alien to Newton's era, still painfully conscious of the long noninevitability of progress, intellectual or material.

4

"I stood upon the shoulders of giants" wrote Newton in an oft-quoted phrase,[3] carrying the implication that the ability to see farther depended on one's ability to scramble to the top of the human pyramid created by our ancestors. Obviously Newton would have understood, as we do, that the scramble was open to all, but the intense individualism of his age prevented him, and his contemporaries generally, from understanding also the correlative, so obvious to us, that many scramblers, more or less successful, more or less sharp-sighted, must glimpse the same new prospect. No one yet spoke of "movements" or "schools," still less of "research programs," all concepts that link intellectual innovation with a sociological interpretation of the evolution of learning or art, and hence the idea that within a given context, and on the basis of a common past experience, the thoughts and experiments of several men must necessarily converge upon the same innovation did not present itself; and it was the less likely to do so when truly creative individuals were (in absolute, not relative, terms) very few and, consequently, disparate in their environment – one educated, let us say, by the Jesuits of La Flèche, another in Cambridge and the Inns of Court, a third in Presbyterian St. Andrews. The phenomenon of convergence, the independent solution by more than one individual of the same problem in identical or closely similar ways, is in historical fact extremely common in seventeenth-century science: Galileo Galilei, Thomas Harriot, and (very possibly) Simon Mayr all turned the newly popularized telescope to the heavens in the same year, 1609; John Napier and Jost Bürgi both invented the idea of calculating by the use of logarithms; Galileo and Christiaan Huygens independently (and successively) devised ways of regulating the mechanical clock by the oscillations of a pendulum; Marcello Malpighi and Jan Swammerdam began the microscopic exploration of the same insects at about the same time, and Malpighi and Nehemiah Grew independently took up the microscopic histology of plants. Such examples are almost innumerable, and it is well known that Newton's own work in mechanics converged closely with that of Huygens and of Robert Hooke, older contemporaries who published long before Newton. Because convergence occurred without its being recognized as a necessary phenomenon of discovery, priority squabbles like those studding Galileo's career were far from rare.

The fact that convergence went unrecognized as a necessary social consequence of an active "research program" – a necessary evil, perhaps, from the point of view of some researchers, like the almost forgotten V. Hensen, who effected the *second* isolation of glycogen from liver tissue – of course did not prevent some examples of it from being perceived. Newton, for example, admitted that Huygens had preceded him in the discovery of the laws of centrifugal force, and it was well known that Wren and Huygens had produced identical laws of collision. Convergence was still more evident in the experimental study of pneumatics during the late 1650s where, however, the individualism of the age is clearly manifest in the universal and eponymous linking of Robert Boyle's name with the fruition of this line of research: not that Boyle's fame was unmerited or that anything much is to be gained by renaming Boyle's Law the Towneley-Power-Hooke-Boyle Law. Because, rightly, it was believed (though perhaps not always as a result of a very judicious examination) that *one* statement only of a truth becoming manifest to several inquirers about the same time was complete and general and richly based on argument and evidence, it was taken not to be unjust to assign credit to this one superior enunciation. So Newton maintained the justice of his claiming the universal theory of gravitation for himself against the pretensions of Robert Hooke.[4]

These various factors – the great value attached to personal merit, the emphasis on innovation as the creation of an individual talent, and the absence of a sociological theory of the growth of knowledge, which are rightly regarded as of a social character – may well suffice to account for the frequency and bitterness of priority disputes in the past, especially when combined with the lack of formalized conventions about behavior in the learned world, conventions that only became settled (and enforced by ostracism) in the nineteenth century. Mathematics, because it readily defines "research fronts," because it offers the possibility of attaining results equally rigorously by different means, and because its logical character virtually necessitates the occurrence of convergence, was peculiarly likely to be troubled by quarrels and priority disputes, just as, at the opposite extreme, natural history was almost completely free of such disagreeable incidents. One might almost venture the generalization that the life of no major mathematician of the seventeenth century was wholly free of such wrangles, except that of John Napier, perhaps. Accordingly, though the dispute between Newton and Leibniz was grander, more dramatic, and more interesting than most, it was far from unprecedented and is merely (as a sociologist might say) indicative in a striking way of those faults in the "reward system" of the period, which were almost every day manifested in lesser quarrels by claim and counterclaim.

It is obvious in particular that the discovery of the methods of the differential and integral calculus was a natural occasion for strife. If we limit the formal honor of discovery to Newton first and Leibniz later, we have to admit that (at the least) very close approximations to discovery were made by Christiaan Huygens, James Gregory, Nicolas Fatio de Duillier, and probably others. Limited mastery of parts of the whole that was to be the calculus had been attained not only by these but by others still – René François de Sluse, Nikolaus Mercator, Isaac Barrow, and again others. The discovery of the calculus was more than a synthesis of previously distinct pieces of mathematical technique, but it was certainly this in part; interest in special cases later subsumed under the calculus – such as the general method of tangents and the quadrature of particular curvilinear areas – had lasted for a generation. We can now see, although this was quite obscure at the time, that what appeared in the 1650s and 1660s as a series of independent problems actually constituted, in fact, a single "research front," and that those who succeeded in making an advance in the solution of any one of these problems were converging upon the concepts of the calculus. Although there were areas of pure mathematics, like the projective geometry of Girard Desargues (himself, incidentally, a victim of charges of plagiarism by one Curabelle), that had no relation to the development of the calculus, one might guess that perhaps a half of all the mathematical activity of the first seventy years of the seventeenth century was more or less contributory to it. So much talent was devoted to this research front that, in relation to particular successes, duplication was commonplace, as with Sluse's and Newton's methods of tangents, Mercator's, Gregory's, and Newton's methods of quadrature by infinite series, the particular series for the circle obtained by Gregory, Newton, and Leibniz, and Newton's and Gregory's formulations of the binomial expansion. With so many men doing similar things successfully, it was not easy for any one mathematician to set his accomplishment apart from that of others. In Hofmann's words:

Infinitesimal problems were being hotly pursued simultane-

ously in France, Italy and England; the improved concept of
indivisibles was being used as a guiding principle by Fermat,
Pascal and Huygens equally as well as by Torricelli, Ricci,
Angeli and Sluse . . . The characteristic triangle – to take up
a particular point – was known already [before Leibniz] to
Fermat, Torricelli, Huygens, Hudde, Heuraet, Wren, Neil,
Wallis and Gregory long before it was made public by Barrow. Each of these predecessors had used it, *but nobody wanted
to expose the jealously guarded secret by which he had found his
results.*[5] [My italics]

Even today, reevaluation by historians of the achievements of
these various mathematicians is occurring: James Gregory was not
recognized for the powerful innovator he was before Turnbull's
researches of the 1930s, and yet more recently, Isaac Barrow's reputation, once so high (at least among English speakers), seems to
be descending toward the status of an elegant codifier. As Hofmann's expressions also indicate, competitiveness produced se-
crecy and envy, obstructing the open and cumulative development
of new methods. Was it likely, therefore, that one man could stand
out from all others as "discoverer of a new infinitesimal calculus"
in that decade from 1660 to 1670 of enormously rapid progress on
the basis of the foundations laid by Cavalieri, Descartes, Wallis,
and so many more? Only if, like Boyle in pneumatics, he could
produce a powerful and persuasive treatise.

Of this, I think, Newton may have been conscious. He put his
new mathematical methods on paper clearly enough, but he
printed nothing, only circulating a part of his work to friends. In
1668 he found himself (as he judged) partly anticipated by Nicolaus Mercator's *Logarithmotechnia*. Three years later, and one year
after the writing of his 1671 treatise on fluxions, Newton received
– with respect to his novel ideas about light and color – a severe
lesson as to the trouble and annoyance springing from ill–considered and incomplete publication of his own work. He toyed at
this time with the idea of an extensive mathematical book, but it
made no progress and in the end was abandoned. If proper publication of one's work required such a vast effort, and hasty publication caused such disasters to one's time and temper, better give
it up altogether.

We cannot, in fact, be confident that the printing of one or more
of the tracts about fluxions and infinite series that Newton had
composed before he put mathematics aside would have guaran-

teed him a swift and easy immortality as "discoverer of the calculus." Newton's friends then and later admired these tracts for their wonderful originality, and they have remained as the public basis of Newton's achievement in the calculus since the early eighteenth century. But who can say that, printed in 1673, they would not have raised up for Newton again those dust devils of incomprehension, misunderstanding, skepticism, and hostility that danced over the field of optics in the 1670s?

It is true that Leibniz in 1684, more than ten years after Newton had renounced pure mathematics for other studies, set his seal upon the differential calculus with only one short paper. But the situation in which that appeared was quite different from that of the early 1670s. Some of the mathematicans in whose shadow Newton had worked were (like Collins and Gregory) dead, and others were no longer interested. The threads so actively spinning and intertwining then had broken off short; Leibniz's own paper was the exposition of ideas he had formed and tested nine years before and then put on the shelf. In violent contrast to the turmoil aroused by Newton's optical paper of 1672, no one now in 1684 challenged Leibniz, or set his paper in its ten-year-old context, or indeed in this flat calm commented at all (for several years) on this contribution to a new and not well-known Leipzig periodical. Several years passed before commentators and expositors appeared, and then fortunately they were friendly and respectful. Newton, when so many voices were shouting against him already, had kept silent; Leibniz had the luck to speak when all else was quiet, to be heard, and to be marked. Hence these tears.

2
BEGINNINGS IN CAMBRIDGE

IN 1718 a French Huguenot refugee in London, Pierre Des Maizeaux, a professional author, was putting the finishing touches to a new book. It was to be a *Collection of Various Pieces on Philosophy, Natural Religion, History, Mathematics etc by Messrs Leibniz, Clarke, Newton and other famous Authors* and was only to appear at Amsterdam two years later, in fact. The pieces concerned aspects of the difference in outlook that had for a number of years divided British scholars from the scholars of the Continent. Among the Europeans, Gottfried Wilhelm Leibniz had, throughout that period of division, stood out as the leading figure, whereas the creator of the ideas that had brought Britain into conflict with Europe was Isaac Newton. Part of the difference in ideas was wittily summarized some years later by Voltaire (for it was to continue considerably longer):

> A Frenchman, who arrives in London, will find philosophy like every thing else very much changed there. He had left the world a *plenum*, and he now finds it a *vacuum*. At Paris the universe is seen composed of vortices of subtile matter; but nothing like it is seen in London. In France it is the pressure of the moon that causes the Tides; but in England it is the sea that gravitates towards the moon; so that when you think that the moon should make it flood with us, those gentlemen fancy it should ebb, which, very unluckily, cannot be proved . . . According to your Cartesians, everything is performed by an impulsion, of which we have very little notion; and according to Sir Isaac Newton, it is by an attraction, the cause of which is as much unknown to us. At Paris you imagine that the Earth is shaped like a melon, or of an oblong figure; at London it is an oblate one. A Cartesian declares that light exists in the air; but a Newtonian asserts that it comes from the sun in six minutes and a half. The several operations of your chymistry are performed by acid, alkalies, and subtile

matter; but attraction prevails even in chymistry among the English.[1]

Far more personally bitter, however, was the dispute between Newton and Leibniz over the discovery of the new infinitesimal calculus, what we now call the differential and integral calculus. This was certainly the most important development in mathematics that had taken place since the early seventeenth century when Descartes had shown in analytical geometry how algebraic equations could mirror geometrical reasoning; arguably – bearing in mind potential future developments that were still unsuspected in the early eighteenth century, including the applications of the calculus in science and engineering – it was the greatest advance in mathematics that had taken place since the time of Archimedes. It was not in question that Leibniz had been the first to present a sketch of the differential calculus in print, in 1684, at which time none of Newton's mathematical writings had been openly published. Newton, however, made two claims against Leibniz's apparently impregnable priority: First, that he himself had made his first discoveries in the calculus nearly twenty years before that publication; and second, that he had shared these innovations with Leibniz (as with other correspondents) during those intervening years. Thus, Newton asserted, if Leibniz had been first in print, it had been with knowledge of what Newton had done previously.

Des Maizeaux had been in touch with Newton about his collection of letters in the hope of securing additional material from him or at least his goodwill toward the printing of what he had already assembled. And Newton had actually paid Des Maizeaux's Dutch publisher to hold back the edition for a while: It had been Leibniz's death in 1716 that had provided the occasion for it. In August 1718 Newton, who had been sent proof copies of part of the proposed book, drafted a long letter to Des Maizeaux about it and the whole calculus dispute. Drafts of this letter, repeating much that Newton had asserted on his own behalf on previous occasions, contain a famous passage referring back to the days when Newton was twenty-three or -four years old, that is, to 1665–6:

> In the beginning of the year 1665 I found the method of approximating series and the rule for reducing any dignity [power] of any binomial into such a series. The same year in May I found the method of tangents of Gregory and Slusius, and in November had the direct method of fluxions and the

next year [1666] in January had the theory of colours and in May following I had entrance into the inverse method of fluxions. And the same year I began to think of gravity extending to the orb of the moon . . . All this was in the two plague years of 1665 and 1666, for in those days I was in the prime of my age for invention and minded Mathematics and Philosophy more than at any time since.

This autobiographical narrative intended for Des Maizeaux in 1718 (though afterward Newton put his pen through the drafts) was far from being Newton's first of the sort; for example, before the violent outbreak of the calculus dispute but at a moment when the date and character of Newton's early work in mathematics was already becoming a matter of national interest, he had written a summary about his formation as a mathematician in one of his old notebooks, on 4 July 1699:

By consulting an account of my expenses at Cambridge in the years 1663 and 1664 I find that in the year 1664 a little before Christmas . . . I bought Van Schooten's *Miscellanies* and Descartes' *Geometry* (having read this Geometry and Oughtred's *Clavis* above half a year before) and borrowed Wallis's works and by consequence made these annotations out of Schooten and Wallis in winter between the years 1664 and 1665. At which time I found the method of infinite series. And in summer 1665 being forced from Cambridge by the plague, I computed the area of the Hyperbola at Boothby in Lincolnshire to two and fifty figures by the same method.

The books mentioned by Newton in this note are all well-known mathematical works of the period, though very advanced for the day and as such not the kind of material that an ordinary undergraduate would be expected to study. The fact that he could, with only slight earlier preparation, work rapidly through such difficult material and press on at once beyond it to innovations of his own, seeing the next steps following from the work of Descartes and Wallis, which those mathematicians had themselves failed to see, is evidence of the power and originality of his mind. "Beyond reasonable doubt," in the opinion of the foremost contemporary student of Newton's mathematics, "he was self-taught . . . , deriving his factual knowledge from the books he bought or borrowed, with little or no outside help . . . Very quickly he advanced to the frontier of existing knowledge and cut a broad path into new mathematical country."[2]

The mass of Newton's surviving papers is very large: Those papers concerned with mathematics are being printed in eight very solid volumes. Hence, it is easy to check on Newton's autobiographical affirmations, though not in all cases easy to verify the dates precisely. The student notes, with criticism and comments, that Newton made on the writings of such earlier mathematicians as the Frenchmen Viète and Descartes, the Dutchmen Van Schooten, Hudde and Huygens, and the Englishmen Oughtred and Wallis still exist and are now available in print; everything in them confirms Newton's later recollections in essentials and reveals his passage from study to discovery, for example in his reading of Wallis's work on infinite series, which led him to the series by which the area under a hyperbola could be calculated (at his first attempt he made two mistakes in the arithmetic) and the general binomial theorem. The technical historian of mathematics can follow the lines whereby, building on the treatment of the geometry of curves by Descartes and his immediate followers (mostly Dutch), Newton came upon the general algebraic handling of curvature, which is "the calculus." An important early step was his mastery of the binomial series expansion as early as the winter of 1664–5; a few months later he was already obtaining derivatives; by the summer he understood that what was later to be called *integration* was the inverse of *differentiation*; and by the autumn he had begun to master the process of integration by means of infinite series. Some of this work was clearly done at Trinity College, Cambridge, where Newton had access to mathematical books; the later and more original part was done in Lincolnshire during 1665–6.

Thus, by about the middle of the year 1665, Newton was able to "set down the standard differential algorithms" – that is, the standard procedures for differentiation – "in the generality with which they were to be expounded by Leibniz two decades later." However, in the autumn of the same year, Newton rejected the idea of *differences*, which he had been exploring in almost the same manner that Leibniz was to adopt, in his turn, ten years later, in favor of what he called *fluxions*, the fluxion being "a finite instantaneous speed defined with regard to the independent dimension of time and on the geometrical model of the line-segment"; that is to say, instead of imagining a variable quantity proceeding by

13

many infinitely little steps from one value to another (the steps being called differences), Newton now chose to think of the variable as flowing from one value to another and to consider its rate of flow, which is a motion or speed. On the last day of October 1665 he began to write – and rewrote the following week more elegantly – a short memoir on "How to draw tangents to mechanical lines" (lines of a more involved nature than the curves represented by algebraic equations in Descartes's system), which opens with a statement of this idea of curve formation, here (as it happens) rather too freely generalized in its development:

> In the description of any mechanical line whatever there may be found two such motions which compound or make up the motions of the point describing it, and by those two motions may the [resultant] motion of that point be found, whose determination is in a tangent to the crooked line.

Newton is generalizing the Cartesian notion of coordinates so as to make the x axis and the y axis both change through infinitesimal intervals of time – electronically, this is just how a curve is traced by a point of light on a television screen. Any curve can be simulated by properly matching a changing flow of x to another changing flow of y; and if at any instant we halt the double flow, the two (now static) rates define a straight line, which is the tangent to the curve at that point.

Two weeks later still, Newton is defining the general rules for finding the fluxions relating to given equations:

> An equation being given, expressing the relation of two or more lines [that is, variables] x, y, z, etc. described in the same time by two or more moving bodies A, B, C, etc. to find the relation of their velocities p, q, r [that is, their fluxions].

The rule follows immediately; it is, of course, exactly the same rule – taking equivalences into account – as was first printed by Leibniz nineteen years later as the foundation of the differential calculus.

This draft, bearing the date "November the 13th 1665" and the title "To find the velocities of bodies by the lines they describe," might be called Newton's first essay in the nascent calculus of fluxions; although the word "fluxion" does not appear in it, the word "velocity" always being used, the increment is symbolized by a lowercase letter *o,* and its use is clearly exemplified by examples.

After this there is a (relatively) long gap in the succession of manuscripts; not until May 1666, apparently, did Newton return to this particular technique, now substituting the word "motion" for "velocity," so that the nascent fluxional calculus becomes "the method of solving problems by motion" – the essence is the same, of course. Finally, all the earlier work was both condensed and developed in an incipient treatise (of forty-eight pages in Whiteside's edition, and it remained incomplete), which at some stage of its composition was dated by Newton "October 1666." Still, as yet, Newton writes of "velocities" and the term "fluxion" does not appear.[3]

For all this, it is evident that (so far as his own early achievement is concerned) Newton's claim to priority in discovering the calculus, as against Leibniz's, is perfectly justified by the ample remaining documents. To quote Whiteside yet again, these show Newton as possessing "a depth of mathematical genius which by late 1666 had made him the peer of" the great Dutch master Christiaan Huygens and of the Scottish mathematician of surpassing originality, James Gregory, whose work was just then starting to appear before the world but of which (in 1666) Newton was necessarily unaware. Newton was indeed probably already "the superior of his other contemporaries," among whom Leibniz, at the age of twenty, still knew nothing of mathematics. To recognize this is not to allow that all the claims made by Newton himself after many years about this age of invention were true; he said things that seem to be mistaken about his own intellectual inheritance and the influence of other mathematicians of note, like Barrow and Fermat, upon his own evolution, and more seriously (perhaps more deliberately also) he made claims about his use of symbols and even the very word "fluxion" that are far from being wholly consistent with the evidence, though, on the other hand, it might be argued that neither are they totally false. And because Newton composed a number of autobiographical recollections at different times, it is not surprising that discrepancies stand out when they are compared. But all these matters are of little weight in comparison with the central truth, which has indeed long been universally recognized, that Newton was master of the essential techniques of the calculus by the end of 1666, almost exactly nine years before Leibniz.[4]

What did Newton do with his mathematical discoveries? We have every reason to believe that no one at this time was aware of the content of his notebooks and papers, and in later life (though publicly affirming in his own name from 1704 onward that his discoveries had been made in the years 1665–6) he regarded everything as private that fell before 1669, when he reckoned that he first became known to others as a mathematician. (This, of course, still left him very well ahead of Leibniz.) Isaac Barrow, soon to become the Master of Trinity, Newton's college, was the first to take note of Newton's unusual abilities, though there is no evidence that he was ever Newton's teacher, as was once supposed; others soon formed an equally high opinion. Through Barrow, a general notion that Newton was an excellent mathematician, and information about his practical success with a novel form of reflecting telescope, reached John Collins in London in this same year, 1669, which was also the year in which Newton succeeded Barrow as Lucasian professor of mathematics at Cambridge, obviously through Barrow's goodwill. Collins, a minor civil servant, had a passionate though rather ill-informed interest in mathematics; he corresponded energetically with mathematicians in the British Isles and abroad in order to learn what was going on and spread news of what was considerable. These two, Barrow and Collins, were cited by Newton long after both were dead in reference to their letters about himself as "these two ancient, knowing and credible witnesses" to his discovery of the calculus in 1665–6. (It is not untypical that in a rejected draft Newton also cited the Oxford mathematician John Wallis as a third witness, though in truth Wallis in the 1660s had no knowledge of Newton's existence.)

Collins lost little time in making direct contact with Newton, now no longer a student but a teacher. He began to urge the young Cambridge mathematician to publish – and it was sound advice – for the benefit of the world of learning and the protection of his fame; here, indeed, we may relate the story in the words Newton used in telling it to Henry Oldenburg, the secretary of the Royal Society, for the benefit of Leibniz in October 1676 (that is, just a year to the month after Leibniz's own first discovery of the calculus idea); Newton starts from a work entitled *On Analysis* written in 1669, which he now describes to Oldenburg as

> a compendium of the method of these [infinite] series, in
> which I let it be known that, from straight lines given, the

areas and lengths of all the curves and the surfaces and volumes of all the solids [formed] could be determined, and con-
versely with these [taken as] given the straight lines could be
determined, and I illustrated the method there outlined by
several series.

Despite the use of the words "method of series" rather than
"method of fluxions" (in the letter quoted Newton made no open
reference to "fluxions" at all), it is obvious from the inversion
(lines to areas, areas to lines) that differentiation and integration,
that is, the method of fluxions, is in question. When Collins
pressed him to go into print, Newton continues in the letter, he
thought of publishing both his mathematical and his optical discoveries in the same volume:

> When by the persuasion of my friends I took up the plan
> some five years ago [that is, in 1671] of publishing the treatise
> on the refraction of light and on colours which I then had
> ready, I began again to ponder upon these [integration] series,
> and I wrote a treatise on them also so that I could publish the
> two together . . .

Now all this was written to Oldenburg for Leibniz (and certainly
read by the latter) *eight years* before Leibniz's first published paper
on the calculus; there is not the least reason to believe that Newton's report of his mathematical history speeded Leibniz's own development of the calculus or influenced it in any way whatever;
but Leibniz (as it is clear from his reply to Newton's letter) understood what Newton meant and saw in fact that Newton was master of some technique akin to his own calculus. Thus Leibniz's
own response partially confirms the existence of the method of
fluxions – though Leibniz was as yet unaware of Newton's nomenclature and algorithm – before October 1676, and indeed at
least as early as 1671, the earliest date positively claimed by Newton in the letter to Leibniz, though by implication even from that
the beginning of the work must have been several years earlier
still, as we have seen it was indeed.[5]

In the narrative Newton recorded two treatises, one a "compendium of the method of these series" written at an unspecified
date (obviously, however, between 1666 and 1671) and the other, hinted at as a revision of the former, prepared in 1671. Both
still exist.

The earlier treatise, entitled by its author "On Analysis by means of equations having an infinite number of terms," was in fact the first of his mathematical papers communicated by Newton to Isaac Barrow and by him, shortly afterward, to John Collins in London. Since Barrow first wrote of it to Collins in July 1669, *On Analysis* must have existed somewhat before that month, though the exact date of its writing cannot be ascertained. (The October 1666 treatise on fluxions was seen by no one.) After reaching London, *On Analysis* was copied by Collins, who was thus able (eight years later) to oblige John Wallis with a copy of his own copy. It was first printed in 1711, more than forty years after its composition, and was reissued by Newton himself in the following year with the perfectly accurate claim that it was a text that had been in Collins's hands in 1669.

The later work is the first of Newton's explicitly to state in its title its concern with the calculus: It is a *Treatise on the Methods of Series and Fluxions.* Not printed in Newton's lifetime, it was first made public in an English translation by John Colson in 1736. Its nature and the (approximate) period of its writing (1671) are well attested not only by Newton's later affirmations but by contemporary correspondence; moreover, as Whiteside has shown, Newton borrowed from it materials he used in his 1676 letters to Leibniz, and in later years it was also examined in private by several British mathematicians. Like so much of Newton's work, the *Methods of Series and Fluxions* was never completed, much less thoroughly revised, and so, although originally intended as a public exposition of the fluxional calculus, it never fulfilled its purpose.

What is the content of these two mathematical studies, of which the latter to a large extent subsumes the former? I have already quoted Newton's letter to Leibniz about them; here is his much later (1712) claim for their originality and scope (Newton writes of himself in the third person here):

> . . . it appears that in the year 1671, at the desire of his friends he composed a larger treatise upon this same method [of series and fluxions], that it was very general and easy without sticking at surds or mechanical curves and extended to the finding tangents, areas, lengths, centres of gravity and curvatures of curves, etc.; that in problems reducible to quadratures it proceeded by the propositions since printed in the Book of Quadratures . . . [*On the Quadrature of Curves*, first

written in 1691, published (abbreviated) in 1704]; that it extended to the extracting of fluents out of equations involving their fluxions and proceeded in difficulter cases by assuming the terms of a series and determining them by the conditions of the problem; that it determined the curve by the length thereof and extended to inverse problems of tangents and others more difficult, and was so general as to reach almost all problems except numeral ones like those of Diophantus . . .[6]

All this is true; Newton's exploration of new mathematical territory was indeed as far-ranging and thorough as he claimed. Outmoded, a mere historical curiosity when Colson finally gave it to the world, the *Treatise on the Method of Series and Fluxions* might have effected a mathematical revolution in its own day, sixty years before its belated publication. Its predecessor, *On Analysis by means of equations having an infinite number of terms*, is, as its title indicates, concerned much more with the usefulness of series than with fluxions, a word that does not appear in its text. Nevertheless, though no one would claim that this little tract could serve as an introduction to the calculus – whereas such a claim could be made on behalf of the October 1666 tract – it does contain (near the beginning) the basic algorithms for differentiation and integration, not of course qualified as such or indeed in any way singled out as constituting a special mathematical technique. Later in *On Analysis*, moreover, Newton explains integration procedures, that is, quadratures or the calculation of the areas enclosed by curves, by calculus methods, even employing a special symbol □, signifying "quadrature," that is, "the integral of," or "the sum of," as indicated by Leibniz later by the long *s* ∫, standing for "summa." Newton in his defensive writings against Leibniz of subsequent years was to cite this passage particularly as providing evidence of his clear priority.

In fact, *On Analysis* was rather the less serviceable for Newton's anti-Lebnizian purposes in 1712 because, at the time of composition, the essay had been directed to the different object of proving that Newton had obtained an independent mastery of the use of infinite series in the solution of compound algebraic equations by extending (with a hint from Viète) the technique used for the binomial expansion to which, in a very simple case, Nicolaus Mercator had seemed to point the way in his *Logarithmotechnia* (1668). Newton was indeed to explain to Oldenburg in 1676 (for Leibniz's benefit) how he had feared, after Mercator's book appeared, that others "would quickly discover the rest before I was mature enough for writing." Against this, *On Analysis* in Barrow's hands was a safeguard. But for that purpose (as a recent writer has stressed), a treatment of series by "a method fully sanctioned by tradition" was more appropriate than one that would have led Newton "into a lengthier treatise embodying a unified – but radically different – approach to the whole problem complex of analysis." In other words, a fully developed matrix of fluxional ideas for the specific nuggets of *On Analysis* would have only blurred the issue for contemporary mathematicians in 1669 and rendered the writing of the essay – hastily compiled – far more difficult. Newton's natural choice of the simpler course of action, however, entailed the consequence that this most "public" of his early mathematical papers was also the least explicit concerning the idea of "analysis by motion," whose significance Leibniz, for example, in his later hasty examination of *On Analysis*, seems to have quite missed. Moreover, the two great letters of 1676 would only confirm the impression that Newton's forte was the study of series and (as the opening of the *Second Letter* autobiographically reveals) that his interest in series did not derive from infinitesimal ideas.[7]

On the Method of Series and Fluxions, on the other hand, addresses itself straightforwardly to "widen[ing] the boundaries of the field of analysis and advanc[ing] the doctrine of curves"; after describing what he calls "computational methods" for the extraction of the roots of equations, Newton plunges into the exposition of his technique – first expressed, as we have seen, in 1665 – for relating curves to the speed of motion of a point tracing them, and then very rapidly to an account of *fluxions*, here openly and specifically treated. In illustration, here is his distinction between the *constants* in an expression, the *fluents* (variables) and their *fluxions* (or rates of change):

> . . . to distinguish the quantities which I consider as just perceptibly but indefinitely growing from others which in any equation are to be looked on as known and determined and are designated by the letters a, b, c and so on, I will hereafter call them [the former] fluents and designate them by the final letters v, x, y, z. And the speeds with which they each flow and are increased by their generating motion (which I might more readily call fluxions or simply speeds) I will designate by the letters l, m, n, and r. Namely for the speed of the

[fluent] quantity v I shall put l and so for the speeds of the
other quantities I shall put m, n and r respectively.

It will be observed that Newton's notation is awkward; much later
– not regularly until 1691 – he made the relationship of fluxion to
fluent more obvious by writing $\dot{x}$ as the fluxion of x, and so on.
It was then also possible to express the fluxion of a fluxion as $\ddot{x}$,
to any desired elaboration. Leibniz's later comment that in the notation of 1671 Newton could hardly express these higher fluxions
was justified, and it is a sign of Newton's weakness that he would
neither concede the lateness of his own "dot" notation nor admit
the superiority of Leibniz's invention in this single respect of notation. Against Leibniz, Newton was apt to claim (as, once more,
in writing to Des Maizeaux in 1718):

> And as the notation used in this book [*On the Quadrature of
> Curves*] is the oldest, so it is the shortest and most expedite . . .

This claim was not true, in that the "dot" notation of 1691 was
certainly not older than Leibniz's "d" of 1684.[8]

This set aside, Newton's claim to have mastered the new infinitesimal calculus long before Leibniz, and even to have written –
or at least made a good start upon – a publishable exposition of it
as early as 1671, is certainly borne out by copious evidence, and
though Leibniz and some of his friends sought to belittle Newton's case, the truth has not been seriously in doubt for the last
250 years. Of course, the fact of Newton's historical priority in
the discovery, and perhaps especially in his very extensive exploration of the method of integration (or quadrature) by infinite series, does not justify or authorize his second claim against Leibniz,
that Leibniz "must have" been aware of Newton's prior work,
which we now know to be as false as the first claim was true.
Which means, of course, that it was unjust of Newton to attempt
to deprive Leibniz of the credit that was due him as an independent discoverer of the calculus and the first in publication. Newton's scornful assertion that "second inventors" deserve no fame
does not apply in a case where (as here) the "first inventor" has
maintained the privacy of his discovery.

Why did Newton ever put himself in the position where doubt
could arise? If his unpublished mathematical essays had been formally deposited within the archives of the Royal Society (as was
done with the first sketch of his *Principia* years later) or in the
Cambridge University Library (as was done with his lectures on

optics), his position vis-à-vis Leibniz would have been far stronger. "If it be asked why I did not publish this book sooner [again, *On the Quadrature of Curves* is meant]," Newton told Des Maizeaux, "it was for the same reason that I did not publish the Theory of Colours sooner, & I gave the reason" in the *Second Letter*, that is the letter for Leibniz written in October 1676. There, indeed, still addressing Oldenburg, Newton reminded him that

> when I had sent to you a letter on the occasion of the reflect-ing telescope, in which I briefly explained my ideas about the nature of light, something unforeseen made me judge it nec-essary to write in haste to you about the printing of that letter. And then frequent interruptions at once were created by the letters of various people filled with objections and other mat-ters, which quite changed my mind, and caused me to call myself imprudent because, in order to catch at a shadow, I had sacrificed my peace, a truly substantial thing.[9]

That was what Newton remembered, but his memory does not quite accord with events as they can be reconstructed. It is true that Oldenburg's publication in the *Philosophical Transactions* for March 1672 of Newton's first "Optical Letter," disclosing his theory of light and colors, had caused him a great deal of labor in replying to criticisms – whether Newton was right to think this *wasted* labor is another question. But his first experience of publi-cation did not, in fact, deter him from planning others in the 1670s, relating to both optics and mathematics, and part of the reason for his plans not maturing successfully was not of New-ton's own volition: It was simply the reluctance of publishers to undertake so serious a work of such small sale. Whether any book of his that might have appeared in the 1670s – for, in fact, a geo-graphical text of which Newton was, in a trivial capacity, editor did come from the press – would have contained *On the Method of Series and Fluxions* or any similar study is obviously impossible to tell: The odds look much less than even in favor. Newton found it almost impossible to regard any of his writings as finished and ready for the reader's eye and was only too eager to seize any excuse for delaying or withholding publication. Had he really wished to print, or deposit on open access, his mathematical dis-coveries of the 1660s he could easily have found the way. At the worst, the relatively small cost of printing a hundred copies of a slender volume would have been trifling to him, for although not yet rich, he was far from destitute or even dependent upon his professorial salary. The fact is that Newton preferred not to publish, for whatever assembly of reasons. He was, clearly, glad to win private merit within the circle of competent mathematicians, particularly those of Britain, and to deserve the admiration of such men as Barrow, Collins, and James Gregory, but he had no wish to contend in a broader field. It was, in a sense, the tragedy of this whole dispute that Newton later, changing his mind and seeking to undo the natural consequences of his earlier inactivity, tried to make this private reputation equivalent to one fully established and recognized by the public.

3
NEWTON STATES HIS CLAIM: 1685

I N THE SUMMER OF 1685, perhaps not long after the defeat of Monmouth at Sedgemoor, Isaac Newton in his rooms by the Great Gate of Trinity College, Cambridge, was absorbed in writing the earliest version of his *Mathematical Principles of Natural Philosophy*, a majestic work whose beginning was still only about a year past.[1] In August 1684 Edmond Halley, one of the secretaries of the Royal Society, a competent mathematician and an astronomer with some years of practical experience, had ridden the fifty miles from London to Cambridge expressly to put to the Lucasian professor a technical question that London mathematicians had failed to solve:

> What he thought the curve would be that would be described
> by the planets supposing the force of attraction towards the
> Sun to be reciprocal to the square of their distance from it.

Newton at once answered – too precisely – that the orbit would be an ellipse. Halley, "struck with joy and amazement, asked him how he knew it; Why, saith he, I have calculated it; whereupon Dr. Halley asked him for his calculation without any further delay." But the paper could not be found then and there, and Halley had to return to London with Newton's promise, soon fulfilled, that the demonstration would be sent to him there.

On 10 December 1684 Halley spoke of the work that Newton was engaged upon to the Royal Society in London, and Newton must by then have embarked already upon a large-scale treatment. He worked incredibly intensely, writing the first draft of a large and densely mathematical book in seventeen or eighteen months, though it took longer to complete and perfect it. By the late summer of 1685 he had finished the first form of what was later to become Book I of the *Principia* and also a great deal of what was to be Book II – the division into Books as we now have it was not yet established at this time. The early version, of which the larger part still survives, was (on some now unknown occasion in the summer of 1685) read and annotated by Edmond Halley. Even if Newton had died prematurely at this moment, the world would have been able to recognize his immortal achievement.

The Lucasian professor was forty-two years old, and he had occupied his professorial chair for sixteen years. Despite a lack of students, he had performed his duties with diligence: Records of his lectures already reposed (unread) in the Cambridge University Library. He seems to have been patient and conscientious with the few, students or seniors, who sought him out and wanted his advice. He was the only active mathematician in Cambridge at that time and almost the only scientist. In this very year an attempt to found a select scientific club in Cambridge had failed (in Newton's words) for "the want of persons willing to try experiments, he whom we chiefly relied on refusing to concern himself in that kind," Newton himself being just as reluctant to promise "any loss of my own time in those things." The writer of the *Mathematical Principles of Natural Philosophy* might well be forgiven his unwillingness to desert his desk in favor of putting mice in an air pump; yet Newton had certainly experimented not long before in optics and in chemistry and would do so again in years to come; below his rooms he had a little laboratory shed with chemical furnaces, and on the staircase above he sometimes stationed a telescope. In 1685 Newton was an academic at ease in his world, comfortable in the sense of achievement, which, at his chosen moment, he might one day lay before the world in proper form. The accident of Halley's question had released in Newton a creative torrent whose vigor must have been unsuspected even by himself, sweeping him in a direction barely hinted at in all his earlier concerns. That Newton's intellectual relations were rather with the outside world of London, the Royal Society, and the *Philosophical Transactions* than with his Cambridge colleagues troubled him not at all. What he had done – even what he was now doing – was for his own satisfaction and not for applause. Through all the past years, though ready enough to share his results when a rare opportunity (like Halley's visit) offered itself, his attempts to address the public at large had been highly tentative. In 1685 none of Newton's mathematical work (amounting in modern form to four very stout volumes) had yet been printed, and little enough was known to any one at all. Much remained unknown, unsuspected, at the time of his death forty years later.[3]

Although he found a few friends when young and drank with them sometimes at the tavern, and although when old he kept a private wine cellar and gave occasional dinners, Newton was neither convivial nor frivolous. His own preferred use of his day was for study, calculation, writing. His addiction to notes, transcripts, and drafts was pathological. He copied extensively from books standing on his own shelves. A young namesake, no relation, his valet and copyist, described him as "very meek, sedate, and humble, never seemingly angry, of profound thought, his countenance mild, pleasant and comely." If never seemingly angry in the Cambridge of 1685, the elderly Newton could be furious, petulant, and even violent in his language – or so his critics report. The Lucasian professor avoided exercise and cared not for food. He was attentive neither to college chapel nor to private prayer and was sometimes haphazard in his conduct and in his appearance. But Humphrey Newton fails to show us Newton managing his Lincolnshire estate with shrewd firmness or to present Newton's barristerlike mastery of detail in arguing a case. If some of Humphrey's account of Newton seems plausible, some exaggerated, and some false, certainly he did not avoid hearsay; he could not himself have observed that, at thirty (ten years before), Newton's gray hair was "very comely, and his smiling countenance made him so much the more graceful." When writing the *Principia* Newton probably looked much as Godfrey Kneller depicted him four years later – not smiling, indeed, but surprisingly youthful (at forty-six now), spare, and intellectual. The head, seemingly with his own long natural hair flowing to the shoulders on either side and ruffled at the crown, appears large in proportion to the body; the eyes are gray, the brows strongly marked and slightly knit, the expression somewhat tense (perhaps from holding the pose). Newton's mouth is full and firm, the lines well marked, the chin large and cleft. It is a powerful but not unkindly face. He wears a loose gown, under which the shirt collar is open; the hands, clasped, show long, delicate fingers like a woman's. The figure is in repose, but does not suggest calm; rather, quiet enforced by the mind on an active body. Indeed, Newton possessed great physical strength and resilience; he worked constantly, he was almost never ill, and he remained alert and active into his eighties.[4]

Why this picture was painted remains so far a mystery. It is not a portrait that would be demanded by vanity, pomposity, or ostentation. His mother was dead ten years when it was painted; he was a posthumous and only child. The portrait can hardly have been intended for Newton's family, and it seemingly remained always in his own possession; it must, therefore, have been painted for his own satisfaction – or Kneller's. Of the man immediately behind the portrait little is obvious either. Newton's lower-gentry origins and his ten years of apprenticeship and introduction, from 1665 to 1675, are clear enough. In the latter year he concluded a phase of communication of his optical researches to the Royal Society and in the year following wrote his two great mathematical letters to Leibniz (of which more later). Then, for the next eight years, he withdrew to Cambridge and (it is facile if not wholly accurate to add) alchemy. The professor whom Halley hastened to visit was a half-forgotten man, whose closest colleague in the outside world at that moment was (ironically enough) the Astronomer Royal, John Flamsteed, Halley's own former mentor and patron. Certainly Halley did not go to Cambridge to seek advice from one of the greatest theoretical physicists of all time; he went to consult a very able mathematical astronomer.

After the stimulus of Halley's visit, mathematical physics demanded all Newton's strength and concentration. Like Sherlock Holmes when summoned to a case, he abandoned (for a time) his chemical experiments and his already declining interest in pure mathematics. During the first period of writing the *Principia* Newton's only extant correspondence was devoted to digging the astronomical data he needed out of Flamsteed; afterward he had to exchange a number of letters with Halley about the printing of the book. He left Cambridge for a couple of weeks between March and April in 1685 and was absent again for a week in June. Thereafter he settled down to his task without a break for nearly two years. It is needless here to consider the *Principia* in order to define Newton's style and originality as a mathematician; this has been most ably done by D. T. Whiteside. In outline, two seemingly antithetical points need to be made. The book is cast very much in the mold of a Greek geometrical treatise, notably in its formal layout of definitions, axioms, propositions, lemmas, theorems, and so on. The nearest ancient model, considering both manner and content, might be found in Archimedes' *On the Equilibrium of Planes* and *On Floating Bodies*, but more immediately the model for geometrical physics had been furnished to Newton's

27

generation by Galileo, Descartes, and their contemporaries. Yet mathematically, the classicism of Newton's *Principia* is illusory.

Not only was Newton's familiarity with the great Greek masters of geometry like Archimedes and Apollonius limited, the essence of his mathematical argument is entirely nonclassical. Does it then reflect the essence of mathematical innovation associated with Newton's name, the method of fluxions? The answer again is no. The method of fluxions is intrinsically algebraic rather than geometrical, and there is not the slightest reason – in the historical evidence or in logic – to suppose that the argument of the *Principia* was ever cast in an algebraic rather than the geometric mode in which it was published. It is true that Newton himself attempted in later life (and with success) to confuse the issue by claiming that he had employed analysis, specifically the method of fluxions, to discover the propositions displayed in his *Principia* and then afterward worked out the synthetic geometrical demonstrations of them, which he had incorporated into the book. Thus in 1712, drafting a (rejected) prefatory note to its second edition, he wrote:

> It has seemed advisable that I should now add the analysis by which I investigated the proposition in this Book of Principles, so that readers who have been instructed in the same [analysis] may the more readily examine the propositions considered in this book, and increase their number by discovering new ones.

In fact, he never attached any account of his method of fluxions to the *Principia*, nor was such a distinct method of analysis used by Newton in writing the book. As Whiteside has said, it is an inescapable fact that Newton wrote the book just as we have it, in a geometrical form, but in accord with the letter rather than the spirit of Greek geometry.[5]

A Greek geometer sought to express a mathematical problem in its most appropriate (that is, simplest) geometrical form and to solve it by establishing relations between such quantities as lengths of lines or areas bounded by lines or even volumes. In many problems the relations under investigation – for example, between two areas *ABCD* and *EFGH* – are constant, though, of course, the areas themselves are arbitrary and indeterminate; but in some problems the Greeks had already dealt with changing magnitudes. To give an instance, the Archimedean spiral is defined as the curve traced by a point moving uniformly along a straight line while that line rotates uniformly about one extremity.

Thus the curve will end or begin at this center of rotation. In general, however, the ancients had found great difficulty in conceiving of relations between *two* changing quantities – hence their extremely cumbersome understanding of what a "constant speed" and a "changing speed" are, because speed depends upon both change of time and change of place or distance. A Greek would have found it still more difficult to conceive of a second-order variable such as acceleration or change in speed or to cope mathematically with the spiral produced in Archimedes' manner but with the difference (say) that the speed of the moving point along the straight line is proportional to its distance from one end of the line.

Now the essence of Newton's mathematical argument in the *Principia* is that he is all the time concerned with the relations between changing quantities. For example, the very first proposition of the book is concerned with a motion in which a body traces an indeterminate curve – in other words, it changes its direction at every instant as it proceeds along its path, but for the purpose of the proposition that is no obstacle; it is not required that the shape of the curve be defined. And – as in this proposition – Newton is concerned with "instantaneous," that is to say infinitesimally small, changes. In this case, and elsewhere, he "models" a smoothly continuous curve by a succession of equal straight-line segments – if these are sufficiently numerous and consequently sufficiently minute, they constitute a line indistinguishable from a smooth curve (note that Archimedes had *precisely* defined the sizes of two polygons, each of ninety-six sides, one outside and one inside a circle in determining limits for the magnitude of a circle in terms of its radius – he had no method, and no impulse, to consider polygons of so many sides as to be indistinguishable from a circle). Whereas the Greeks had operated (in general) with unvarying geometrical quantities, Newton, like Kepler and Huygens before him, operated with varying quantities but handled infinitesimal limit increments of the second order as well as the first; that is, problems were to be solved not by considering change in a quantity but rather the rate of change of different quantities.

Newton's techniques for solving such problems need not concern us; in general they involved the use of the methods of series, which he had discovered in the 1670s. What is important to observe here is that the advanced geometrical procedures involving infinitesimal line segments displayed by Newton in the *Principia* are to be considered as equivalent to other procedures in the direct and inverse method of fluxions and, accordingly, in the differential and integral calculus of Leibniz. That the *Principia* was a book founded on "the calculus" was a truth understood by Newton's contemporaries and successors, as well as by modern historians, it being appreciated that "the calculus" in his sense means an informal and general mathematical process, not the particular algebraic algorithms of Leibniz or of Newton himself. Indeed, a famous, and to Newton a deeply irritating, criticism (arising from a mistake occurring in the treatment of Proposition 10 in the first edition of Book II, where Newton is concerned with resisted motion) was precisely that in this place his mistake had arisen from a failure to obtain second differentials (the differential of a differential) in the proper way – this (mistaken) accusation could have had no force had it not been plausible to argue that Newton was relying on the calculus for his solution to the problem. There is thus no question but that in the *Principia* – though not employing fluxions – Newton could formulate and resolve problems by the integration of differential equations, and, in fact, he anticipated in his book many results that later exponents of the calculus regarded as their own novel achievements.[6]

There was nothing in principle either strange or unprecedented in Newton's proceeding in this way rather than in developing the science of motion by analytical algebraic methods, as was soon to be done by other mathematicians including Leibniz, the Bernoullis, and Pierre Varignon. Particular instances of differentiation and integration had been handled by earlier mathematicians geometrically as well as algebraically. Both older contemporaries, like the Dutch nobleman Christiaan Huygens, and young ones, like Newton's Swiss friend Fatio de Duillier, were familiar with analogous geometrical procedures. To the few skilled mathematicians in Europe around 1685 who were capable of understanding Newton's mathematical arguments at all, however expressed, the form he actually adopted in *Principia* was far more convenient and familiar than either the method of fluxions – known to no one but Newton – or the Leibnizian differential calculus first described, rather obscurely, in print only in 1684 and equally as yet adopted by no other mathematician. Leibniz's calculus, indeed, was still too undeveloped to have served Newton.

Accordingly, one will look in vain in the earlier pages of the book for any apology for the use of mathematical methods unfamiliar to the reader or for a preliminary explanation of them. The *Principia* was notoriously a book that only the expert could follow on every point of its argument, but this was not because its argument depended on some wholly new branch of mathematics, such as the calculus would have been. In the opinion of Whiteside, "Newton did not set an impossibly high estimate on the mathematical competence and technical expertise of his reader, who is assumed to be familiar with the Euclidean *Elements* of the geometry of the straight line and circle but otherwise expected to know only the simplest Apollonian properties of conics, the rest being proved *ab initio* as needed in the progress of the argument or (in rare instances) justified by an appeal to the general algebraic 'quadrature of curves' which Newton had expounded [but not yet published] some fifteen years before." The difficulty of the book lay not in its technical novelty but rather in its concision and density. Only in Lemma II at the beginning of Book II does Newton explain in elementary form a new mathematical idea, which, in fact, we now name (after Leibniz) differentiation. Why Newton chose to introduce the "fluxions lemma" at this point is a mystery; it would have been at least as appropriate, or more so, for him to have done so in Book I, where he had been happily differentiating without troubling to give a particular explanation of his procedure. A possible reason for the lemma will appear in a moment.[7]

Differentiation is simply a way of finding out how mathematical quantities change. Thus if one quantity y changes in some way that is dependent upon change in a first variable x, differentiation establishes the manner of the change. Or consider a geometric curve, whose form is defined by an algebraic equation; the direction of the curve at any point is determined by the slope of the tangent at that point and this in turn can be discovered after the equation to the curve has been differentiated. Newton spoke of flow (flux) rather than change in relation to mathematical quantities, and accordingly he spoke not of finding the differential but of finding the fluxion, or rate of flow. (Although there are important differences between a differential and a fluxion, a statement in differentials can be systematically rewritten in fluxions.) However, in the lemma in Book II of the *Principia* he uses still different language again, though the content of the lemma is based on an unfinished draft treatise, "On the Geometry of Curved Lines," written perhaps six years before, which was devoted to fluxions.

In the lemma the argument hinges on the notion of a *moment*, which Newton defines as the instantaneous increase or decrease of a changing quantity. It is not itself a very small *finite* quantity, however (an "atom" of number-ness, as it were), for finite quantities are the end products of moments: "We are to conceive them [the moments, Newton adds obscurely] as the just nascent principles of finite magnitudes," that is, as the germ cells of finite quantities. Here, as always, Newton finds it very difficult to express his conception: The moment is not nothing (zero), for if so it could not be multiplied as he proceeds to do; it is a changing thing, yet capable of being seized upon at an instant of time; it is not a fixed minute quantity, for to allow this would be to deny the perfectly smooth transition of a changing magnitude from one value to another.

We can see (as Newton explains) that the moments of any number of related quantities are proportional to their fluxions, because in any short length of time (during which these quantities are supposed to change in value) the sum of the added or subtracted moments will be proportional to the rate of change, that is to the fluxion of each quantity.

The upshot of the lemma is that Newton provides a highly unsound demonstration of the perfectly correct result that the moment of any quantity x^n is proportional to nx^{n-1}; in Leibnizian terms Newton is saying that the moment of x^n is $d(x^n) = n \cdot dx \cdot x^{n-1}$. This is the beginning of the calculus that everyone learns, and the fact that Newton's demonstration in the lemma is self-nullifying by no means negates its operational utility. Nor is it surprising that, for all his profound intuitive grasp of the rightness and good sense of what he was doing, Newton was unable to give complete logical validity to an infinitesimal calculus, leaving cracks into which Berkeley and other critics could insert powerful destructive wedges; Leibniz was not very much more successful in this task, and the problem was only settled in the nineteenth century. Meanwhile, the philosophical problems of allowing mathematicians to handle quantities that were neither zero nor finite magnitudes did not inhibit the development of the calculus, nor its application to science and engineering. However, as we shall see, Newton's uncertainty as to the definition of his fundamental concept (was it moment, or a fluxion, or something else?) and his diverse justifications for its usefulness in mathematics were to be made to appear as weaknesses later.[8]

Moments make very little further appearance in the *Principia*, and as Newton had decided to leave his readers in ignorance of his methods of fluxions and of series and to allow them to take for granted his use of limit increments, he might as well have omitted this "fluxions lemma" also, but for one reason, that he followed it with a scholium, or explanatory comment, of an autobiographical character:

> When, in letters exchanged between myself and that most skilled geometer G. W. Leibniz ten years ago, I indicated that I possessed a method of determining maxima and minima, of drawing tangents and performing similar operations which served for irrational terms just as well as for rational ones, and concealed the same method in transposed letters [which, when correctly arranged] expressed this sentence: "Given any equation involving flowing quantities, to find the fluxions, and vice-versa"; that famous person replied that he too had come across a method of this kind, and imparted his method to me, which hardly differed from mine except in words and notation.[9]

One can hardly doubt that the lemma was written so that this scholium could be added to it, or so that (not to mince words) Newton could claim his mastery of a powerful new method in mathematics as early as 1676 (and by an inference, which Newton would not hesitate to make and substantiate by evidence, as long a time again before 1676). The historical claims of the scholium are clear and precise: Newton then already possessed a true method of some generality – a method as we would say of differentiating expressions – going beyond mere rules for determining tangents or other ad hoc successes, and this method extended to all algebraic expressions, not merely easy ones. This was a very important claim because, as Newton well knew and was ready when necessary to concede, there was a long story of partial success in differentiation extending far beyond his own time; therefore, the only significant claim he could make at this stage was to an *unrestricted* success in this procedure. At the same time, Newton did not assert that he had informed Leibniz of his counterpart success in integrating expressions or in obtaining quadratures (the determination of curvilinear areas); integration, as Newton already knew and some of his mathematical contemporaries were learning, is the inverse of differentiation; but this is only indicated (by the two words "vice-versa"!) in the so-called anagram, which

Leibniz could not, of course, have deciphered. Nor does Newton in the scholium claim that Leibniz had learned anything from himself, or indeed deny that Leibniz's discovery of his own method – assuming that to be equally general and valid with his own – might have antedated his own letters in which the method of fluxions had been so darkly hinted at. And it positively concedes that at the time of his reply in 1676 Leibniz possessed something – some art of differentiation – of considerable importance, of which he had imparted examples to Newton.

It is easy enough to imagine why Newton, conscious of the weight and extent of his innovations in mathematics, should not wish to see his discoveries passed over as they were overtaken by the subsequent discoveries of others. But why, after maintaining silence through nearly twenty years, should Newton at last choose this moment and this awkward manner of expression, in which his methods were not allowed to stand alone but were put in challenging juxtaposition with some analogous accomplishment of a German philosopher nowhere else mentioned in the *Principia*? Why suddenly resurrect here a brief correspondence that had been dead for ten years?

The reason, in fact, is clear: Leibniz's own belated emergence in print. We now are quite sure – though Newton never knew it – that Leibniz had the beginnings of his differential calculus in October 1675, shortly before his correspondence episode with Newton. But his first account of it – when the method was very considerably matured – was published in a Latin periodical produced in Germany, the *Acta Eruditorum* (Transactions of the Learned), for October 1684. The article was headed, in part, "A new method of maxima and minima and also tangents, and a singular kind of calculus for them"; as with Newton, Leibniz's discovery had remained almost unknown since its inception until this first announcement, which mathematical readers found quite difficult to assimilate. Nevertheless, from this article (and others with which Leibniz followed it) the whole history of the calculus has stemmed. Newton at once realized its importance, and as soon as he read it his mind flew back to Leibniz's letters of 1676 and his own to which those of Leibniz had been replies. There can have been no doubt in Newton's mind that the article represented the fulfillment of the promise of Leibniz's letters, whereas he himself had nothing on record, save in letters to one or two friends. Hence his eagerness to state his claim and to state it in a way that made

34

his independence of – not so much yet his priority before – Leibniz absolutely plain and solid. For, of course, Newton understood at once that the differential calculus and the method of fluxions were different expressions of the same mathematical idea. It is not surprising, on reflection, though at first it may seem so, that Newton waited a year to react to a paper put on sale in, say, November 1684 at Leipzig. The transmission of periodicals, of all postal packages indeed, across Europe was at that time extremely slow. Only if deliberately carried from Germany to Cambridge by a friend could it have reached Newton's hands in a few weeks; by normal commercial means very few copies of the *Acta Eruditorum* can have reached England at all. Nor had Newton any special reason to look out for one; the *Acta* was not primarily a mathematical or scientific periodical and was but recently begun. Finally, virtually no one alive in 1685 could have foreseen Newton's passionate interest in Leibniz's article. All these factors combining together, the delay of a year or less before Newton reacted is explicable; and he may well have taken further time to decide how he could decently react in a scholarly and gentlemanly fashion. In the end we may well feel that what he chose to say was excessively diffident.[10]

At this time Newton could so very easily have announced himself in the pages of *Philosophical Transactions*, for example, by means of his new friend Edmond Halley, or afforded to issue a small separate pamphlet of his own. The expense could not have been a serious object to him, not so much as it had been in the early 1670s when he came nearest to an early revelation of his mathematical researches. Only in recent years has Whiteside put into print the already almost finished treatises that Newton had to hand, such as *The Method of Fluxions and Infinite Series* going back to 1671. Or he could have issued an effective autobiographical narrative. It is nowadays beyond all challenge that Newton had begun to formulate original ideas in mathematics in 1665 and 1666, the time he had described as the "prime of my age for invention," and that he had then written his first account of his method of fluxions, which he had perfected in 1671. And there had been much more mathematics since, notably Newton's sophisticated study of series, which was the key to the inverse method of fluxions or integration. In the scholium, though tabling his claim to consideration as the prime inventor of a new brand of mathematics, Newton was too scrupulous (or too cautious?) to allude to this private record of achievement or to give more than the barest and bleakest hint – not without its possible chances of misleading the reader – of the vast intellectual apparatus of which the mathematical arguments of his *Principia* are offshoots or exemplars.

It really seems as though Newton – who could justly have written so extensively and creatively in expressing his mathematical thoughts since 1666 – was most eager not to make for himself the maximum just claim (as any one may morally do) but the minimum claim that would still leave him, as it were, in the game. The language of the "fluxions lemma" is so stilted and inadequate that (as Whiteside has remarked) the Leibnizians were deceived as to the real quality of Newton's achievement:

> In hindsight we may see how a little more forthrightness on Newton's part when he first presented his fluxional method to the world would have saved him a great deal of the bitterness and sense of frustration he experienced twenty-five years afterwards when he fought tenaciously to safeguard his priority of discovery.

Whatever one's doubts about the appropriateness of the lemma, and the related scholium referring to Leibniz, it is obvious that Newton would have been well advised to elucidate their puzzling obscurity and reinforce his position in major mathematical publications after 1687. He seems to have contemplated no such move, and in fact his next major step toward a mathematical treatise – this time not on differentiation at all but on integration, the "quadrature of curves" – was taken not in response to Leibniz's mathematical innovations, but to published work coming from the Scottish mathematician David Gregory, who had sought Newton's acquaintance and support not very long before.[11]

This was the second time that Gregory, professor of mathematics at Edinburgh, had seemed likely to step forward for a share of the fame that Newton might expect to receive if his unpublished work were ever to be appreciated. Starting from some old private papers of his uncle, James Gregory, who had held the Edinburgh chair before himself, David had advanced far in the techniques of series, as could be seen in a fifty-page pamphlet, *A Geometrical Essay on the Measuring of Figures*, published in 1684 and sent by its author to Newton in June of that year. In an accompanying letter David Gregory claimed (justly) that the treatise contained "things new to the greatest part of the geometers," though

he recognized – from his late uncle's correspondence – that Newton himself had "of a long time cultivate[d] this method and that
the World have long expected your discoveries therein." Newton
would have gathered what Gregory was too cautious or modest
to state outright, that Gregory's explorations came close to his
own. How close must have surprised him; the scene has been well
reconstructed by Whiteside:

> Thus forewarned, Newton was well prepared for Gregory's
> following sketch of the principles of exact algebraic integra-
> tion and their exemplification in problems of the quadrature
> and rectification of curves and the mensuration of their solids
> and surfaces of revolution. But, as he read on, a growing
> feeling of the *déjà vu* must have come over him, for Gregory
> devoted the remainder of his tract to elaborating (with nu-
> merous well-chosen examples) two of the three methods of
> reducing quantities to series . . . which Newton himself had
> set down in "Reg. III" of his *De Analysi* fifteen years before;
> in addition, he gave, without proof, two instances (seemingly
> taken from his uncle's papers) of extracting the roots of a lit-
> eral equation . . . and declared his intention of publishing a
> full explanation of the method "on another occasion" unless
> John Wallis, . . . should forestall him by printing his "enu-
> cleation" of the equivalent Newtonian doctrine. The chal-
> lenge thrown out to Newton himself was unspoken but no
> less real: publish or be published.[12]

Newton's response was strange and typical. Although believing
himself (as he wrote in a draft paper) less able "to resist the entrea-
ties of my friends" that he should publish, and that he would be
"better advised to be swiftly acquiescent rather than have to submit to annoyance [!] at a later, lesser opportune time," as Gregory
had already declared his hand, he decided, nevertheless, rather
than publish any systematic exposition of his own or write any
piece wholly new, that he would more oblige the public ("after
what Mr. Gregory has done") by publishing his letters to Leibniz
of ten years before, "especially so since in them is contained Leibniz's extremely elegant method, far different from mine, of attaining the same series – one about which it would be dishonest to
remain silent while publishing my own." And so a work – which
never left Newton's hands – took shape under the title "Specimens
of a Universal System of Mathematics," on which Newton must
have labored (perhaps quite briefly, for the sketch was never com-

pleted) in June or July 1684, in any event shortly before Halley's trip to Cambridge, which plunged Newton into the composition of the *Mathematical Principles of Natural Philosophy*, to the abandonment of his interest in pure mathematics. [13]

The choice of policy by Newton is really very extraordinary. David Gregory, relying pretty heavily on his uncle James's unpublished ideas, had lifted the veil from one section of the Newtonian edifice of mathematics. Was Newton's reaction to say to himself that the time had really come at last to put all his cards on the table and set out properly that whole of which others might reveal parts piecemeal? No; he was conscious of no debt to learning, no ambition for fame; publication was an "annoyance." All he thought it worthwhile to do was to prove that he had known all this (and more) before Gregory. His reaction seems almost pathologically historicist; he is not eager to get on with the business of adding his searchlight to Gregory's torch; he thinks only of citing documents ten years old to prove that he himself was first to shine the light. It is hardly surprising, therefore, that many years later his defense against Leibniz took an equally historicist form and indeed had at its core exactly the same documents, the Newton-Leibniz letters of 1676. It is true that as Newton went on with the preparation of the *Specimens* his native interest in the subject began to take over; but in the end this dried up and he abandoned the matter altogether. The final outcome was that Newton allowed John Wallis, the Oxford doyen of English mathematics, to wave the Newtonian flag by summarizing Newton's two letters to Leibniz in his *Algebra* of 1685.

To complete this aspect of Newton's relations with Gregory and to introduce the last of Newton's attempts to prepare a full exposition of his fluxions, we must go forward again into the post-*Principia* period. About two years after the last incident, in 1686, Gregory succeeded in reconstructing, from information innocently brought to him by another Scotch mathematician, John Craige (of whom more in Chapter 5), Newton's general theorem for squaring curves which had already been imparted to Leibniz in the *Second Letter* of 1676. Not long thereafter (despite Craige's showing Gregory that his result was identical with that obtained by Newton long before) the theorem was obscurely published as Gregory's, without a word about Newton, by the former's physician friend Archibald Pitcairne. Fortunately no report of this came to Newton who, like everyone else, was soon deep in the excitement of the English Revolution. When that was done, Gregory, who in 1691 was seeking the Savilian professorship of astronomy at Oxford (which he was to win), wrote asking for Newton's assistance in publishing "his" theorem with examples and a proper acknowledgment of what Newton had previously accomplished. Newton, starting to draft a formal letter of reply for Gregory to print, passed rapidly into the composition of a regular treatise on the quadrature of curves. *De quadratura curvarum* rapidly grew into a work of considerable extent and by early 1692 greatly impressed Fatio de Duillier. But alas, as always before, Newton's enthusiasm waned after a few months and it was never finished. Here, for the first time, Newton used his dotted fluxional notation (which thence, in extracts, went to Wallis and so appeared in print in Wallis's Latin *Algebra* of 1693). As Gregory's theorem appeared in the same book, the circle was now closed, and, indirectly, Gregory had helped to bring to light at last the first effective statement of the Newtonian calculus. As for the original *De quadratura curvarum*, it remained forgotten until Whiteside rediscovered it; the version printed by Newton himself with *Opticks* in 1704 – too late now to advance current mathematical knowledge – was a revised and shortened text.[14]

Even under provocation, even when fully aware of others pressing up behind him, Newton could not bring himself to polish and publish his own work in mathematics. Even his two great books, the *Principia* and *Opticks*, were not (in their first published versions at least) truly completed works. The *Principia* lacked in 1687 a vital theme, a dynamical theory of the moon's motions, which Newton could not yet supply. Newton went on rewriting the book for the rest of his life. The *Opticks* breaks off in its third book and turns into a series of *Queries*; there is evidence that Newton had intended to continue in due form but was defeated by the nature of his remaining material and his ideas about it. Endless drafts upon everything he wrote, numberless unfinished sketches prove that Newton, a perfectionist, found it very difficult to take his hand from any paper. In mathematics his foible was strongest – may one guess, perhaps, both because his original powers were here most strongly and intimately at work and because he was most critically conscious of the weaknesses in what they wrought for him – so that in all his long life no major publication in mathematics came from Newton's hands: What he did give out, as late as 1704, were mere tidied extracts of larger books. But two further remarks, one biographical, one psychological, ought to be made concerning Newton's impotence in this respect. First, in 1684, when contemplating his response to Gregory, which was almost Newton's last chance to make his mathematical researches public in a decent, noncontroversial way, he was still unaware of Leibniz's emergence in print, which was (fateful irony!) to happen only two or three months later. Leibniz, at that moment, was chiefly remarkable to Newton as someone who had made by no means trivial criticisms of his own ideas, which ought to be answered. For the rest, Leibniz had kept as quiet through ten years as had Newton himself and had probably given up mathematics – Newton thought he had become a politician or civil servant. As for Gregory, his position vis-à-vis Newton was perfectly clean and innocent, and he seemed to have said the proper things so far as his uncle was concerned. Newton had no wish to stand in his light. I think Newton's modesty at this time, and his anxiety to help Gregory and Leibniz get all the praise to which they were entitled, were perfectly genuine and highly moral. It is true that (as so often when people behave morally) Newton thought of himself as master of the situation. He thought it was in his power to let Gregory (or Leibniz) have just as much fame as he judged really due to him, for he believed that he possessed the means (through the drafts and letters of which he had copies) to cut other mathematicians down to size if any one should claim too much. Although one might have supposed that earlier experiences of the truth, that all is not simple and rectilinear even in the realm of the intellect, would have taught him otherwise, Newton seems still to have believed that very simple and honest actions on his part would always make the republic of learning perceive that what he, Newton, understood to be right and true really was right and true. Again, throughout a long life Newton seems always surprised that the minds of men whom he knew to be able and intelligent, and in every other respect just and moral, could be capable on some point or other of fixing immovably upon ideas that in Newton's eyes were base and false.

And finally, as regards Gregory in 1684, Newton was the less concerned because series were at stake, not the method of fluxions. Of course, Newton was aware that for all the great importance of his work on series – the binomial expansion, his "prediscovery" of Taylor's series and so forth – in this part of mathematics he was in a continuing fertile tradition. In his own early working lifetime, the German-born London mathematician Nicolaus Mercator had published *Logarithmotechnia* (1668), covering a great deal of new ground, and the study of series had also been advanced by James Gregory, the uncle of David, by Christiaan Huygens, and by young Leibniz, not to mention some other English contributions. It was almost inevitable that what Newton had accomplished in this area would soon be repeated by others; as long as he could make his own just priorities understood – and note that these were *unpublished* priorities – Newton was quite content to let others proceed with publication.

But in the spring or summer of 1685, as he was writing the first draft of the *Principia* before placing it in Halley's hands, Newton must have discovered that Leibniz had published a description of a new form of infinitesimal calculus. Whether Newton himself read Leibniz's *Acta Eruditorum* article at this stage with due care and attention, or whether he had some indirect information about it, we cannot tell. Newton always believed (rightly) that the progress he had made in the infinitesimal calculus, the method of fluxions, was far more important than the work he had done on series; the former swelled from a profoundly original idea, the comparison of flowing quantities, of which previous anticipations were remote and slight, and which, when developed in systematic form (and here it must be remembered that notation, though significant, is still a lesser element), proved of tremendous power and potential. Hence the blow he received from Leibniz, though the record of its impact in the scholium seems slight enough, was more severe than the blow he had received two years before from Gregory. Hence Leibniz became in time (he was not yet) an enemy, whereas David Gregory, even though he had to be put right twice as to Newton's priority rights – in 1691 as well as in 1684 – remained always what one may call a "friend" of Newton's. Nevertheless, and perhaps it is not surprising, Newton reacted to the challenge from Leibniz much as he had a year earlier reacted to Gregory, that is to say that even so far as the calculus was concerned he was prepared only to offer rudimentary evidence of the nature of his progress with his own ideas and to follow a laissez-faire policy: "Let Leibniz get on with it if he chooses, for the historical evidence that *I* was first to gather this harvest is irrefutable." Perhaps one may see that the Gregory incident induced Newton into a fundamental error in appreciation of his new situation vis-à-vis Leibniz – he was himself, after all, plunged deeply into quite different work. As a rejoinder to Gregory's claims, publication of Newton's 1676 letters to Leibniz (which Wallis actually undertook), with their copious evidence of Newton's mastery of the use of series, was good evidence of Newton's priority. But in response to Leibniz's assertion of this discovery of a unique and powerful new area of mathematics, the same letters contained no similar abundant evidence of Newton's mastery of developed infinitesimal calculus expressed in a systematically formulated system or algorithm. The "anagrammatic" sentences of the *Second Letter* (of which one, "Given any equation involving flowing quantities, to find the fluxions and vice-versa . . ." was now quoted by Newton in plain language in the scholium) could mean much or little. They could hardly be regarded by any dispassionate observer as testimony to their writer's possession of a great new mathematical system; there was nothing in the letter for Leibniz's eye but "bare enunciations" of the fluxional concept, "which would have told him little."[15]

If Newton in 1676 had gone on – as Whiteside conjectures he had some notion of doing – to explain fluxions to Leibniz in some detail, or even if now in the immediate post-*Principia* years Newton had produced even the work he had already by him, the whole future agony might have been avoided. It appears to me that Newton's failure to perceive the frailty of his *Principia* scholium is the turning point of the whole story. Newton did not understand that the historicist argument, which in 1684 had worked so well in his own interest against David Gregory, was almost totally ineffective when turned against Leibniz; not only because Leibniz had no reason or inclination to be deferential toward Newton as Gregory had, but because, as a sheer matter of fact, the evidence of the 1676 anagrams offered by Newton was extremely thin. Why should Leibniz be expected to take it seriously? Actually, for a number of years Leibniz was perfectly happy to concede Newton possession of some procedure approximating to his own calculus, thus accepting the assertions of the scholium at face value, although neither he nor anyone else knew what the method of fluxions entailed.

It is also possible that just as Newton overestimated the ease with which he could dispose of any threat to his own prestige – while surely greatly underestimating the psychological importance of that prestige to himself – so he may not at this time have perceived the full magnitude of Leibniz's own achievement. It was, he knew and he conceded, different from his own. Newton was very properly eager that he should seem – while making the historicist claims for his own priority set out in the *Specimens* and the scholium – to allow full merit to the independent though posterior researches of Gregory and Leibniz (here again, one might think that Newton imagined the two cases to be more alike than they really were); but he was well aware that he himself had done much more in series than what Gregory had published as new and interesting in 1684, and he must have believed that similarly he had done much more in the methods of differentiation and integration by 1685 than had Leibniz. Indeed, Newton's judgment to this effect may well be historically just, but it is in any case irrelevant, and Newton should not (if this was his view) have relied upon it. Newton knew that in 1676 he had been far ahead of Leibniz, who was in his eyes little more than a beginner in advanced mathematics; was it not likely that even nine years later Leibniz was still far behind him in the rich deployment and complex unfolding of this new calculus of infinitesimals?

If Newton made any such judgments to the effect that Leibniz, like Gregory, was a competitor whose challenges were still easily in Newton's power to control, by his own great resources and historical evidence, he was making a bad mistake. And if he also supposed that no fame won by Gregory or Leibniz could ever disturb his own peace of mind, he made a worse mistake concerning his own nature. For certainly he should have either brought his own unpublished work into the light of day in the late 1680s or remained silent thereafter – and enforced silence on his friends. That is, if he wished to avoid a fight.

4

LEIBNIZ ENCOUNTERS NEWTON:
1672–1676

I F IT TAKES two to make a quarrel, it takes two men of genius to make a famous quarrel. If Newton is one of the half-dozen mightiest figures in the history of science, Gottfried Wilhelm Leibniz enjoys an equal eminence in the history of philosophy. And though in the folk culture of the Germans Leibniz may stand as a lesser man than Goethe, just as in Anglo-Saxon eyes Newton must bow to Shakespeare, by more formal standards each appears as the dominant figure of an aspect of European intellectual life. It is perhaps not accidental that they were contemporaries, Newton's life-span exceeding Leibniz's by a few years at either end, for this was the moment when European intellectual development toward freedom and maturity offered the highest opportunity for creativity. This was the point of flexure on its growth curve. Of course, both Newton and Leibniz were men of transition, thinking for the future with minds conditioned by the past. It might be imagined, perhaps, that Newton the scientist, the man of numbers rather than words, belonged more decisively to the new age than did Leibniz, of whom it has been written that "he is, in relation to the new scientists [of the seventeenth century], a man sunk deep in medieval conceptions, a weaver of metaphysical systems, a believer in the necessary unity of theology, philosophy and science."[1] What the writer of these words meant was that Leibniz was, like Aristotle, Aquinas, and Descartes, a rationalist: He regarded the essential truths of the physical (as well as the moral) universe as being what the human intelligence must see as necessarily true. Thus Leibniz assures us that because God aims at perfection, there can be no gaps in his living creation that could possibly be filled by any extra animal or plant; this is the basic law of natural history. Yet is this Leibnizian law so much more metaphysical, or so much less accessible to proof or disproof, than Newton's celebrated first law of motion:

Every body continues in its state of rest, or of uniform motion in a straight line, unless it is compelled to change that
state by forces impressed upon it.

Who has examined "every body" in the universe? And where, in
this universe full of matter and of the forces associated with matter, is any one body found that is in even an instant of time free
from all force acting upon it?

There is, of course, a genuine contrast between scientist and
philosopher, which, in a broad way, the two individuals Newton
and Leibniz represent. The philosopher seeks to "reconstruct" the
universe (or history) by pure thought; the scientist acquires facts
and makes mathematical models. But it would be too simple to
suppose that Leibniz (professionally both a diplomat and a jurist,
a man who worked on both the mechanical calculator and the
clock and was much interested in the progress of German chemistry) was a mere dreamer, or that Newton (who spent thousands
of hours reading alchemical authors and elucidating the prophecies
of Daniel and John, who believed that to discourse of God did
certainly belong to natural philosophy, and who supposed that
only divine intervention saved the universe from fairly rapid dis-
integration into chaos) introduced some new epoch of hard-
headed, empiricist science in which ideas played no part.[2]

Anyone living since Darwin, Freud, and Marx might, in fact,
suppose that Leibniz and Newton had far more in common than
in dispute between them; that their differences in philosophy and
metaphysics are, after all, minute, sectarian distinctions compared
with the gulf that separates seventeenth-century thinkers from
those of the modern world. They were at one in seeing the problem of God's relation to his creation as one of supreme importance
to men; the notion of an evolutionary universe would have been
as appalling to both of them as the Freudian concept of an unconscious level of the mind filled with repressed emotions. Accordingly, both viewed man as the centerpiece of the universe, the
climax of God's enterprise in creation and redemption, the posses-
sor of a unique intellectual capability of apprehending the universe
like that of the divine mind itself, though infinitely inferior. In all
such ways Leibniz and Newton resembled each other because they
were men of the same age, just as they also resembled each other
in a variety of personal ways, in intense seriousness and wide
range of interest, in their never marrying, in their reluctance to
"dazzle the intellectual world, and perhaps acquire a personal fol-

lowing, by publishing an apparently coherent and all-embracing philosophical system."³ These words written of Leibniz are as true of Newton. But for all that, in the fine-grain structure of the process of change in thought to which they contributed so profoundly, the differences between them cut deep. Leibniz was not Newton's equal as an experimental scientist, and he rebelled at Newton's instrumentalist attitude to gravitation; Newton was not Leibniz's equal as a philosopher, and he rebelled against Leibniz's metaphysics and theoretical mechanics; indeed, when the theoretical mechanics of Leibniz and Newton have been accurately juxtaposed and compared, then it may be that the gulf between them will be best defined.⁴

Although the most recent German student of Leibniz the mathematician has taken the opposite point of view, there is really no evidence at all to indicate that up to the time of writing the *Principia* Newton felt any dislike or distrust of Gottfried Wilhelm Leibniz. He knew very little of him; perhaps the most important thing he knew was that during his longish visit to England in early 1673 Leibniz had favorably impressed Newton's most intimate acquaintances in the Royal Society, Henry Oldenburg and John Collins. Both these men were half a generation older than Newton. Oldenburg, like Leibniz, was of north German birth. He had lived continuously in England since 1653 and had been the Royal Society's highly efficient secretary since the society's formal foundation in 1662. Particularly, Oldenburg conducted the society's extensive correspondence at home and abroad, seeking to encourage foreigners by making them aware of the society's interest in their researches; in fact, he brought most of the distinguished scientists of the Continent into the society as Fellows: such men as Christiaan Huygens, Giovanni Domenico Cassini, Marcello Malpighi, R. F. de Sluse, and Antoni van Leeuwenhoek. But to find men in Germany who really shared the Royal Society's vision of the advancement of science and worked actively in the same direction had been very difficult. Oldenburg spotted such a man in Leibniz, and it was natural that he should try to add Leibniz to his circle of correspondents. In replying to Leibniz's first letter introducing himself to the Royal Society, Oldenburg praised him as revealing "a degree of advancement in the physical sciences remarkable for your years" and expressed interest in the new ideas Leibniz's letter had hinted at. "You will prove yourself to be a true philosopher," he wrote, "if you bring a project of such importance to completion, and you will perform an act most welcome to the Royal Society if you will take the trouble to explain the gist and foundations of that hypothesis [of yours]." As the correspondence developed, Leibniz clearly found it highly rewarding, and it was this that induced him to inspect the scientific milieu in England. If Oldenburg had not died in 1677, the break between Leibniz and England would not have occurred as soon as it did.[5]

Collins was a minor government servant, an accountant, his acuity of vision fallible. His penchant was toward the study of higher algebraic equations rather than toward advanced analysis and geometry. Collins acted as Oldenburg's mathematical adviser but also maintained an active, independent correspondence with mathematicians in England and abroad, including Newton at Cambridge (from 1670 onward) and James Gregory at Edinburgh; indeed Collins was the only man in the world who knew something of what both these highly original mathematicians were doing and indirectly provided a weak link between them. Collins, unlike Oldenburg, was inclined to be chauvinistic in his estimations of mathematical merit. It is to the everlasting credit of these two men of great industry but mediocre talent that they recognized the great powers of Gregory, Leibniz, and Newton; and that as regards the two younger mathematicians they gave them every spur to further achievement and publication.

Leibniz's third most important contact in England was Robert Hooke, ten years older than himself, curator of experiments to the Royal Society, who was unreasonably disagreeable about the mechanical calculator that Leibniz had devised and brought with him to London during this first visit. Hooke made Leibniz unhappy, but as he had also made Newton very unhappy, his critical opinion of the young German could hardly have affected Newton himself.

Leibniz was in London for more than two months, from January to March 1673; as he visited neither Oxford nor Cambridge, unlike most foreigners, he met neither Wallis nor Newton, nor, as it happened, did he come across John Collins, who was in poor health at that time. The chief point of interest to the English was his calculator (which would multiply and divide as well as add and subtract), and his interest in serious mathematics only emerged rather slowly. He had been in correspondence with Oldenburg for the best part of three years on matters of philosophy, without mathematics cropping up as a theme; nor is this at all surprising, because Leibniz had received little training in mathematics at school or university. His occupation up to the age of twenty-five or so had been with linguistics, theology, philosophy, and law – he bade fair to become a polymath, but a literate rather than a numerate polymath. It is true that he had some years before written a piece, "On the Combinatory Art," concerned with a (then) unusual aspect of mathematics, but his most recent interest had been in a "New Physical Hypothesis" of a highly speculative character in which, adopting an improved variant of Cartesian physics, Leibniz tried to find the origin of all effects of nature in a highly active universal ether. The *Hypothesis Physica Nova* had been received in London with only bare politeness, but Leibniz did not know this.

He came to London having just spent nine months in Paris in the service of the Elector-Archbishop of Mainz, and to Paris he returned, remaining there until 1676. Then, in October, he spent a week in London en route from that heavenly city of philosophers to employment as librarian to the Duke of Hanover – the best job he had been able to find. His time thereafter was much committed to that Court, to its diplomacy, and to unraveling the ancestry of his noble employers. Somehow he was also to find time for an enormous correspondence with scholars in all parts of Europe on a great variety of topics, for his own studies in philosophy, and (still) for mathematics, which occupied him greatly in his letters, though Leibniz's mathematical publications continued to be sparse and brief. The Paris sojourn had an enormous effect on Leibniz's intellectual life, and possibly (though not probably, perhaps) he would have done nothing in mathematics without it. For there he met Christiaan Huygens and plunged almost at once into an exciting world of discovery – flashes of ingenious perception came to him so rapidly and brilliantly that he became proudly conscious of powers of insight and innovation of which he had been hitherto unaware.

Yet Leibniz was by no means a modest man. He was really different from Newton in this respect. Newton had intellectual arrogance in that he knew the worth of his own work in both mathematical and experimental science and, accordingly, had little patience with those who offered blundering objections against it, but he had no inclination to parade it before the world, to pick up patronage or (through all the first half of his life) do other than what pleased him in his college rooms and laboratory. After the publication of the *Principia* and his engagement in public affairs he became tired of his professorship, but even then was so unconvinced about the wisdom of his decision to become a principal officer of the mint in London that he wrote to friends denying its truth.[6] Although it is unlikely that the warden (nominally the chief officer) of the Royal Mint in London would have been widely regarded as lower in prestige than the Lucasian professor in Cambridge, one may suppose that Newton's doubts about his change of career sprang from his own internal sense of values; he realized that he was abandoning his free and whole commitment to intellectual life in favor of the mundane world of the king's business where (as he once angrily retorted to Flamsteed) he ought not to be thinking of mathematics.

So far as possible (no man who publishes anything can be utterly without fault in this) he tried to avoid giving concrete authority to ideas that were merely speculative, however tempting and plausible. The metaphysics to which Leibniz (and Newton, too, in a different way) subscribed required that these pictures and explanations should be mechanical, that is to say that effects should be produced by the motions of the particles of which material bodies (like the sun, or a magnet, or the heart of a man) are composed, and that these motions of the component material particles or, indeed, of whole bodies like planets should be produced by some other cause, likewise mechanical, which in the last resort was taken (though never by Newton) to be the ether filling all space and the intervals between particles.[7]

Leibniz was confident that he understood all such philosophic notions extremely competently and that his ideas were suitable for "fanning the true philosophic blaze." He believed that he had so mastered the true theory of movement that he could disclose "the true cause of the cohesion, bending and hardness of bodies, hitherto accounted for by none," and also demonstrate that the laws of motion as hitherto understood were "neither primary nor absolute nor evident, but arise by accident from a certain state of the globe [we inhabit]," which was, in fact, the occurrence everywhere of a restless ether. Thus he could, he goes on, "give the theory of all the motions in bodies that puzzle us," including the phenomenon of light, and "you will be astonished, of the three chemical principles . . . by means of a familiar and almost mechanical way of reasoning." Not only were Leibniz's early intellectual ambitions extremely far-reaching, but he was quite complacent about his prospects of success in attaining them; not merely in science alone, but in such matters as linguistics (he thought he knew how to work out a perfect language and symbolism) and in law also. Leibniz knew he possessed great gifts, and even the reformation of engineering seemed within his compass.[8]

At his first arrival in England the only name of an English mathematician known to Leibniz was that of Thomas Hobbes, whose mathematical arguments (unlike his philosophy) have proved little more interesting to posterity than to his contemporaries. Probably, in fact, Leibniz had little acquaintance as yet with Parisian mathematicians either, and one of the impressions he picked up in London was of the vigorous new work being done in both countries. On his return to Paris he was able to make new learned friends and report their activities to Oldenburg. News of the experimenter Mariotte, of the death of the Jesuit mathematician Pardies, and of the activities of others in summation of series fills out the first letter he wrote to London on 16 April 1673, after regaining Paris, and soon mathematics became the dominant theme in the letters. Already too Leibniz had received from the pen of John Collins (but translated by Oldenburg into the style of a Latin letter) the first of a series of "mathematical reports" that came to him from London – largely but not wholly devoted to British activities. Here Newton's name appeared three times, first as promising a discourse on the formation of equations and their construction, second as the inventor of graphical and mechanical methods of solving equations, and third

> As to solid or curvilinear geometry [Collins wrote] Mr Newton hath invented (before Mercator publish't his Logarithmotechnia) a general method of the same kind for the quadrature of all curvilinear figures, the straightening of curves, the finding of the centres of gravity and solidity of all round solids . . . with infinite series for the roots of affected equations, easily composed out of those for pure powers. Which doctrine, I hope Mr. Newton is a publishing . . . [9]

Possibly this was Leibniz's first acquaintance with Newton's name, but it is far more likely that he would have heard a good deal of the wonderful Cambridge professor while he was in London. Only eighteen months previously Leibniz had printed a tiny four-page pamphlet on optics, which he sent to the Royal Society and about which he wrote Oldenburg a letter. The pamphlet (and the letter) are concerned with the improvement of lenses for optical instruments, so it is really inconceivable that in the winter of 1673 Leibniz would not have heard of (and indeed seen) Newton's remarkable little reflecting telescope, which had created such great excitement and enthusiasm only a year before and was then in the possession of the Royal Society. Indeed, it is quite possible too that, indirectly and without Newton's ever learning his name, Leibniz's pamphlet and letter caused Newton to send his little reflector up to London to show what he had already done – at least, if we imagine that Leibniz had succeeded in starting some fresh discussion of practical optics.

Unfortunately, a break in the surviving correspondence between Leibniz's letter about applied optics of October 1671 (which was certainly received by Oldenburg) and January 1673 conceals from us all the exchanges between the two men during Leibniz's first period in Paris. Surely Oldenburg would have replied to Leibniz's October letter, in which specific questions had been put:

> what is the state of optical science with you; how big does the best of your telescopes make the moon without loss of clarity; what is the extreme size of a louse seen through your best microscope? These inquiries I was ordered to make by his Eminence the Elector of Mainz, who is most deeply interested in optical matters; he is very well aware of your design, as befits so markedly intelligent a Prince. Also, at what distance can the largest printed letters be read, like those of service books? We also wish to learn what was the result of the schemes of Wren and D'Esson, celebrated in the grinding of conic sections.[10]

Whatever influence Leibniz may have had in bringing Newton's optical researches to light, we may be sure that Oldenburg would have responded responsibly to such precise questions from such a source, and equally that Leibniz would have informed Oldenburg of his departure on mission to Paris (February–March 1672), at which time too Oldenburg would have been most anxious to tell Leibniz of Newton's marvelous experimental work in optics. Finally, Leibniz must by letter or message have advised Oldenburg of his impending visit to London before January 1673.

When so much is surely lost at a critical moment we cannot be quite sure that the secretary of the Royal Society and his mathematical friend Collins were totally ignorant of Leibniz's delighted plunge into mathematics under the guidance of Christiaan Huygens during the latter part of the year 1672. He could hardly have had a more eminent or more intelligent mentor. The Huygens family was rich, able, and noble. Christiaan's father was one of the most eminent literary and diplomatic figures in Netherlands life and had been an intimate friend of the former Stadholder; after the violent overthrow of the De Witts, Christiaan's brother Constantijn the younger was equally close to the new Stadholder, who was to become William III of England. Christiaan, though Dutch, was honored under the personal patronage of Louis XIV as the leading member of the Royal Academy of Sciences in Paris. He had made brilliant discoveries in astronomy and solved the problem of Saturn's ring; he had introduced the pendulum clock and thoroughly investigated the dynamics of the pendulum; he had made many other important studies in mechanics, in optics, in pneumatics, and he was above all a very fine pure mathematician. He was master of the latest work of the mathematicians of Italy, France, Britain, and the Netherlands; in 1687 he was one of the half-dozen (or fewer) men in Europe who were fully qualified to probe Newton's arguments in the *Principia*. Moreover, Huygens, a Fellow of the Royal Society of London, knew England well; he deplored the xenophobia of the English but admired the institution of the Royal Society and trusted the men in it.[11]

Early on, Leibniz convinced himself – and, typically, made much of his idea – that it must be possible to obtain the sum of any series of terms (each successive term, of course, being formed by some stated rule), even if the series be infinite in number, provided that the terms decrease toward zero. To test his idea, Huygens asked Leibniz to find the sum of the infinite series of terms

$$\tfrac{1}{1} + \tfrac{1}{3} + \tfrac{1}{6} + \tfrac{1}{10} + \tfrac{1}{15} + \ldots$$

which he knew already to be equal to 2. And at the same time he suggested that Leibniz study series in the recent work of the French mathematician Grégoire de Saint-Vincent.[12] Leibniz, not worrying very much about the reading, not only came back with the correct answer but with a general method for obtaining the sums of series of fractional numbers of this sort, for example

$$1 + \tfrac{1}{3} + \tfrac{1}{7} + \tfrac{1}{14} + \ldots = \tfrac{5}{3}$$

Elated by his successes and by his sense (too optimistic as it turned out) that he had already probed far deeper into the nature of series than earlier mathematicians, Leibniz pressed on with methods of summation. By the end of the year 1673 he had in all probability discovered the equivalence

$$1 - \tfrac{1}{3} + \tfrac{1}{5} - \tfrac{1}{7} + \ldots = \tfrac{\pi}{4}$$

Thus, that long-sought quantity, the square equal in area to a circle, could be expressed as an infinite series (he was surely unaware of Brouncker's continued fraction for the same purpose and other such anticipations).

Neither Leibniz nor Huygens nor any one else, of course, knew how to bring this last infinite series to a sum, or even how to find an algebraic expression for a finite partial number of terms. So the problem of "squaring the circle" had only been rewritten. But equally (at least in the opinion of Huygens, which was probably shared by Leibniz) there was as yet no valid proof that the circle series, or perhaps some other version of it, could not be brought to a finite sum. True, the Scottish mathematician James Gregory claimed that he had proved the impossibility of squaring the circle algebraically, but Huygens and Leibniz thought they had found fallacies in his demonstration. In any case, Leibniz's new series for the circle was far and away the most simple and elegant of its kind yet produced; in a letter to Oldenburg he wrote of "that most wonderful [theorem] by means of which the area of a circle or some given sector of it may be exactly expressed in a certain infinite series of rational numbers." And Huygens praised it (in words Leibniz never forgot) as "very beautiful and very successful. And in my opinion it is no small thing to have discovered in this problem which has exercised so many minds a new approach which seems to give some hope of arriving at its true solution." For a while, at least, Leibniz seemed to have his hand on the veil shrouding one of the greatest mysteries of mathematics, beyond which so many had despaired of seeing. For even Huygens went on to comment (vainly, as we now know) that it "would not appear impossible to obtain the sum of this series and consequently the quadrature of the circle, when you have shown that you have determined the sums of many others that seem to be of the same kind." And, in any case, should this fail, Leibniz would have discovered a very remarkable property of the circle, which would be forever celebrated among geometers.[13] No wonder Leibniz found mathematics delightful and took immense pride in what seemed to him (mistakenly) the original discovery that series were related to curves. As Leibniz himself put it referring to the particular case of the circle: The problem of quadrature, "which up to now has been examined in vain, has now been transferred from geometry to the arithmetic of infinitesimals. Therefore it only remains to perfect the theory dealing with the sums of series or progressions of numbers."[14]

That was still nearly two years in the future when Leibniz discovered the circle series. Meanwhile, he had much to learn. As was made clear to him early in his acquaintance with Huygens, in 1672, Leibniz was still ill read even in those areas of mathematics that interested him; he had no notion of how much had been done in Italy on the handling of series, or in France (by Blaise Pascal, for example) in the analysis of the method of differences, and everywhere in the development of the infinitesimal calculus, to which Leibniz had as yet hardly addressed himself. When he was first in England, his lack of familiarity with the state of the art caused him an uncomfortable moment. This happened at the house of Robert Boyle's sister, Lady Ranelagh, in Pall Mall, where the great natural philosopher resided when in town; in Leibniz's own words

> When I was yesterday at the very illustrious Mr. Boyle's, I met the famous Mr. Pell, a notable mathematician, and the topic of numbers chanced to come up; I remarked, under the stimulus of the conversation, that I possessed a method of forming the terms of any continually increasing or decreasing series whatever from a certain sort of differences that I call generative . . . [To explain: taking the series of numbers given earlier, 1, 3, 7, 14, 25 . . . the *third differences* obtained by subtraction of successive terms form a new series 2, 4, 7, 11 . . . ; repeating the successive subtractions gives the *second differences* 2, 3, 4, 5 . . . of which the *first* or *common difference* is unity.] Mr. Pell answered that this was already in print, reported by Mr. Mouton, Canon of Lyons, as the discovery of the very noble François Regnaud of Lyons. As for myself . . . I was unaware that the book had appeared, for which reason picking it up at Mr. Oldenburg's [who lived almost next door to Lady Ranelagh] I ran through it hastily and found that what Pell had said was perfectly true.[15]

He was deeply embarrassed and so wrote about the nature and development of his idea of differences up to this time in order that no suspicion might gain ground of his trying to claim "for myself credit for the thoughts of others." There is every reason to sympathize with the young man who found himself thus placed. John Pell, a Puritan whose real achievements in mathematics never matched the reputation built up around him by his friends, considerably older than Leibniz and with a career permanently blasted by the restoration of the monarchy in England in 1660, did not have the generosity of spirit that Collins and Oldenburg consistently displayed. And Leibniz was particularly sensitive to any shadow upon his integrity. We may confidently believe that he had never seen or heard of Mouton's book, yet it was typical of him that he strove to exaggerate the advance of his own ideas beyond those of the French priest.

One need not imagine that this trivial incident did Leibniz any harm in England. Yet, curiously enough, a copy of his letter describing it and made by Oldenburg exists among Newton's papers. Did Oldenburg merely send it to Newton as evidence of his young protégé's temper of mind? As both Newton and Oldenburg took much trouble to satisfy Leibniz's desire for mathematical enlightenment in 1675 and 1676, we need place no ill construction upon Newton's retention of his copy of the letter.

Enlightenment was certainly needed. Leibniz had still very much to learn. It is needless to go into detail here, save perhaps to indicate some likenesses and differences between Newton and Leibniz, who was, after all, roughly ten years behind Newton in his reading. Newton was far more influenced in his development by older mathematicians such as Descartes (with all the improvements and additions incorporated by Frans van Schooten in his great edition of Descartes's *Geometry* in 1659) and John Wallis (*The Arithmetic of Infinites*, 1656). Like Leibniz, Newton studied the *True Quadrature* and the *Universal Part of Geometry* by James Gregory, which were written during his stay in Italy (1664–8), from which he returned to take up the chair of mathematics at St. Andrews. The latter book especially was a very important synthesis of the most modern work in infinitesimal analysis; Newton, more mature, was far better qualified to appreciate the subtleties of such a book than, as yet, was Leibniz. During his first visit to England Leibniz also bought and cursorily examined the *Geometrical Lectures* of Isaac Barrow, Newton's predecessor in the Lucasian chair of mathematics, but as he did not read it at all carefully till much later, it had little effect on him, as is probably true of Newton also. Leibniz was more influenced by the ideas of such Continental mathematicians as Sluse, Pascal, and of course Huygens than was

Newton. But what was crucially important for later events was that Leibniz did not know what the two outstanding British mathematicians, James Gregory and Newton, had accomplished since 1668; indeed, it was impossible that he should know this at the time of his first visit to England because nothing of it was in print.

Instead, he had to find out the hard way. As soon as he was back in Paris from the first London visit he began to learn of this British work (and a good deal more, such as the *Algebra* of John Kersey published in 1673–4, which did not prove to be a book of permanent importance) from the mathematical reports and letters sent him by Henry Oldenburg, who used information provided by Collins and later directly by Newton; this phase ended with Newton's last great letter of October 1676. Collins was seemingly incapable of writing a clear, systematic, logically ordered report; certainly he could not take one interesting theme (like the use of series for quadrature) and develop and illustrate it. He hopped about in his sketches of British mathematics from this name to that, from one topic to another. And although he staked a lot of claims for this or that achievement, he rarely made it exactly clear what the achievement was or how it had been effected. Hence, a good deal less was to be learned from Collins's letters than might appear at first, but on the other hand "a word to the wise" is often quite enough, and the slightest hint might give an acute mathematician an entrance to a new investigation of major interest. We know how James Gregory took (and "such was the acuity of his mind," as Newton said, developed) just such a hint from Collins about Newton's work on series. Of course Collins was not trying to instruct Leibniz, only to show what had been done and to defend British prior achievements. Leibniz, focusing on his own private objectives, does not seem to have appreciated how much relating to his own lines of mathematical research might be contained in these documents – especially the two 1676 letters from Newton; he realized after a time that Gregory and Newton had pressed on far with series, but not how much they knew of the processes of differentiation and integration in the infinitesimal calculus (naturally, without the as yet unformulated Leibnizian algorithm).

It was, indeed, at first with astonishment and chagrin that Leibniz learned from London that his discovery of the circle series had been anticipated in Britain. Oldenburg's letter commenting on his first announcement of this result (delayed until July 1674 by the need to work out an adequate presentation of his result for publication) was lost, so Leibniz, after waiting a while, repeated the news in October.[16] Oldenburg made cautious answer in December, advising Leibniz that he was far from first in the field, and that he should not press his hopes too far. The method and procedure for measuring curves had been extended by both Gregory and Newton "to any curve whatever, even the circle itself," and, in relation to Leibniz's claim that the area of the circle could be *exactly* determined by the sum of an infinite series, he told him that Gregory claimed to be able to prove that this could not be done.

> I do not mean this statement to be a check on your talent and enthusiasm [Oldenburg concluded], but to express caution out of my affection for you, so that indeed you may properly reflect upon it and turn it over in your mind before you deliver it to the press.

This advice was surely given in good faith; if Leibniz rushed into print with a half-developed and exaggerated discovery he would make a fool of himself. Oldenburg was in a difficult situation; he did not know at firsthand what Newton and Gregory had already done, but he knew enough (and was obviously pressed by Collins) to feel that he could not let their achievements go unstated. On the other hand he was neither ethically at liberty nor in practice able to communicate them to Leibniz in order to make it clear that he was not bluffing. The issue of national priorities had been made more acute recently by a quarrel between Huygens and John Wallis over the question: Who had first defined a curve whose straight-line equivalent length could be stated (that is, rectified the curve) – was it an Englishman or a Dutchman? Only a few months before, Oldenburg had had to remonstrate with Huygens for cutting himself off from the Royal Society because of this quarrel. Huygens (who was, in spite of these incidents, the mildest of men) had also quarreled with James Gregory over the possibility of squaring the circle algebraically, which Gregory denied. It was unfortunate that Leibniz's chief mentor and patron in mathematics was at odds with English mathematicians – and therefore the more important that Oldenburg should record their discoveries.[17]

This exchange was to provide much later the first item in the list of accusations against Leibniz devised by Newton. Then Newton claimed, rightly, that the circle series had been known to him since 1669, and that Collins had imparted it to other mathematicians soon thereafter. Leibniz, however, had known nothing about such series when he came to London in 1673. It is in fact true that Collins had written generally to Sluse at Liège about Newton's work in series and had passed on Newton's circle series to James Gregory. The implication of Newton's later narrative is clear enough: Leibniz could easily have learned of the circle series through Collins, directly or indirectly; indeed, Newton emphasized the fact that though Leibniz *claimed* the series in the letter to Oldenburg, he had not actually *expressed* it until after Oldenburg had done so first. For, in the event, Leibniz let three months go by without answering Oldenburg's cautionary letter of December 1674, and then wrote much more modestly, asking if he might see something of the English work on series in greater detail. Having developed his arithmetical calculator to an almost perfect state, he had gone on to more work in mechanics, this time an improved clock. His request produced a long account from Oldenburg (prepared as ever by Collins) reporting numerous series already discovered by Gregory and Newton, including the same circle series that Leibniz himself had found, but no hints as to methods (or proofs of the various equivalences stated). Although some of this work on series he could take as an extension of what was in Gregory's published books or in Mercator's *Logarithmotechnia*, a good deal must have been far from obvious to him. Some copying errors in the paper that came to Leibniz's hands did not help his understanding of it, but, in fact, these bare results were a good deal beyond his attainment. In methods of interpolation, in the use of power series, and in expansions, he had still much to learn, but he was learning fast. His rapid progress was indeed a contributory cause of the later tragedy, for Newton would never in later life believe that Leibniz had, by his own unaided efforts and without any improper "borrowing" from himself or others, by the end of 1676 got so far ahead of his earlier rather elementary position.[18]

For the moment, Leibniz wrote back rather cheekily that he had not yet examined "the series which you sent and compared them with mine . . . For it is now a few years since I found mine, by a route which is pretty unusual." Again, Leibniz had laid himself open to later Newtonian accusations, because in truth he was not yet master of Gregory's and Newton's series, nor was his own beginning in the field so very remote. Leibniz, no less than Newton, was consistently guilty of underrating the opposition. And again like Newton he could never *quite* accept the notion that his own marvelous ideas, those tremendous illuminations, had been shared separately by others. To the end of his life, Leibniz could not reconcile himself to his loss of priority in the study of series. At the height of the dispute over the first discovery of calculus, after the publication of Newton's *Commercium Epistolicum* (*Correspondence*) in 1712, he wrote to his friend Johann Bernoulli:

> I now hear for the first time that my discovery of the magnitude of the circle is to be attributed to [James] Gregory also . . . Newton himself praised my discovery at the time when it was imparted to Oldenburg, and admitted that my own way of discovering it was an original one. Therefore he did not then know of Gregory's discovery.

And shortly afterward, in drafting the notorious *Charta Volans* (Flysheet), Leibniz attempted to rebut the false and unjust charge that he himself had stolen the circle series from the British mathematicians with the allegation that the British themselves – Wallis, Hooke, Newton, and the younger (David) Gregory, James's nephew – had remained ignorant for thirty-six years of what James had achieved and had acknowledged Leibniz as the sole discoverer of that series. Indeed, Leibniz went on, no doubt relying upon memory's blandishments, Newton himself had "admitted in a letter that (so far as he knew) this method of series was not yet employed by others." Alas again for the elderly philosopher, Newton had made no such admission in the letters of 1676, and on the contrary had taken some trouble to display before Leibniz at that time the rich accomplishments of the British.[19]

To go back to 1675, what must be emphasized is that the story has now almost reached the point where further contacts between Leibniz and the British mathematicians could no longer affect the development of his own mathematical ideas. Further letters were exchanged in the early summer of 1675, but these are rather concerned with algebra than with series or calculus. When Leibniz asked in July whether the British mathematicians could determine the length of arc of the ellipse or the hyperbola, Oldenburg waited three months before sending a reply to the letter, when he admitted that this could only be done approximately, to "err by less than any assigned quantity." And he went on to say that another German who had spent some time in England but had moved on to Paris could inform him of Gregory's method of rectifying any arc of a circle. The traveler here mentioned was Ehrenfried Walter von Tschirnhaus, whose interest in mathematics was keen but idiosyncratic; for a short time, before Tschirnhaus continued his travels into Italy, he and Leibniz were close acquaintances in Paris. Tschirnhaus remained always a mathematical amateur, rather notorious for his bad behavior than famous for any solid achievement, and Newton is hardly to be blamed for failing to guess that Tschirnhaus and Leibniz were men of such very different weight and temper. One important piece of work of a different sort should, however, be recorded to Tschirnhaus's credit: He had a large share in the European rediscovery of the secret of the true Chinese porcelain, and so with the start of its manufacture at Meissen.

It was at one time supposed – probably Newton supposed – that Tschirnhaus, after a stay of several weeks in England, had conveyed a very useful picture of British mathematics to Leibniz. The supposition was mistaken. Leibniz's own notes indicated that he had only casual conversations with Tschirnhaus early in October and discussed no mathematical topic with him in any detail before the end of November 1675, when his own ideas on a new infinitesimal calculus had crystallized. Moreover, in the opinion of the best authority, J. E. Hofmann, Tschirnhaus was too opinionated and too locked up within himself to have formed a useful impression in England; thus he would "never have been capable of transmitting accurate reports of English mathematical methods to Leibniz; not even if he had in fact been told anything in detail." True, Leibniz's conversations with Tschirnhaus, like their joint study of the unpublished papers of Blaise Pascal, further advanced his mathematical education; he began to see something of the use of "imaginary" numbers – like $\sqrt{(-1)}$ – in algebra, for example. All this contributed to the synthesis forming in Leibniz's mind. But again he borrowed nothing from Tschirnhaus.[20]

It was a gloomy summer and autumn for Leibniz, who saw his stay amid the intellectual delights of Paris drawing to a close. His appointment in the diplomatic service of the Archbishop-Elector of Mainz had virtually run out and for a time he lived on such commissions as he could pick up while postponing his acceptance of a firm offer of employment from the Court of Hanover. His friendly relations with Huygens seemed to cool. In any case, Huygens, as a Netherlander uncomfortably placed in the French Academy of Sciences in the middle of a French invasion of the Low

Countries and doing what he could to mitigate the atrocities of war inflicted by French troops on his own country, could do little to help Leibniz, who would have liked to join him as a salaried foreign member of the academy. As there were already three influential foreigners so employed, his chances of being appointed were never good. The final word rested with Louis XIV's great minister Colbert, to whom Leibniz desperately tried to make an approach in order to awaken his interest in the calculating machine. He failed. Equally, the chair of mathematics in Paris, which he would have liked, went to an "inside" candidate.

This check, and his consequent departure from Paris and unavoidable immersion in the affairs of Hanover, were the more unfortunate because this was the supreme creative moment in Leibniz's life as a mathematician. The calculus came decisively within his grasp. By early November he was talking of publishing his latest mathematical work in the form of letters addressed to influential persons (to interest their patronage, of course); but, as he said, "to produce something clear and elegant you must have a free and uncluttered mind." He would have been able to get on rapidly had he remained as a scholar in Paris. A month later his next letter to Oldenburg also hinted at great things: Promising to send Oldenburg his mechanical device for constructing algebraic equations geometrically and also his circle series, he went on, "I have recently found a successful approach to another geometrical problem whose solution hitherto has been almost despaired of, about which I shall say more when there is leisure for perfecting it. From which you will recognize, I think, not only that I have solved some problems, but also that I have discovered new methods, and to this I attach a unique importance." And Leibniz makes it plain to Oldenburg that his recent successes are not merely tactical victories; they have strengthened and clarified his conception of the unity of knowledge and the possibility of proceeding to the solution of all the deep problems that confront the human mind by means of a superior science, a kind of metamathematics, which itself (though he does not say so here) was to have a highly flexible and general symbolic character:

> This algebra (of which we deservedly make so much) is only part of that general system. It is an outstanding part, in that we cannot err even if we wish to, and in that truth is as it were delineated for us as though with the aid of a sketching-machine. But I am truly willing to recognise that whatever

algebra furnishes to us of this sort is the fruit of a superior science which I am accustomed to call either Combinatory or Characteristic, a science very different from either of those which might at once spring to one's mind on hearing these words. I hope to explain the marvellous strength and power of this science some time by means of rules and examples, given health and leisure. I cannot here describe its nature in a few words but I am emboldened to say that nothing can easily be imagined which is more effective for the perfection of the human mind, and that after this way of philosophising has been accepted the time will come, and come soon, when we shall have no less certainty about God and the mind than about figures and numbers, and when the invention of machines will be no more difficult than the construction of geometrical problems.

In these words Leibniz places himself, and his work on the calculus, in the long and noble line of those who have devotedly though vainly sought for some universal symbolic logic, an algorithm of nature and ideas no less certain and formal than the algorithms of mathematics. This was an ambition Newton did not share with him; the algorithms of mathematics had no transcendent significance for his mind as facets of some universal system. The method of fluxions was a means to an end, a key to handling some difficult problems of mathematics – which Newton was often, however, equally happy to treat in alternative ways. The differential calculus was, to Leibniz, an integral part of the universe of the mind. The fact that, in late 1675 and continuing right through to the time of his departure for Hanover via London in the autumn of 1676, this new mathematical insight continued to yield him tactical successes was not so much significant in itself as indicative (to Leibniz) that he now possessed the key to a far wider realm of ideas.[21]

It was in 1676 that, after renewed exchanges with Oldenburg and Collins, Leibniz at last came across mathematical products of Newton's mind. (Whether or not Leibniz was already familiar with Newton's thought on optics, as printed from time to time in the *Philosophical Transactions*, it is impossible to know; probably Huygens would not have thought it essential for Leibniz to read these pieces.) It was again the English who took the initiative, for Leibniz had been too busy to keep up his correspondence with them. Collins and Oldenburg took advantage of the journey to

Paris of a young Dane, who had spent some time in London, to send Leibniz for the second time Newton's pair of series for determining the arc from the sine and the sine from the arc; they had been first communicated to him a year before, or less, but by May 1676 Leibniz seems to have forgotten all about that – which may indicate that they did not seem significant to him or (more likely) that they were mysterious to him, for in his reply Leibniz asked for the proof of the two series to be sent to him. Presumably he could not see how they could be obtained, even now. The two Englishmen immediately did as Leibniz wished.[22]

If planning his future was a problem for Leibniz, in London the situation was not altogether without difficulties either. Collins was in danger of losing his job (which did happen in the summer of 1676), and Oldenburg was caught in the controversy between Newton and some Continental critics of his theory of light. Moreover, he had been under severe attack from Robert Hooke as one far too partial to foreigners and negligent of the interests of Englishmen. It was a depressing time for English mathematics; Gregory died in October 1675, and Newton (as Collins sadly reported in his very last letter to the Scottish mathematician) now, like his predecessor Isaac Barrow, found mathematics "dry and barren"; he had taken to chemical studies and had failed to write to Collins for nearly a year. In any case, the efforts of these and other lesser English mathematicians seemed likely to be as sterile as Newton feared, as none of their work was likely to be printed: After burning their fingers on Barrow's writings and others of the like sort, the London publishers would not take mathematical books without a dowry; when books that cost five shillings each to publish were remaindered in Little Britain for ninepence, the trade looked around for more lively goods.[23]

Nevertheless, Leibniz could get private information, and did. It is really rather extraordinary that Leibniz's English correspondents should have been portrayed as surly, suspicious, and reluctant to give Leibniz what he wanted. On the contrary, they took much trouble on behalf of this young foreigner; Oldenburg was already praising him in most enthusiastic terms to his other correspondents, and after he met him in the autumn, Collins wrote to another mathematical friend that although illness had prevented his seeing much of the "admirable Mr. Leibniz," yet "I presume I perceive him to have outtopped our mathematics as the moon's brightness dims that of a star." This, in fact, was putting Leibniz (as yet) too high. As soon as Collins learned of Leibniz's renewed interest in May, he had begun assembling from the letters and papers in his own possession an account of Gregory's mathematical discoveries (later called the "Historiola") through which Leibniz was to rummage while in London in October, and into an "Abridgment" of this he brought (with Newton's permission) some allusions to Newton also. It was not Collins's fault that he was, as Hofmann rightly remarked, a pygmy between two giants; he was neither an ignoramus nor a fool; he did his best, not foreseeing (how could he?) the crucial role that the infinitesimal calculus, rather than half a dozen other topics in mathematics, would assume in the future. Collins's lack of judgment and occasional mistakes had only a minuscule effect in creating the gulf between Newton and Leibniz. The same is true of Oldenburg who, meanwhile, tackled Newton himself; not much more could be done to enlighten Leibniz concerning Newton's work without his direct cooperation. Not only, as a result, did Leibniz receive in return a great deal of material prepared for him in London, he also had a copy of a letter specially written by Newton.[24] There were sufficient riches here to last a mathematician half a lifetime.

The documents came to Leibniz's hands on 16 August, having been finally dispatched (again by hand) from London on 26 July, and Leibniz at once, on the very next day, sent off a carefully composed answer. Through one of those stupid historical accidents that so often tend to exaggerate the logical trend of events, this exchange of letters (which clearly caused Newton no offense or alarm at the time) was to be erected by him into a cause of complaint against Leibniz. When Newton's *First Letter* was initially put before the world by John Wallis in the third volume of his own *Mathematical Works*, the date of the dispatch – presumably as given by Newton to Wallis – was mistakenly given as 6 July, and Newton forever after perpetuated this error, making it appear that Leibniz had mused over his *First Letter* for three weeks or more before in turn putting pen to paper. To make matters worse, when he came to reconsider these events years later, Newton also convinced himself that the collection of mathematical memorabilia assembled by Collins from his correspondence (which in fact never left London) had been sent to Leibniz in Paris on 26 July and so had been available for Leibniz to study throughout the summer before his arrival in England for the second time. All this slow poring over other men's work, so untypical of Leibniz, who actually was more inclined to rediscover than to read, was in fact a product of Newton's imagination and an injustice to Leibniz's character and abilities.

The so-called *First Letter* from Newton is formally addressed to Oldenburg; Newton immediately opens it with a compliment to Leibniz:

> Although Mr. Leibniz's modesty in the extracts which you lately sent me from his letter attributes to our countrymen a great deal in connection with some views on infinite series which have begun to be talked about, yet I have no doubt but that he has not only come across a method of reducing any quantities whatever to series of that sort (as he affirms) but also speedy methods, perhaps like my own or better.

However, Newton went on, as Leibniz was eager for information about what the English had done, and as he himself had entered into the business a few years before, he was sending some of the things that had occurred to him in order to satisfy Leibniz's wishes, at least in part. Nothing could be more cordial. It is, in fact, a strange irony that the *First* and *Second Letter* to Leibniz – the latter finished in October – were to constitute Newton's final communication on mathematical topics for several years. By then (1684) Oldenburg and Collins were both dead. The *First Letter*, eleven substantial pages of Latin, is no trivial document, even though Leibniz had already heard in outline of everything contained in it, and though only results (with explanatory examples) were conveyed without demonstrations, and though Newton left a great deal for his reader to explore by himself.

Leibniz's historian, Joseph Hofmann, has complained of obstruction on Newton's part: "Everything was done," he writes, "to prevent Leibniz from, as it were, improperly penetrating the world of Newton's thought" and in particular "nothing was said of the central problems – nothing of [Newton's] method of fluxions or the differential equations into whose solution by power series Newton already possessed considerable insight." One may wonder, in the first place, whether Newton *could* have put so much mathematics into the compass of a single letter, and whether he would have been likely to write a complete treatise (in a few weeks) for the benefit of a total stranger. In the second place, Newton understood Leibniz to be interested in the sort of series about which Collins had informed him slightly before; it was not in the request that he should also give Leibniz an account of his own advances in the infinitesimal calculus (of which Oldenburg indeed, who had written to Newton on Leibniz's behalf, knew nothing). And in the third place, though Newton could (as Hofmann indicates) have given Leibniz a fairly elaborate account of the process of differentiation, he perhaps could not have given Leibniz such a full view of integration by power series as Hofmann supposes; the treatise "On the Quadrature of Curves" was only to be written by Newton much later. The implication that Newton was ungenerous in his response to Leibniz seems therefore groundless; and it is even worth remembering that Leibniz, who had made quite large claims for himself, had as yet sent far less than this to England in the way of real mathematics.[25]

At any rate, Leibniz seems to have been well pleased with what he received – which indeed included what (after Pythagoras's theorem) many nonmathematicians will recognize as a landmark: the binomial theorem. This was one of the points on which, in writing back and describing some of his own work, Leibniz desired further explanation. Just as Newton's *First Letter* seems genuinely open and honest, so too was this reply from Leibniz; he did some fairly obvious masking of his methods, but Newton's later charges against Leibniz's sincerity of conduct in thus maintaining the correspondence with himself were quite unjustified, the miasmic products of the evil atmosphere then existing. But it is perhaps illuminating to note that the discussion of series – the topic long before of letters between Leibniz and Oldenburg, now enriched by Newton – went back to an earlier, pre-calculus stage of Leibniz's mathematical development. He had by now begun to be master of the calculus. It was, in a curiously paradoxical way, fatal that Leibniz should – could – see no sign of anything similar in Newton's *First Letter* or even in his *Second*. Newton's July letter spoke to his earlier, pre-calculus self. He was indeed interested in what Newton told him, but it in no way seemed to him to relate to his own recent great discovery. As Newton might have noticed, and as he was perhaps originally aware until overcome by jealous rage, Leibniz never *asked* a single question about the infinitesimal calculus, partly because he did not need to (doing so well unaided) and partly because he had it fixed in his mind, all his life, that Newton's great expertise was in the manipulation of series – and nothing else. And it must be said that the natural effect of Newton's *First Letter* – and to an overwhelming extent of his *Second* also – would be to confirm him in that view.

This great *Second Letter*, which Newton completed in October, was a small treatise of nineteen pages. Again, it opens with a generous compliment that Leibniz was fond of quoting to his own credit in later days:

> Leibniz's method of obtaining convergent series is certainly extremely elegant, and would sufficiently display the writer's genius even if he should write nothing else. But there are other things scattered through his letter most worthy of his reputation which also arouse in us the highest hopes of him.

Most commentators on these exchanges have seen this attitude toward Leibniz as permeating the whole of Newton's letter, which is very rich in technical content and merited immediate printing, if Newton would have allowed such a thing. The most recent of them, Dr. Whiteside, writes of Newton's "friendly helpfulness" to Leibniz. On the other hand, the leading student of Leibniz's mathematics has written of Newton's evident impatience with Leibniz, his haste to put a stop to their correspondence, and his belief that Leibniz was an unworthy opponent. He reconstructs Newton's view of Leibniz's last letter to Oldenburg as being: "That there was here no independent invention, nor even a rediscovery, but simply an attempt at plagiarism seemed certain to Newton when Leibniz asked him to explain again more explicitly the decisive points in his letter, namely his methods of series-expansion and series-inversion." It is hard to understand Hofmann's view. Certainly this was how Newton viewed Leibniz's conduct at a much later time, when his view of the German had completely altered, but there is nothing explicit in the *Second Letter* at all to justify the attribution of such an opinion to Newton in 1676. Nor is there anything implicit either. I myself cannot find a word in it to upset the most tender recipient, unless it is Newton's reluctance to describe the content of a treatise he had begun to write, not greatly concerned with infinite series but with other things, such as a most general, natural, and perfect method of drawing tangents, of determining maxima and minima, and other topics "of which I am not now speaking" (that is, to Leibniz). And then Newton concealed even the mention of *fluxions* and *fluents* in the famous "anagram" (frequency table of letters) already mentioned. Was this action merely grudging, mean-spirited? Again, Whiteside comments that "What prevented Newton on this occasion from being more explicit on the subject of his fluxional insights was, almost certainly, lack of self-confidence and the memory of the hail of criticism he had had to endure when he made public his equally novel theory of light a few years before." And, in fact, there really was no occasion (or room) in this letter for Newton to embark on a very different mathematical subject.[26]

Some historians of mathematics have tended to see the priority dispute between Newton and Leibniz as beginning in 1676, though remaining dormant for twenty years more. They assign Newton the role of suspicious, distrustful, watchful opponent of Leibniz's mathematical evolution.[27] This seems to me a mistake. I think that Newton was cordial to Leibniz in 1676 when he wrote the two letters, though finding him a little obtuse and inexperienced in certain areas, and I feel confident that if Newton had discovered in Leibniz signs of ill behavior or a propensity to plagiarism he would have left his opinion on record at the time in no uncertain terms. I cannot believe that Newton would have written his *Second Letter* for the benefit of a man he had judged as a potential criminal, and it would have been indeed extremely foolish of him to do so. As for Newton's silences in these letters, they were reasonable and justifiable. No man is bound to pour out all his developing ideas for the benefit of someone else's friend. After 1676 Newton probably forgot about Leibniz for eight years. And then, when the relationship between Leibniz and England revived again, there was renewed cordiality. We have already seen that Newton was eager to put his own case fairly forward by 1686, but no less intent upon being fair to his old German correspondent. Even in 1686 it is a gross anticipation of events to make Newton an opponent of Leibniz; all the more so ten years before.

Even if it be granted that in the last years of his life, say from 1709 onward, Newton became proud, overbearing, and tortuous, it by no means follows that he had always been so. It is difficult to find more than perhaps the faintest signs of such characteristics in all his early correspondence, down to the publication of the *Principia*. The evidence that he did *not* make great public claims for himself, did *not* rush to put each great idea on record, did *not* seek to make a noise in the world, is overwhelming. He more than once – as he does in his *Second Letter* for Leibniz – speaks of his willingness to keep his work to himself when he sees that others were making known things that he might have published first. No doubt – as I have said – this was an unwise policy, one that could bring troubles about his head; but it was genuinely his, a product of his introverted desire not to have his peace of mind disturbed. There is no reason why this modest, reticent scholar, who had patiently imparted his work to Collins and Oldenburg (and, through the former, indirectly to James Gregory, whom Newton must have perceived as a dangerous rival to himself if his mind had been working that way), should suddenly develop violent suspicions of Leibniz's motives in seeking further enlightenment on technical matters of mathematics. And why, if he had such suspicions, should he nevertheless prepare for Leibniz, drawing once more upon his own storehouse of materials, another nineteen pages of mathematical riches? The idea that Newton had "rumbled" Leibniz at this time does not make sense.

5

THE EMERGENCE OF THE CALCULUS: 1677–1699

EWTON'S *Second Letter* was retained in London for lack of a safe means of conveyance until May 1677 and only came to Leibniz's hands in late June, eight months after it was written and about as long after Leibniz had settled in Hanover, in the new world where he had to make his career. It was, he acknowledged to Oldenburg, a "truly excellent letter . . . I am enormously pleased that he has described the path by which he arrived at some of his very elegant theorems." And he reiterated his praise of Newton's results throughout the letter – no evidence here of a sense that anything had been begrudged him. Before turning to series again, Leibniz took up Newton's brief allusion to the method of tangents, leading on to the anagram (which Leibniz does not mention), and in doing so expressed "publicly" for the first time (unless in informal private communications) his calculus notation; dx, he said, is the difference between two closely related values of any changing quantity x, and dy the corresponding change produced in a second variable y, which is related to x by some mathematical expression. Then if dx is constant, dy will define the slope of the tangent at x. Leibniz showed in rather detailed steps how this could be worked out when applied to a curve whose shape was defined by some algebraic expression, arguing rather naïvely – as Newton also sometimes did – that though a product such as $x \cdot dx$ was not negligible in quantity, the product of two infinitesimals – that is, $(dx)^2$ or $(dy)^2$ – was too tiny to count and could be ignored: Thus he could demonstrate that $d(x^2) = 2x \cdot dx$. Further, as he showed in his examples of this new method and notation, they were "useful also at a time when irrational quantities intervene, inasmuch as these do not in any way impede it . . . " There was, in fact, a good deal more basic calculus in this letter of June 1677 than Newton was to make public in his *Principia* lemma or anywhere else before 1704. It is indeed very impressive, even though Leibniz's exposition was quite elementary and naïve in its basic idea of an infinitesimal. From its calculus, many of the special results in the geometrical analysis of curves obtained during the last generation followed easily, and Newton did not (like Huygens later) reject it as ungeometric but rather was bound to perceive that Leibniz had hit upon a general conception analogous to his own method of fluxions.[1]

In short, the significance of Leibniz's mathematics must have been crystal clear to Newton, even though the letter contained some mistakes and incomprehensions that Newton (at any rate, later) judged inexcusable. For he was never able to understand the patchiness of Leibniz's performance as a mathematician, which puzzled Huygens also. Meticulous, thorough, methodical himself, Newton could not sympathize with Leibniz's impatient and pragmatic way of tackling problems or perceive how Leibniz's mind could be so penetrating in some contexts, so seemingly obtuse in others.

Because we have no notes by Newton on Leibniz's first "public" essay on the calculus, we cannot tell how he reacted. In the *Principia* scholium to the fluxions lemma (Chapter 3) – his first recorded comment – Newton recognized Leibniz impartially as making an independent, though subsequent, discovery. Even when his anger against Leibniz was fully aroused, in 1712, he could still allow as much, writing anonymously (of Leibniz's argument that $2x \cdot dx$ is the differential of x^2) and generalizing further:

> That is, if the second term of a binomial is the difference of the first term [e.g. $(x + dx)$], the second term of a power of the binomial $[(x + dx)^3 = x^3 + 3x^2dx + 3x(dx)^2 + (dx^3)]$ will be the difference of the power $[3x^2dx]$. This is the foundation of the differential method already laid down by Leibniz. And Newton had in the year 1669 [in *On Analysis*] laid down this same foundation of his method. By very similar calculations Newton inferred moments and Leibniz differences, and they differ only in the names they gave things.

All this is true, without accusation of plagiarism, and repeats the *Principia* claim. But Newton was also in 1712 to say far harsher things of Leibniz's June letter; for example, in a bitterly ironical passage Newton wrote of the help he had given to Leibniz in the

understanding of reciprocal series consisting of fractional terms:
>when he [Leibniz] received it in this year [1676] he had diffi-
>culty in understanding it, and when he had understood it he
>soon discovered from his old papers that he had formerly dis-
>covered it himself . . . Thus he discovered first (or at least
>independently, by his own efforts) the method which he had
>long desired, had asked for and received and almost failed to
>understand.

Newton now dismissed the idea that Leibniz had entered upon the
study of such series by an approach different from his own. He
made other comments such as "Even the calculus expounded in
these examples differs from Newton's calculus only in the for-
mulae of notation, but is made more obscure by a less apt notation" to indicate the derivative nature of Leibniz's achievement.
For by 1712 Newton was convinced that the Leibnizian calculus
was essentially not new, not an independent discovery but an imitation of his own method, and he imagined (falsely) a number of
stages through which, successively, Leibniz had followed in his
own footsteps. First, he had learned generally of Newton's work
from the letters sent him by Collins and Oldenburg; second, he
had learned more from Tschirnhaus, who was (in this scenario)
supposed to have received a complete picture of British mathematics while in London; then followed third, the benefit of the
First Letter; and fourth, Leibniz had had access to all the materials
in Collins's possession during the summer and while making his
October visit of 1676. Finally, he had carefully picked the gold
from Newton's *Second Letter*, and having studied all this, he
was – on the basis of the published work of Gregory and Barrow – able to "reconstruct" Newton's method of fluxions in the
form of the differential calculus.[2]

However plausible this might seem to one wholly ignorant of
the other side of the story – as any criminal accusation presented
by a clever counsel may appear to be – we now judge from the
documents (and Newton himself helped to make these available)
that the intuition from them of the calculus would have been virtually an independent discovery anyway. Further, because we now
know that the calculus was born in October 1675 – precisely nine
years after the method of fluxions – we also perceive that some of
the documentation that Leibniz had from England came too late
to affect the genesis of his ideas. There remains, however, the possibility that Leibniz could have gained important aid from New-

ton's work during his second visit to London or through Newton's letters in a more subtle way. For the discovery of the calculus
was not a single act, like turning a key in a lock. When the first
principle of differentiation (or obtaining the fluxion of a quantity)
has been mastered, one has to find out how to differentiate pro-
gressively more complex algebraic expressions. Then the mathematician has to discover what to do with derivatives, which are
not much use by themselves without further operations. Then
there is the problem of second differentials (a change in acceleration is the second differential of a velocity). And most difficult of
all, there is the frequent necessity of reversing the business, that
is, integrating; it is very difficult to devise systematic procedures
for obtaining integrals, which indeed are commonly tabulated
from previous differentiation results. Now it is perfectly conceiv-
able that Leibniz could have taken the first independent steps to
differentiation, and then, on seeing Newton's work and having
appreciated its value, gone on to "borrow" the development of
calculus in his own notation. This would have been a modified
theft, involving less than the whole cloth. And the most likely
source would have been in the Newtonian treatises that Leibniz
examined in October.

Unwittingly, Johann Bernoulli, who was Leibniz's most eminent pupil in the calculus, demonstrated how this might have
been. Bernoulli claimed, and he said that Leibniz had conceded
him as much, that the integral calculus as it was being developed
in the early eighteenth century was virtually his own independent
creation. Of course, he did not deny that Leibniz had properly
understood integration as the inverse of differentiation and had
given some examples of how it was to be done, but he maintained
that the development of this basic notion into a large algebraic
apparatus that made the calculus workable was his own. Could
not Leibniz, already blessed with his own insight, have in a similar
way taken over a developed apparatus from Newton?

There are difficulties in such a suggestion. The Newtonian
writings were not so elaborately developed as all that – and this
was particularly true (at that time) of the study of integrals. But
the best evidence that nothing like this happened is provided by
Leibniz's own notes of his Newtonian reading, recently published
by Hofmann. This private record must, in Whiteside's words, finally "clear Leibniz of any lingering suspicions still felt by any
ardent Newtonian supporter that he made good use of this chance

to annex for his own purposes the fluxional method briefly exposed there," that is, in the Newtonian transcripts that were then in Collins's hands. Leibniz did indeed take quite long notes (thirteen printed pages) from Newton's *On Analysis by means of equations having an infinite number of terms*, written in 1669, but his notes deal exclusively with the formation of series, their use in quadrature and in the extraction of roots. He made a brief annotation also on Newton's letter to Collins of 20 August 1672. In all this, as in Leibniz's notes on the *Historiola* (which in any case contains very little mention of Newton's work), there are no signs of fluxions or of infinitesimals of any kind; Leibniz passed without remark over Newton's brief and obscure allusions to what is tantamount to differentiation at the opening of *On Analysis*, because there was in them nothing new for him.[3]

Two obvious questions about Newton's early opinion of Leibniz's calculus – whose power he could not but acknowledge – cannot be positively answered. Did he at any time accept Leibniz as an honest independent discoverer? If so, when did he change his mind? My own answers – clearly subjective – are that in all probability he did not at first judge Leibniz any more indebted to his own discoveries than James Gregory had been; and like Gregory, it is worth remembering, Leibniz had not yet made any open claim to an exclusive discovery of the calculus. I think he remained unchanged in his opinion until after the *Principia* was published, that is until the 1690s. Then, as we shall see, things altered.

The lapse of correspondence between Leibniz and England might seem to indicate a collapse of confidence, but it was largely accidental. Writing to Leibniz in July 1677 Oldenburg still did not know whether Newton's "treasure" of a letter had reached Hanover or not, but he sent a budget of scientific news. On the ninth of August the secretary of the Royal Society addressed Leibniz for the last time to tell him that his two letters announcing the arrival of Newton's *Second Letter* in Hanover had been safely received in London; he advised Leibniz not to expect a speedy reply from either Newton or Collins, as both were out of town and much involved in other business. He sweetened this with an account of recent observations of the great red spot in Jupiter made by the astronomer Giovanni Domenico Cassini at Paris. It is a very friendly letter, also informing Leibniz that Oldenburg had sent him a parcel of books from London by a traveler. Oldenburg could have had no reaction from Newton by this date, for Collins seems to have sent Newton a copy of what Leibniz had written to Oldenburg only on the last day of August. By then Oldenburg had gone to Kent for his summer holiday, and there both he and his wife contracted a fever and died within a day or so of each other. There was now no one left in England to correspond with Leibniz. Even Collins, who lived a few years longer, did not attempt the task, not possessing the contacts through whom Oldenburg had sent his letters abroad in days when most of the Continent was beyond the reach of the public postal services. It is not surprising that Newton made no further response to Leibniz.[4]

Lacking any contemporary word from Newton (who vanishes from sight, practically, from the summer of 1677 to the spring of the following year), it is worth quoting in full John Collins's appreciation of the situation, as related to the Oxford mathematician John Wallis:

> Concerning Leibniz, you say he fell into the method of Mr Newtons infinite series which the said Mr Newton himself grants, by a new method of transformation of curves by which the extraction of roots of simple powers in species is avoided; whether this was not learnt or may not be derived from Dr Barrows Geometric Lectures is the question. Also the said Leibniz avoids the extraction of the roots of adfected equations in species by an improved method of tangents.

Collins was commenting upon a large narrative of British mathematics drafted by Wallis, which was eventually to figure in Wallis's *Algebra* (1685); particularly, Collins corrected and added to what Wallis had written of the recent advances made by Gregory and Newton, with whom, of course, he had been intimately acquainted during the last few years as Wallis had not. Now Wallis was the last of Englishmen to play down the importance of his countrymen's work, and it was certainly not Collins's intention in his comments to do so either. He would not have spoken of Leibniz as "a most learned ingenious man" if he had thought there was anything shady about his conduct, nor would he have gone on to suggest that it "would not be unpleasing either to [Leibniz] himself or to the Royal Society" to explain Leibniz's system of differentials "more largely than himself could have leisure to do" and that the Royal Society could quite properly publish as Leibniz's work the content of Leibniz's private letters to Oldenburg, its secretary. Clearly Collins felt (wrongly, as a matter of fact, but his mistake was a natural one) that Leibniz belonged to the "English

School" of mathematics because he had begun from the procedures of Isaac Barrow. Virtually all Newton's extant mathematical
work in the calculus had passed through Collins's hands, and he
knew better than any man alive what Leibniz knew of it. Yet he
hinted at no indebtedness to *Newton* on Leibniz's part.

James Gregory, however Collins went on to explain to Wallis,
had been a bit peevish with him because he had failed to publish
Gregory's solution of Kepler's Problem in the *Philosophical Transactions,* "which I would not doe as knowing Mr. Newton's series
were made use of therein and yet he [Gregory] had a good right
so to do for he really advanced the doctrine [of series]." If Collins
was obtuse so far as any possible grievance that Newton might
have against Leibniz was concerned, he was obviously sensitive in
general to the question of property in mathematical work.[5]

In fact, as nothing emerged at present from Wallis's efforts to
create a perspective on English mathematics (including his own
innovations), Wallis did not become seriously involved in the relations between Newton and Leibniz for several years more.

Nothing further is to be said about Newton and mathematics
for several years, and Leibniz too did little more. In 1678 he published in the Parisian *Journal des Sçavans* the quadrature of a particular area of a cycloid, which he had found when working with
Huygens three or four years earlier; but as long a time again
passed before he printed in the *Acta Eruditorum,* the new journal
that German scholars had begun to publish from Leipzig, his
study of the quadrature of the circle by means of infinite series,
which also belonged to his work with Huygens. In this paper –
the nub of which is the series

$$\tfrac{\pi}{4} = 1 - \tfrac{1}{3} + \tfrac{1}{5} - \tfrac{1}{7} + \dots$$

(Chapter 4) – although Leibniz mentions a number of earlier
mathematicians who had attempted to evaluate π (or else improperly "square the circle"), he refers nowhere to the various series of
Gregory and Newton also serving to evaluate π, which he ought
to have believed were prior to his own. Certainly Leibniz was
entitled to publish what were indeed his independent discoveries,
whose significance he assessed in his own way, but it was discour-
teous to make no mention of what he knew others had done in
the same direction; indeed, he could well have mentioned (as he
did not) those like James Gregory and Nicolaus Mercator, who
had dealt with such kinds of series *in print* before Leibniz touched

on them at all. Finally, in October 1684, he sent to the *Acta Eruditorum* his first published exposition of the calculus. Here there is no historical introduction; without a word of the development of methods of tangents and of determining the quadrature of curves through the mid-seventeenth century, without naming any other mathematician, Leibniz plunges directly into his own method.

This first calculus paper was restricted to differentiation and its geometrical uses; Leibniz waited another two years before going on to integration. Six pages long, it is pathologically compressed and, as Boyer remarks, must have repelled most readers. Some found it an enigma rather than an explanation of what Leibniz had in mind. The second paper (1686) is rather longer.[6] It opens with a curiously boastful apology:

> As I gather that several things which I have published in these [Leipzig] *Transactions* relating to the advancement of geometry have won no slight approval from certain learned men and are gradually indeed being introduced into general use, but that either through mistakes made in writing or for some other cause certain points [in my papers] were not clearly enough grasped, I have on that account thought it worth while to add here some matters that can serve to illuminate the former papers.

To whom could Leibniz have been referring? Who, before 1686, had made his difficulties with the 1684 calculus paper – for that, of course, was the one at issue – known to Leibniz? His next sentence answers the question and explains his anxiety to put on record his own investigations of integration and quadrature, when he refers to the one mathematical publication in Europe that had already given immediate and favorable recognition to Leibniz's differential calculus. But, gratifying as this must have been to its inventor, the effect was marred by extensions of the calculus that Leibniz, naturally, wished to retain for himself.

The new author was a Scot, John Craige, recently (in all probability) a pupil of David Gregory at Edinburgh but living in Cambridge and in acquaintance with Newton when he wrote and published *Methodus figurarum . . . quadraturas determinandi* (The method of determining the quadratures of figures) in 1685. Craige is not well known. Presumably he was still a young man at that time – an older one, more set in his ways, might not have reacted so positively to Leibniz's ideas. He was to live until 1731. Very possibly he first heard of the mathematical papers coming out in the

Acta Eruditorum from David Gregory, who certainly had some familiarity with recent mathematical work in Germany. Unlike Gregory, Craige saw great possibilities in Leibniz's new algorithm and adopted it. Several years later, when he knew more of Newton's discoveries and methods, he would still write in his second little mathematical book of 1693:

> In order not to seem to assign too much to myself or to detract from others, I freely acknowledge that the differential calculus of Leibniz has given me so much assistance in discovering these things that without it I could hardly have pursued the subject with the facility I desired; how greatly the very celebrated discoverer of it has advanced the solid and sublime art of geometry by this one most noble discovery cannot be unknown to the most skilled geometers of this age, and this treatise now following will sufficiently indicate how remarkable its usefulness has been in discovering the quadratures of figures.

To go back to 1684–5, Craige was interested in the quadrature of figures, confident that in many cases it was only because of the impotence of existing methods that areas defined by recalcitrant curves were said to be not quadrable. He acknowledged the extreme kindness of the distinguished Mr. Newton in permitting him to examine his manuscripts (this is almost the first reappearance of Newton in mathematical history since 1676) and was skeptical of the value of what Tschirnhaus had recently printed on the subject in the *Acta Eruditorum*. He felt far otherwise, however, about what Leibniz had claimed in the same journal, about the treatment of irrational terms in a mathematical expression:

> For to this difficulty the outstanding geometer G. W. Leibniz has furnished the best of remedies. For that very famous person shows a neat way of finding tangents even though irrational terms are as deeply involved as possible in the equation expressing the nature of the curve, without removing the irrationals. How that method is to be applied to our present purpose I will show by one example . . . [7]

Clearly Craige had gathered – from the inversion implicit in the two operations of integration and differentiation, or quadratures and tangents – that if the tangent to a given curve can be defined by any method, including Leibniz's for the curves expressed by more complex equations, then the mathematician might aim "to invert this problem generally, that is, having the Tangent to find the Curve whose Tangent it is," which process "would give us a General Method of determining the Quadrature of any Curvilinear space." Of course Craige understood that in obtaining more complex integrals mere inversion of differentiation would not suffice; he understood the method of approximate quadrature by infinite series, and when tackling the hyperbola obtained the series published long before by Mercator, and for the circle sector he found the series discovered by Newton and first printed by David Gregory.

No wonder Leibniz was eager to increase his stake in the game. But to a modern reader, the amazing fact is the very slight allusion to Newton's work, even on the part of a mathematician who had personally consulted him at Cambridge. It is true that in an "Addition" to the *Philosophical Transactions* review of his first book (1686), Craige described an additional method of finding a curvilinear series by means of an arbitrary equation, and improves it by a "little dodge" (*observatiuncula*), which Newton taught him, but there is no sign of a major indebtedness, even in his second book of 1693 after Craige had read the *Principia* with enthusiasm. As for the reviewer of the *Method of Quadratures* in 1686, he simply (following Craige himself) lists those who had advanced the method of tangents (differentiation) as being Descartes, Fermat, Sluse, Barrow, Wallis, Tschirnhaus, and Leibniz. No mention of Newton (who had indeed, as yet, published nothing on mathematics). The omission of Newton's work as analogous to that of Leibniz and antecedent to it is clearly the product of Newton's silence and the writers' inevitable unawareness. Many years later (1718) Craige issued yet a third tract, "On the Calculus of Fluents," pursuing his old line but now, as the title shows, converted to Newton's method of fluxions. In this tract he wrote that in Cambridge, when he was working on the *Method of Quadratures* in 1685, he had asked Newton to read through the tract; Newton, such was his kindness, had done so and had even furnished Craige with some ammunition to use on behalf of Leibniz against Tschirnhaus, as well as other insights into integration by means of converging series. (It was in this way, through Craige, that David Gregory learned of Newton's study of series, as mentioned in Chapter 3). And yet, there is absolutely no indication that Newton dropped any hint to Craige of his own work on differentiation and all that went with it and still less on integration as the inverse of a method of differentiation.[8]

Thus by silent implication, Craige's evidence also testifies to Newton's having no feelings of criticism toward Leibniz in 1685 or any desire to proclaim to the world that *he* had been the first to discover a general process of differentiation and exploit its applications. He left Craige's warm feelings for Leibniz undisturbed. Craige's knowledge of Newton also led him to form – however mistakenly – the same impression of Newton as a mathematician that Leibniz had formed ten years before: He believed Newton to be a great master of the use of infinite converging series for the quadrature of curves – such as, classically, the circle and the hyperbola; he was as yet, and was to be into the 1690s, totally ignorant of Newton as a master of the process of differentiating functions, as were also at this time, still, David Gregory and almost all other British mathematicians. They saw no reason not to award Leibniz's discoveries the highest praise. Nor did Leibniz, for all that he could have known of Newton.

Although Craige was first in the field, and Leibniz himself was to praise the discoveries he had made by means of the calculus, he was not to play a major role in the diffusion of the new calculus. The real development of Leibniz's initiative occurred on the Continent and centered upon the activities of two brother mathematicians, Jakob and Johann Bernoulli. The family was one of prosperous merchants and bankers in Basel, Switzerland; the elder brother Jakob (b. 1654), having studied science and mathematics against his father's wishes, spent a long period abroad during which he made a visit to London and then settled as professor of mathematics in the city of his birth. Johann, his father's tenth child, was thirteen years younger than Jakob (under whose direction he studied) and more ambitious. Perhaps the two facts help to explain their subsequent fraternal quarrels. It is sometimes implicitly assumed that the Newton-Leibniz dispute was unique of its kind; it was only so in its duration and the majesty of its leading participants. Throughout this period, intellectual life was punctuated by bitter wrangles and struggles for supremacy, in which nationalist sentiments also played no small part. Leibniz was involved in other debates besides that with Newton, a relatively friendly though serious debate with his former friend Tschirnhaus over mathematical issues going back to their years in Paris together and another more rancorous dispute with the Dutch philosopher and mathematician Nieuwentijt. The Royal Society itself split into factions for a time under Newton's presidency. The quarrel of the Bernoulli brothers was produced by the trouble-provoking practice of the age whereby a complacent mathematician might challenge his fellows to produce the solution of some problem which he had himself (supposedly) mastered; this had once been done to great effect (and some damage to friendships) by the great Pascal and was now revived again in the excitement and personal rivalry accompanying the early development of calculus.

Jakob, who worked on physics, logic, and probability theory during the 1680s as well as algebra, came to the infinitesimal calculus through his study of the writings of the English mathematicians Wallis and Barrow. He did not reach Leibniz until 1687, two years after Craige. Apparently from the first the brothers Bernoulli worked at the calculus together, for despite the terseness and obscurity of Leibniz's 1684 paper, Johann reported, "It was for us only a matter of a few days to unravel all its secrets" in his *Autobiography* long after. It was Johann who soon coined the modern term "integration" to describe the process inverse to differentiation, which had previously been generally known as "quadrature." Jakob first proved his mastery of the new calculus by his solution (1690) of the problem posed as a challenge by Leibniz three years earlier: to assign the curve along which a body would descend (without friction) at the earth's surface at a constant speed, that is, without, as normally, accelerating in its fall. Huygens had produced an immediate solution to this problem by conventional means, and Leibniz published his own solution in 1689. Shortly after Jakob's success, in 1691, Johann too placed himself in the top rank of mathematicians in his first independent work, his study of the catenary (the curve formed by a loaded chain, like that of a suspension bridge). Very soon too the brothers began their correspondence with Leibniz, which went on for many years; Jakob, however, died in 1705 before the open outbreak of the dispute between Leibniz and Newton.[9]

Unlike Leibniz and Newton, neither of whom had direct pupils to whom they taught the calculus in an organized way, the Bernoulli brothers were active and capable teachers who rapidly diffused the new calculus and its notation. Their pupils included, besides Germans and Swiss, the French, the Dutch, and even a few young Englishmen. And, of course, they began to extend, develop, and apply the calculus even more rapidly and spaciously than Leibniz himself. The Bernoullis and their pupils, a consistently able group of men with a dozen major names among their leaders, dominated European mathematics for a generation. They were fertile, inventive, and close-knit in an association that was familial in both a literal and a metaphorical sense; the group controlled the university teaching of mathematics over Europe from north Italy to north Holland. Such a group could not have come into existence had not Leibniz originally given to mathematics a new direction and a new power of great significance and had he not himself (in active correspondence with its leading members) continued to inspire and guide its work. But it was in a sense luck that gave Leibniz such a brilliant team of partisans, luck that in the true Napoleonic spirit he knew how to exploit to the best advantage. Newton was less fortunate. None of his adherents was a Bernoulli, though men like Roger Cotes, John Keill, and Brook Taylor were more than ordinarily competent mathematicians. Newton himself did nothing to advance his method of fluxions after 1691 and hardly anything to make it publicly known before 1704. It would be meaningless to argue whether the Newtonian notation was inherently superior or inferior to the Leibnizian or whether the Newtonian concept of a fluxion was more logically correct than the Leibnizian differential at a time when, as a matter of simple historical fact, Leibniz's calculus was firmly established, widely taught, and proved highly successful in paper after paper all over Europe, whereas even in England Newton's fluxions were still barely known. Not until after Newton's death was the systematic teaching of the method of fluxions in England begun, by indifferent mathematicians for the most part.

In France, possibly the most notable early convert to Leibniz's calculus was the Marquis de L'Hospital, an officer retired early from the army, whose mathematical talents had been evident since boyhood. He had figured for several years among the rather few men in Paris who took a serious interest in mathematics and had corresponded with Huygens since 1690. He first directed a highly flattering letter to Leibniz at the end of 1692, indicating (as was to remain the case) that although he had already mastered the differential calculus he was finding constant difficulties with the inverse problem of integration. In a subsequent letter he told Leibniz how, six years before, "when the *Transactions* of Leipzig came into my hands, I found therein your method of tangents, which delighted me so much that I have since that time composed some papers wherein I explained the method at greater length, and provided proofs of all your rules." This letter would put the Marquis's beginning in calculus to the year 1688.

Whether this date was correctly recollected or not, it is certain that the Marquis had not proceeded entirely by his own unaided efforts before making his first personal contact with Leibniz, who was of course delighted to instruct and indeed flatter so notable, not to say, influential a pupil. For Johann Bernoulli had arrived in Paris in the autumn of 1691, where he was welcomed by the group of intellectuals associated with the neo-Cartesian philosopher Father Nicolas Malebranche, on whom he made a great impression by his mathematical prowess in the manner of Leibniz, particularly his treatment of the curvature of curves (actually borrowed from his brother Jakob). In Paris Johann was engaged by L'Hospital, in what has been called "the most extraordinary contract in the history of mathematics," to act as his salaried teacher in the calculus, under a promise to impart his discoveries and writings to no one but L'Hospital, who for greater privacy had Bernoulli live with him for a time at his château at Oucques. Before leaving Paris, however, Bernoulli seems to have given Malebranche and his close associate, Father L. R. L. Byzance, a first introduction to the methods of calculus, which they extended in subsequent years, so that, in the last decade of the century, all of the French mathematicians trained in Leibnizian methods (L'Hospital himself, who died in 1704, Malebranche, Byzance, Charles René Reyneau, Pierre Rémond de Monmort, and Pierre Varignon) were of Malebranche's circle, and all of them were directly or indirectly pupils of Johann Bernoulli. The Swiss had continued his instruction to L'Hospital even after his departure from France, by a kind of correspondence course, thereby enabling the Marquis to compile (in French) the first textbook of differential calculus, his *Analysis by infinitely small quantities* of 1696. This book, which was rendered very successful by its own merits and the considerable interest in its new branch of mathematics, was reprinted for many years. It gave L'Hospital a high reputation, which was partly merited and partly owed to his intimacy with Bernoulli and Leibniz. Only in the present century has it become clear that the *Analysis by infinitely small quantities* was based on a text by Johann Bernoulli, now at Basel, and that even its Preface originated with another ghost, in this case the great stylist and recording angel of French science, Bernard de Fontenelle.

L'Hospital had intended to write a second book on the integral calculus, but this he never did, though again it is certain that he and certain of the Malebranchists received instruction in it from Johann Bernoulli, and in this case his material was published under his own name – but only half a century later, in 1742. It is not perhaps unnatural that Bernoulli, in later life, tended to feel that he had never been given sufficient credit for naturalizing the calculus in France.[10]

The most notable expositor of calculus methods and their applications to physical science in the French Academy of Sciences, though a relatively late convert – for he did not begin his studies until about 1693 or 1694 – was Pierre Varignon, a Malebranchist. Naturally, he used (after 1696) L'Hospital's textbook and a few years later was energetic in defending the Leibnizian calculus from those within the Academy of Sciences who sought to attack its logical foundations. Varignon was an older man who came late in life to academic pursuits (his family were stonemasons), and it is a further remarkable testimony to the power and excitement that Leibniz's work held for contemporaries that, at the age of forty, Varignon set himself to master and teach these new methods and later to embark upon his most considerable achievement, the setting out of the whole of the science of mechanics up to that time (including the work of Newton) in the new mathematical language.

In the last years of his life Varignon became an admirer and friend (by correspondence) of Newton's, whose *Opticks* he helped to popularize in a French edition issued at Paris. He was in these years heavily courted by both sides – both the English and the supporters of Leibniz – as one who could control the allegiance of France to one party or the other.

Considering the slow acceptance of the theory of quanta and relativity in the early years of the present century, not to mention the earlier resistance to Darwin and Freud, the enthusiastic and swift adoption of the differential calculus, and the complementary evolution of the integral calculus in several quarters, appears astonishing. For as critics asserted, its logical foundations were insecure. Geometry had prided itself through the centuries (with illusory complacency, as the nineteenth century was to discover) on its absolute certainty. A properly proved proposition in geometry was universally understood to be the most secure truth – after those of religion – that the human mind could entertain. About algebra, mathematicians felt less certain – among other things, the imaginary root and the imaginary number presented conceptual problems not yet then fully resolved; nevertheless insofar as much algebra could be verified by geometrical reasoning, its sound logical status seemed assured. The essence of mathematical proof, then, was that it should stand – through long chains of reasoning via antecedent theorems – upon simple, self-evident truths about number, space, and their relations.

However, since the early seventeenth century the calculus of infinitesimals had presented logical difficulties. Geometrically speaking, a line is continuous, not an infinite row of points, and a surface cannot be *exactly* represented by a multitude of little rectangles, however great. Yet its techniques could yield accurate results. Leibniz's calculus seemed open to the same objection that in logic – whatever the end product – it was approximate rather than precise; moreover, the very concept of the infinitesimal or differential was vague. To get around the objections, Leibniz often adopted a pragmatic or operationalist stance: The calculus was justified by its fruits. It worked. "There is no need," he wrote, "to make mathematical analysis depend upon metaphysical controversies" – a rather curious statement, perhaps, from one who criticized Newtonian physics for the inadequacy of its metaphysical foundations. In order to avoid problems in discussing the near-infinitely small and the near-infinitely large, it was sufficient, he believed, "to explain the infinite by the incomparable," that is to say, one can always imagine a quantity incomparably smaller (or larger) than any finite small (or large) quantity. Thus one can always argue that the error in dealing with infinitesimals can be made less than any assigned arbitrary quantity. Or as he puts it again, one could regard the differential as an "ideal notion" or "well-founded fiction" (an operator, in modern terms) like $\sqrt{(-1)}$. Leibniz was much impressed by such identities as

$$(1 + \sqrt{(-3)})^{1/2} + (1 - \sqrt{(-3)})^{1/2} = \sqrt{6}$$

where the sum of two incomprehensible quantities is readily intelligible. If the differential was as incomprehensible as $\sqrt{(-1)}$, it could nevertheless be equally handled according to rigorous rules. Accordingly, Leibniz provided no deep philosophical justification for it, preferring to stress the algorithmic nature of the new method.[11] The calculus was a device, not a new philosophy of quantity and number. Newton was far more concerned about the concepts of the infinitesimal differential, or fluxion than was

Leibniz and (as we shall see) judged his own fluxion to be the concept far more securely rooted in logic.

It is true that considered as a *language* the calculus was, for Leibniz, an offshoot or branch of the universal philosophy of linguistics and symbolism that he wished to develop, and in this way the calculus related strongly to the whole body of learning and was not merely a mechanical process. But, unfortunately, Leibniz's explorations on these lines do not, of themselves, serve to clarify the logical difficulties involved in operating mathematically with a quantity that is sometimes finite, sometimes equated to zero.

It was largely because of this lack of rigor that Leibniz's chief mathematical mentor displayed indifference and resistance to the calculus. Christiaan Huygens was by now elderly, in poor health, and separated from the city and the Academy of Sciences to which he had given his middle years. Illness had compelled him to return home to the Netherlands in 1681, and thereafter the anti-Dutch, anti-Protestant policy of national aggrandisement pursued by Louis XIV ensured that he was never invited to return to Paris or even granted an honorable retirement. So Huygens spent his last years at The Hague or in the family's country house just outside the capital. To him, the use of nonrigorous shortcuts in mathematics in order to find an answer so that a formal proof of it might be devised afterward was by no means new. Archimedes' "Method" (unknown at that time) had been an analytical process of this sort; and the infinitesimal calculus had been generally regarded in the same light during the seventeenth century. No one had supposed its procedures to be rigorous. Indeed, in analyzing a problem, anything went − up to a point − because analysis was concerned with investigation, exploration, and discovery. For in the eyes of sound mathematicians of the seventeenth century, analysis was but the prelude to synthesis, to the rigorous proof of whatever result analysis had disclosed. The so-called synthetic geometry of the Ancients provided the highest standard of authenticity; it was the standard claimed (rather curiously, really) by Newton, and the standard demanded by Huygens of his own published work. Analysis, therefore, and most particularly an approximating method of analysis, was, in the judgment of a traditionalist, of relatively less significance and weight than synthesis.

Although there had been a few trivial exchanges between the now elderly Dutchman and the still young German during the fourteen years since Leibniz had left Paris, the latter waited until

July 1690 before writing directly to Huygens to set before him briefly, in a private letter, the principles of his calculus.

> I do not know, Sir [he wrote] if you have seen in the Leipzig *Acta* a method of calculation that I propose in order to subject to analysis that which M. Descartes himself had excepted from it. Instead of those exponents of quantities which have hitherto been used in calculation being restricted to roots and powers, I now employ sums and differences, such as dy, ddy, $dddy$, that is to say the differences and increments or elements of the quantity y, or else the differences of differences, or the differences of differences of differences. And just as roots are the reciprocals of powers, in the same way sums are reciprocals of differences. For example, as $\sqrt{y^2} = y$, and $\sqrt[3]{y^3} = y$, so also $\int dy = y$ and $\int\int ddy = y$. By means of this calculus I presume to draw tangents and to solve problems of maxima and minima, even when the equations are much complicated with roots and fractions . . . and by the same method I make the curves that M. Descartes called mechanical submit to analysis . . .

Huygens's reply began coolly: He had indeed seen Leibniz's papers but had not gone into them deeply. He believed he had developed equivalent methods of his own and had found Leibniz's treatment very obscure. Leibniz's letter, however, had led him to suppose that the new calculus might be worth looking into further, especially if Leibniz would explain its methods clearly; he set Leibniz a problem to test its capabilities, which Leibniz easily solved. Writing to him again, after devoting some weeks to the new calculus, Huygens now reported that be understood it and found it "good and useful," though not going beyond his own methods. These analytical procedures Huygens never formalized into an algorithm, and he would never have dreamed of publishing the analysis leading to a certain result in mathematics as a *proof* of that result. And, therefore, he told Leibniz that one of his new mathematical processes "admitted things which cannot be regarded as self-evident," reminding him that "although such kinds of reasoning might be allowed sometimes in the process of discovery, the mathematician must make use of others afterwards to attain more certain demonstrations." Huygens seems to doubt whether Leibniz was justified in making such large claims as he did for his new calculus and the results he had obtained with its aid, especially as Huygens was at this moment interested in the investigations of another young Swiss mathematician, Nicolas Fatio de Duillier, who was to move on to England and become a friend of Newton's. Although Fatio was by no means of Leibniz's circle, he nevertheless seemed to have equally promising methods, and at one time Huygens endeavored to promote what he obviously thought would be a fertile exchange of secrets between the Swiss and Leibniz.

In short, whereas Leibniz himself, the Bernoullis, and members of their circle were confident that Leibniz had started a completely new chapter in the history of mathematics – and their judgment proved correct – to Huygens, Newton, Fatio, and others too, it seemed rather that, though Leibniz's invention of calculus was highly important, it was only one among several promising new developments. Although Leibniz could not wholly deny the evaluation, he did retort against Newton, for example, that Newton's notation as indicated in *Principia* was very imperfect; at the same time, however, he did admit that in one respect he had been forestalled by Fatio. It was, in fact, only after Huygens's death in 1695 that Leibniz's calculus emerged decisively as the outstanding new method of analysis – partly, of course, because Newton's relevant mathematics was still unpublished and partly because of the great amount of development effected by Leibniz and his disciples.

Because Fatio, like Craige, L'Hospital, Johann Bernoulli, and others, was heavily engaged on the general theory of integration, for him and others, Huygens devised a number of expressions to try their teeth in finding the integrals. Gradually Huygens became convinced that Leibniz's new calculus offered no infallible and general method for the resolution of such problems, any more than other processes did, although he admired generously the facility with which Leibniz himself employed the differential calculus. When his study of the catenary (suspension-bridge) curve came out, and Huygens had painstakingly verified Leibniz's results, he told Leibniz: "I went on to consider why several of your discoveries had escaped me, and judged that this must be a consequence of your new calculus which offers to you, as it seems, truths which you have not even looked for . . . " Nevertheless, Huygens was still able, by his own means, to recapture what he had formerly missed, once Leibniz had thus indicated the way, and he still could not agree that Leibniz had invented a uniquely miraculous key.[12] Huygens's letters, courteous despite some provocation by Leibniz, indicate a certain disappointment that Leibniz did not somehow find time to set out fully and plainly his own full achievement in the integral calculus, and when they fell to the discussion of problems in physics, Huygens's sympathies were in many respects closer to Newton than to Leibniz. In what was to be almost his last letter to Leibniz, Huygens paid him an outstanding tribute:

> When I received your letter with the solution of what I had proposed to you . . . I saw that you had resolved the problem very elegantly by an unusual approach, which I should be very glad to learn one day. These are the master-strokes which you keep in reserve for yourself, even though you have modestly said that by the use I and others make of your new calculus you have already received the reward you sought. You could put together an excellent treatise on the various uses of this calculus, and I exhort you to do so as the work would be both fine and useful, and come better from you than from anyone else.

Clearly, though Huygens would not agree that calculus opened a new era in the history of mathematics, he would not either have withheld from Leibniz the highest praise for his personal achievement.[13]

In the very next sentence after those just quoted appears Newton's name: "Mr. Wallis has sent me the new Latin edition of his big book on Algebra, enlarged by something new in the way of series from Mr. Newton, in which there are differential equations very much like yours, apart from the symbols." Newton is mentioned many times in the correspondence between Huygens and Leibniz, nearly always as the physicist author of the *Principia* rather than as a mathematician, though Huygens named him (along with L'Hospital and Leibniz himself) as one who might be supposed to be far advanced toward the general solution of problems of integration as Huygens saw them.[14] On Huygens's side there is no allusion to any suggestion that Newton might have an independent claim to be considered a discoverer of the calculus, though Huygens was (as we shall see) made aware of a pro-Newtonian case by Fatio, now in England and close to Newton; Huygens could well allow Newton as much a mastery of an analogue to the differential calculus as he possessed himself; indeed, to such a profound student of the *Principia* as Huygens, it must have been perfectly evident that Newton employed there geometrical ideas (at least) equivalent to differentiation – like others before him, of course. And equally he could recognize (though he did not wish to emulate) Newton's integration procedures. But all this, powerful as it might be – and non-calculus procedures had proved and were still proving extremely powerful in Huygens's own hands – did not constitute a new branch of mathematics such as Leibniz claimed to have begun, nor even evidence that Newton felt the need for such a new branch of mathematics, any more than Huygens himself did.

Perhaps it is worth stressing the point that the difference perceived and to be examined here is not mathematical but historical. Considering just these three principal figures, Huygens, Newton, and Leibniz, all pure mathematicians of the first order, there were no discrepancies in their technical understanding of the calculus. It was by no means the case that there was some power, actual or even potential, in the new Leibnizian method of analysis to which Huygens and Newton were blind – except (and the reservation is far from trivial) to the extent that Leibniz valued the notational and (as it were) mechanical characteristics of his calculus far more highly than the other mathematicians did, and rightly. The main divergence between them occurred with respect to the evaluation of the calculus: Was it (like the methods of Huygens and Newton also) simply a continuous development from the methods of analysis known before, a progressive step, or was it a mutation, bringing into existence powerful methods of analysis of a quality totally unlike any that had previously existed? Huygens did not see either Leibniz's calculus or Newton's fluxions as such a mutation; nor probably, did Newton, though he was more conscious, naturally, of the innovatory character of his own achievements. It is, one may be sure, no small ingredient in Leibniz's contemporary success and ultimate fame that he did perceive calculus as a mutation, a progressive step so great, like the introduction of algebra, that mathematics would never be the same again. If Leibniz could not see that fluxions deserved an equivalent recognition, he is not altogether to be blamed in the sense that Newton himself, seemingly, did not appreciate the true historical significance of his own mathematical concepts and powers until (in a sense) instructed by Leibniz. He was always conscious of standing on the shoulders of giants and often humble before the grandeur of historical continuity.

In later, far more bitter, years Leibniz was to claim that Huygens, the great acknowledged master in European mathematics at the time of his death in 1695 (when Newton was over, and Leibniz slightly under, fifty years of age), had virtually endorsed his own claim to the discovery of the calculus and had known nothing of any basis for a rival claim by Newton. By and large this is true. Newton replied that Huygens's testimony was irrelevant, because Huygens possessed no such knowledge (as Leibniz, he claimed, did) of what Newton himself had done long before. Newton's reasoning is not perhaps perfectly well founded, for with the exception of *On Analysis* Huygens could have seen in print in 1695 virtually all that Leibniz could have read before his departure to Hanover in 1676. Huygens's recognition of "differential equations" in the documents published by Wallis and of Newton's excellent if heavy-going work on finding integrals by the use of series is (indirectly at least) contemporary testimony to Newton's priority in discovery, though Huygens is making no assertion to that effect; certainly he understood Leibniz's crucial role in introducing the calculus as a living and thriving technique in European mathematics.

Thus, despite a slow start in part occasioned by his move to Hanover and the historical work that occupied him there (and caused him to travel restlessly about Europe in search of archival material bearing on the history of the House of Brunswick), Leibniz had by 1695 established himself as the natural successor to Huygens (who died in that year); he became the doyen of mathematics on the Continent. In the eleven years since the publication of his first paper the calculus had become a new branch of mathematics, which almost every aspirant was bound to seek to master, with its first textbook about to appear, with its vehicles of publication in the *Acta Eruditorum*, the *Nouvelles de la Republique des Lettres*, and so on. By his tremendous energy, by his preeminent mastery of the new mathematical methods devised by himself, by the power, authority, and versatility displayed copiously in the letters he had exchanged with leading colleagues like the Bernoullis, L'Hospital, and Huygens, Leibniz enforced respect and perhaps veneration – and he was not loathe to welcome such homage. In 1695 a Continental mathematician not sensitive to the vibrations of the geometrical spider's web whose master was Leibniz was simply out of date and out of context.

In England it was far otherwise, but not because Newton's glory eclipsed that of Leibniz. Indeed, as we have had occasion to see again and again, Newton's work was hardly better known to his countrymen than that of his German analogue. Leibniz's work was presumably not more widely appreciated because of English insularity, in particular English lack of interest in foreign journals. John Wallis of Oxford was to admit (and say he was not ashamed to do so) that he really knew only the 1682 Leibniz paper, which had been translated into English in Robert Hooke's *Philosophical Collections* (1682). There had been no *book* yet that could be imported for the benefit of English mathematicians – and little profit in such a trade anyway. Far more Englishmen read Latin than French, which the Marquis de L'Hospital chose for his text. No Collins or Oldenburg was active. And the number of Englishmen capable of making a start (unaided) on Leibniz's ideas was in any case tiny. For a quarter of a century now Continental universities had been modernizing, slowly perhaps, but at least adopting Cartesian science, modern textbooks, and even, here and there, experimental demonstration of scientific principles. Mathematical courses were being taken far beyond Euclid and Ptolemy. New mathematical chairs were founded, and lively young men were seated in them. Of such an academic renaissance there was in Britain little manifestation, and Newton himself (though he had pupils) trained no one to stand beside him. The single-handed effort of Craige to interest his countrymen in Leibniz's calculus proved futile; the only sequel to his first tract of 1685 was published by himself eight years later. When Wallis came to publish a collected edition of his own mathematical works in the same year (1693), he still knew virtually no more of Leibniz as a mathematician than he had when first issuing his *Algebra* in 1685.

Wallis was by now seventy-seven years old; failure on his part to keep up with the literature is readily excusable. Following the attention to Newton's work on infinite series (especially as used in integration processes) in Chapters 85 and 91 to 95 of the 1685 *Algebra* (as mentioned earlier), Wallis had written simply of Newton's 1676 letters from which his account had been derived:

> There is a great deal more (in these papers [of Newton's]) of like nature, and somewhat of the same kind hath been done by Leibnitius and Tschirnhaus abroad, and Mr. James Gregory and Mr. Nicholas Mercator with us, which are most of them but particular cases within the compass of Mr. Newton's general rules.

He then very briefly describes Leibniz's circle quadrature. Now this is already a quite unfair or uninformed remark. Mercator had published (though less generally) on infinite series before Newton's name had been heard outside Cambridge. And both Leibniz and Tschirnhaus had published their particular cases, whereas Newton's "general rules" had remained unknown until Wallis himself gave, as he admitted, only a sketch of them.

When Wallis reissued his *Algebra* in somewhat enlarged form and in Latin as the second volume of his collected works (but the first to appear) in 1693, he went even further and wrote of the doctrine of infinite series as "long ago introduced by Mr. Isaac Newton and pursued by Nicolas Mercator, Mr. Leibniz and others" as though Newton had made his methods known and the others had adopted them, which (for Mercator at any rate, and as we now know for Leibniz also) is palpably false.[15]

This was not the only change made by Wallis in Newton's favor. Once again, it must be reiterated that in the *Algebra* of 1685 there is no mention of fluxions or of differentiation or of any equivalence between them. Nor was there allusion to a reciprocal relationship between, for example, the method of tangents and the method of quadrature. Wallis had both of Newton's 1676 letters, "full of very ingenious discoveries," to Leibniz in his hands when writing his English *Algebra*, but he chose there to make use largely of the *First Letter* alone, in the belief (as he was to explain in 1693 in the revised, Latin version of the book) that Newton meant to publish the material in the *Second Letter* himself:

> . . . This [that is, the content of the immediately preceding paragraphs] is what I inserted in the English edition of 1685, taken from the letters of Newton noted above. Many other things there worthy of note were omitted, because I was aware that [he] wished to publish this piece and that which he had kept by him. However, as he has not yet done so, several of these points may be touched on here, lest they be lost to sight.

This indicates that Wallis's discretion was responsible for his initial silence, but one is bound to wonder whether fluxions (to which the emphasis now shifts) had struck him as so very important in 1684, when he ignored them, even though the same letters were later supposed by Newton to have been so readily instructive to Leibniz.

However that may be, Wallis now (and for the first time in history so far as print is concerned) embarked upon a brief essay

on the calculus of fluxions, following initially the text of Newton's *Second Letter*. But one soon sees that Wallis is working not
only with this letter but with material that must have been put on
paper by Newton since October 1676, and indeed must be quite
recent. Wallis knows what the sentences about fluxions were that
Newton had sent to Leibniz simply as strings of letters, and quotes
them. He refers to letters, now lost, from Newton dated 27 August and 17 September 1692. And he launches into a comparison
of fluxions and differentials that can only have come from Newton's own mind:

> By fluents [the given functions] Newton understands indeter-
> minate quantities which in the generation of a curve by the
> local motion [of a point] are perpetually increased or dimin-
> ished, and by their fluxions he understands the swiftness of
> their increase or decrease. And although at first glance fluents
> and their fluxions seem difficult to grasp, since it is usually a
> hard matter to understand new ideas; yet he thinks the notion
> of them quickly becomes more familiar than does the notion
> of moments or least parts or infinitely little differences; . . .
> Although he does not neglect the use of such [least] parts but
> uses them only when by their means the work is to be done
> more briefly and clearly, or leads to the discovery of the ratios
> of the fluxions.

It is certain that these words or some very similar were written by
Newton himself, embodying as they do the apparently contradic-
tory double claim that Newton considered fluxions to be easier to
grasp than differentials but, not ignorant of the latter, used their
equivalents occasionally when it was advantageous to do so – thus
admitting that the differential method, if less "geometric," was
certainly in some uses more convenient.[16]

Now it should be emphasized that Wallis's narrative in the Latin
Algebra is perfectly fair and truthful. It is made plain (at least to a
properly attentive reader) that nothing of fluxions was openly or
intelligibly communicated to Leibniz in October 1676. Wallis here
uses the Newton dot notation for the first time in print ($\dot{x}$ is the
fluxion of x, $\dot{y}$ that of y, $\ddot{z}$ the second fluxion of z, etc.), and again
the reader would see that this and much of the explanation is being
given *for the first time* (though indeed and veritably taken from
Newton's now very old mathematical drafts). Wallis made no hint
of a claim that Leibniz had taken anything from Newton or had
been helped by him in the least; he merely put it on record that

Newton had long possessed and devised a notation for the differential and integral calculus recently emerging on the Continent.

So also, two years later still, in the Preface to Volume I of his *Mathematical Works* (appearing *second*, in 1695) Wallis wrote:

> Here [in Volume II, that is in the Latin *Algebra*] is set out Newton's method of fluxions, to give it his name, which is of a similar nature with the differential calculus of Leibniz, to use his name for it, as anyone comparing the two methods will observe well enough though they employ different notations . . .

But here Wallis now began slightly, probably innocently, to mislead his reader, for he says he has taken Newton's method from the two famous letters of 1676 "which were then communicated to Leibniz in almost the same words, where he explains this method to Leibniz, having been worked out by him more than ten years previously." In no significant sense had Newton "communicated" the method of fluxions to Leibniz in 1676, even if we omit the total concealment of the words *fluents* and *fluxions* in the anagrammatic passages; and insofar as the method was explained and exemplified by Wallis himself, this was done with authentic material *not* in the *Second Letter*. Different senses of words are becoming confused in Wallis's Preface: If a person declares, as Newton did in his *Second Letter*, "I have two methods of solving that problem," his words may afterward be taken as evidence that at the time of his statement he did possess these methods, provided that we believe the speaker to be trustworthy, but they give no evidence as to the nature of the methods, nor do they *communicate* anything (except perhaps inspiration, for one man may strive to do what another has already done). Indeed, the *Second Letter* contained many mathematical treasures, but not the concept of a fluxion, nor one example of an expression involving fluxions, nor any hint as to how they might be used. Inadvertently, therefore, Wallis was beginning a process of public deception, and it is not surprising that Leibniz was to write to Thomas Burnet: "I am very satisfied with Mr. Newton, but not with Mr. Wallis who treats me a little coldly in his last [volume of] works in Latin, through an amusing affectation of attributing everything to his own nation."

Almost throughout his life as a mathematician Wallis had fought for the claims of Englishmen against foreigners. He had asserted the merits of Thomas Harriot against Descartes, the priority of Neile against Heuraet. He had offended almost every foreign mathematician with whom he came in contact by his robust, unabashed xenophobia – yet his candor, rugged charm, and ability kept his friends faithful. Once he had begun to suspect that Newton's typically English inventiveness was likely – as in so many earlier instances! – to be overlooked, he longed to take up his case. As he was unembarrassedly to admit to Leibniz at the end of 1696, he knew only two of Leibniz's lesser publications, nothing of his infinitesimal analysis, nothing of his differential calculus save the bare name, which he had picked up just as the Preface to his first (1695) volume was being printed. *But* – and the expression is unconsciously revealing – Wallis when he heard of it "did not want it to be said that nothing was here [in the *Algebra*] written of the differential calculus." It almost seems as though Wallis is saying, It really isn't clear to me what Mr. Leibniz has done but Mr. Newton certainly did it first! Surely, in fact, Wallis must have known a little more of what was happening among Continental mathematicians, especially since the arrival of Johann Bernoulli in Paris, than he was willing to confess, pleading his great age.[17]

Unfortunately, the letters exchanged between Wallis and Newton in the summer of 1692 are for the most part lost, but it is clear that Wallis asked Newton if he might not amplify the account of his work given in the English *Algebra* of 1685 as that work was to be reissued in Latin in Wallis's projected *Opera Mathematica*. The reason for Wallis's request had nothing to do with Leibniz at this stage, but with David Gregory's overwrought claims for his own quadrature series; Wallis wished to assert that Newton had forestalled him with a more general achievement. Newton's answer indicates that Wallis did also ask him about Leibniz's work, or at any rate about his circle quadrature published in 1682 (and reprinted in England), for Newton in reply plunged at once into a report of what was contained in his *Second Letter* (of which, as we have seen, Wallis probably already possessed a copy), explaining its enigmas and adding flesh to the very bare bones of the fluxional calculus presented in the *Second Letter* (when decoded) from the rich materials in *De quadratura curvarum*, which Newton had begun to write in 1691. There is no doubt that what Wallis published on fluxions in the revised *Algebra* of 1693 came from Newton's pen in this way, as indeed Wallis noted himself in his own copy of the book.

Finally, at the end of the few pages on fluxions, Wallis remarked

> Analogous to this method is the differential method of Leib-
> niz and that other method, older than either, which Barrow
> expounded in his *Geometrical Lectures*; and this is acknowl-
> edged in the *Leipzig Transactions* (January 1691) by a writer
> making use of a method similar to that of Leibniz . . .

So much for the old man's protestations of feeble ignorance until
the last leaf of the Preface was set up by the compositor! On page
396 of his *1693* volume he had already known of the Leibnizian
method and of the work employing it that was published in Leipzig in the *Acta Eruditorum*. And there is no reason why one should
suppose Wallis's interest in recent Continental mathematics had
revived again only as late as 1691, the year cited from the *Acta*,
though that would be time enough to account for his approach to
Newton concerning the reprinted *Algebra*. Moreover, it is very
likely that he received a warm hint from Craige or David Gregory
that something more should be said to do justice to Newton's
achievements.[18]

For in May 1694, no doubt also at Craige's instigation, David
Gregory had visited Newton at Cambridge and under his tolerant
eye had rummaged freely through Newton's mathematical papers;
from the notes then made he put together a first sketch for a treatise on "Isaac Newton's Method of Fluxions, in which the Differential Calculus of Leibniz and the Method of Tangents of Barrow
are explained and illustrated by many examples of all kinds."
Gregory's draft, though never printed, was in turn examined and
copied by other mathematicians later; it was wholly derived from
Newton's manuscripts, supplemented by the compiler's study of
the published papers of Leibniz and Jakob Bernoulli. Though
Whiteside has qualified these as "feebly wrought endeavours" and
lamented Newton's failure to "attract no more able and gifted a
disciple" than the younger Gregory, the point here is simply that
by this time outline knowledge of Newton's discoveries was sufficiently widespread for Wallis to have known of them.[19]

When the learned men of Leipzig came to review his book, particularly it would seem Volume I rather than the Latin *Algebra*,
which had appeared two years before, they were amazed to find
so much space and praise given to Newton's mathematics, of
which nothing had yet appeared in print. Admittedly Leibniz had
praised him in their own journal as "that geometer of most pro-

found intelligence," but that hardly justified Wallis's making him a precursor of Mercator or his calling Mercator an Englishman. And an *Algebra* of 1693 should reasonably have given space to the most modern improvements that were generally known and in current use by many mathematicians. When Leibniz, with indefatigable intellectual energy, sought through some intermediary to sound Wallis out further on some mathematical point Wallis had raised years before, the Oxford mathematician took the opportunity to relieve himself, by pleas of ignorance, from any suggestion that he had been actuated by prejudice. Not at all: He had wished to print Leibniz's letters to Newton along with the replies he had used, if Newton had only been able to produce them. He made it clear that (in his view) Leibniz had a perfect right to an independent and distinct, though posterior, discovery and was worthy of the highest esteem.

With this Leibniz seems to have been perfectly content, perhaps not believing that any serious competition could come from the (apparently) inactive and silent Newton, who (as Leibniz knew) in 1696 gave up the academic life for government service in London. At any rate, the correspondence between Wallis and Leibniz begun in that year went on to almost the time of Wallis's death, and fills eighty printed pages. It is gravely cordial in the style of the times and not by any means wholly mathematical, for Leibniz asked Wallis about his early code-breaking activities and had him do a number of small scholarly errands. When Wallis came to the third volume of his *Mathematical Works* he decided to print there not only Newton's 1676 letters but those of Leibniz also, which had by then come into his possession. Leibniz acceded to his request, and took in good part an elaborate commentary from Wallis on his own mathematical accomplishments of twenty years before. In these exchanges of 1698 there is no word of Newton, no allusion to any rivalry between him and Leibniz, between England and Germany. All was peace and amity. Wallis could now see the new infinitesimal calculus in perspective as a descendant of the Greek geometers' "method of exhaustion," and tells Leibniz:

> You readily profess – such is your fairness of outlook – that your differential calculus has many things in common with purport of others, even that of Archimedes, yet it is not for that reason to be the less highly esteemed. For there are many things of which the bases were not unknown to the Ancients but which are so intricate and full of difficulties that they have

been, in our own time, rendered far more plain and fit for use (such as, to name only a few, the algorithm of the Indians using numerical symbols and the analytic calculus or specious arithmetic of the moderns, also the formation of conic sections by a plane and a cone and many other things which the present century has added to the discoveries of the Ancients). And thus, though I do not mean to deprive the Ancients of their praise (for we have built much on their foundations) so also I do not mean to flatter the moderns lest they cease to make fresh progress, but rather to urge them farther on, and yourself above the rest.[20]

Such passages and the whole tone of the correspondence, the deference on one side of a mathematician whose powers have withered to a younger master, on the other of present authority to a living fossil, a survivor from the long past age of Cavalieri, Descartes, and Fermat, indicate that there was nothing in these exchanges with a most patriotic Englishman to trouble Leibniz's self-esteem. Rather Wallis seems to have accepted (in these letters) that very same estimate of Leibniz that the whole Continent, with rare exceptions, accorded to him and which he now took for granted. His service to the house of Hanover and above all the charm and greatness of his personality had earned him the warm respect and confidence of the Electress Sophia, heir apparent to the throne of England in the probable event of the Princess Anne's failing to have children to survive her; and in 1700 (some three years before Wallis's death) Leibniz moved to Berlin at the behest of the Electress's daughter Sophia Charlotte, now herself Electress of Brandenburg. There his plans for the Berlin Academy, the first institution of its kind in northern Europe, were approved, and he himself was appointed its first president. Eminent in history, law, and philosophy as well as in mathematics and theoretical mechanics, he could claim attention as Europe's foremost intellectual.

It is pleasant to record that during these years of his increasing fame and achievement Leibniz had one cordial correspondent in England, besides John Wallis, whom he could regard as less of a xenophobe than most of his nation seemed to be. This was Thomas Burnet, son of the Royal Physician for Scotland and a distant relative both of Gilbert Burnet, Bishop of Salisbury and historian of his own times, and of his own namesake, who wrote the *Sacred History of the Earth*. This Thomas Burnet took his M.D. degree at Leiden in 1691 and in the course of extensive traveling made a considerable impression in 1695 at the Court of Hanover, where he became acquainted with Leibniz, who used him as a source of news about English intellectual life and also as a means of conveying letters to Newton, Wallis, and Locke. Toward the end of 1696 Burnet wrote to Leibniz that he had, during the summer, placed in Newton's hands a note (now lost) from Leibniz: "He told me to thank you on his behalf, because he honours your merits extremely. He added that the new office of Warden of the Mint conferred upon him by the King will compel him to retard his thoughts upon colours, so that he does not know when he will have leisure to resume that investigation; yet he had it almost finished, apart from some kinds of colours which he had not yet enough experiments to examine." In reply, Leibniz, while commending Newton's appointment to the Mint, very reasonably regretted its diverting him from more important thinking. Leibniz would have preferred Newton to continue the physical researches that he had begun, and frequently returned to this point in subsequent letters – he could not, he said, accept Newton's excuse for dropping his intellectual task, and he wished him ten times the £1,000 received by Dryden for his translation of Virgil so that he might be free to work at it without distraction. From Burnet, Leibniz also heard of Newton's solving the brachistochrone problem in two hours "when he was quite overwhelmed by business of a different kind," and of Nicholas Fatio de Duillier's return to England in the summer of 1698.

The last ten years or more of this correspondence show more interest in the discussion of books on religion, philosophy, and politics and the problems they raised than in matters of science and mathematics. There are few allusions to Newton after 1700, but the letters from Leibniz are remarkable as containing several kindly references to Nicolas Fatio *after* he had characterized Leibniz as only the second inventor of the calculus, which speak well for Leibniz's humanity. When Burnet made a second European tour (in the course of which Leibniz helped to secure his release from the Bastille, where he had been imprisoned as a spy) Leibniz reminded him that the two Fatio brothers were excellent mathematicians residing at Basel, and when he learned some years later (in 1708, in fact) of Nicholas Fatio's involvement with the Cevennois and the troubles thereby brought about his ears, he wrote of this annoyance "because of my love for Mr. Fatio, for he is a man excellent in mathematics, and I do not understand how he could have got involved in such an affair." Was there not some mistake, Leibniz asked, in his condemnation, for surely one could not doubt of Fatio's good faith?[21]

At the time of the incident, however, Leibniz was exceedingly angry, and he wrote to John Wallis demanding an explanation of Fatio's discourtesy. Incapable as so aged a man as Wallis was of appreciating the issues involved in the development of the calculus during the 1690s, he nevertheless certainly understood that Leibniz was not a man to be attacked lightly. Like most academics in that age as in our own, he felt a deep respect for authority and the ranks of hierarchy and reward, all of which Leibniz might now seem to represent. Receiving Leibniz's account of a seemingly malicious libel put about in a book bearing the imprimatur of the Royal Society, Wallis responded with signals of distress and sympathy.

In making his complaint, in a letter of August 1699, Leibniz was particularly riled because his accuser, Huygens's former acquaintance Nicolas Fatio de Duillier, had been a Fellow of the Royal Society since 1687 (soon after his arrival in England), whereas Leibniz himself had been a Fellow far longer. Had Fatio's attack the backing of the Royal Society, Leibniz wanted to know, and Wallis was able to reassure him: The attack on Leibniz had not been sanctioned by the society (even though a vice-president had inadvertently approved Fatio's book for publication as a technical geometrical treatise) and the society had no thought of putting such a one as Fatio ahead of Leibniz in its esteem. Moreover, Wallis added, Fatio was in any case not an Englishman but Swiss! The burden of Fatio's onslaught upon Leibniz, according to Leibniz himself, was that Leibniz had no respect for any mathematician who was not of his own school and thought that no others were capable of resolving the more difficult questions of geometry then widely under discussion. This the Swiss had resented, proclaiming that there were others, like himself, just as capable, who owed nothing to Leibniz and his calculus.[22]

This sounds remarkably trivial, and so from the Wallis-Leibniz letters it appears. But there was a good deal more to Fatio's *Geometrical Investigation of the Line of Quickest Descent* (1699) than the charge that Leibniz was an intellectual snob. Fatio had become an ardent partisan of Newton's, just as he had for a time been an ardent partisan of Huygens's, and the apparent slight against his own talents that he had detected in not being considered (abroad) one of Britain's leading mathematicians gave him the opportunity of firing off far more serious accusations.

The portrait of Nicolas Fatio conveys the impression of one who is highly volatile and excitable rather than shrewd and steady. Yet he made a strong impression on two men of genius, partly by actual achievement of his own, partly by promise and receptivity. Like the Bernoullis, Fatio was born in Basel (where his elder brother, Jean Christophe, studied mathematics with Johann Bernoulli); he grew up in Geneva, however, where his father removed the family. At first astronomy was his passion; he worked for a time with G. D. Cassini at the Paris Observatory and even hoped (like Leibniz before) for a place as pupil or assistant in the Academy of Sciences. This failing, he went at the age of twenty-two to the Netherlands, where in September 1686 he sought out Huygens at The Hague, who took a liking to this active young man and encouraged him in pure mathematics. They worked together on the demonstration of errors in a publication by Tschirnhaus. How much of a mathematician Fatio was before he encountered Huygens, how far he advanced in association with Huygens, is not clear. At this stage Huygens seemed to find Fatio's mathematical ideas more familiar and assimilable than those of Leibniz, and (as mentioned earlier) endeavored to promote exchanges between them. Leibniz too was persuaded of Fatio's abilities by Huygens's praise, though the two never met. Fatio's major success at this time was with a process for discovering the equation of an unknown curve, given the slope of its tangent at some value of x; in tackling this inverse tangent problem it appears that Fatio was independently treading the path previously taken by Newton, Huygens, Wallis, and others. This was at a moment when Leibniz's calculus was barely known to, or studied by, anyone in Europe, and Fatio's finished statement of his method, written to Huygens from Oxford in October 1687, is no contemptible piece of mathematics even if he did not get as far as Newton had some fifteen years earlier.[23]

It is likely that Huygens's well-meant efforts created the first dislike of Leibniz in Fatio's mind, for the older man made it clear that (not unreasonably) he thought himself likely to be the loser in any exchange of mathematical "secrets," and that he was really not at all keen on the idea. After all, Huygens had admitted that he and Fatio had not yet mastered the calculus, and sometimes he spoke of Leibniz rushing far ahead into realms of mathematics where he himself could not follow. Perhaps Huygens was a little blind to the point that while Leibniz was very ready to extend the deepest respect to his old Parisian mentor, one of the greatest European figures, he did not feel at all the same about the great man's latest protégé. This is a situation that the protégé resents. In England, where he arrived in the spring of 1687, Fatio was soon absorbed in the atmosphere of incipient triumph preceding the publication of Newton's *Principia* (of which the first book had been received in manuscript in London almost a year before). The English teased him, as a newcomer from the Continent, with sticking to old-fashioned Cartesian principles. He quickly wrote a letter to Huygens about the wonderful new physics of forces, which the recipient, naturally, failed to understand, even though Fatio expressed a tactful hesitation concerning Newton's "principle of attraction" (gravitation) between the heavenly bodies. In Newton (though he had not yet met him) Fatio was to find the third, last, and greatest of his scientific heroes. (However, he was to return and work with Huygens again for the greater part of the year 1691.)

If Fatio had not already sought Newton out at an earlier date, he must have first met him early in 1689 when Newton was attending Parliament at Westminster. By October of the same year Newton was writing to him in that tone of urgent familiarity and concern that is found among his earlier letters to this young man and nowhere else in Newton's correspondence. The letter (from which portions have been excised) is worth quoting in full:

> I am extremely glad that you [have met with your] friend and thank you most heartily for your kindness to me in designing to bring me acquainted with him. I intend to be in London the next week and should be very glad to be in the same lodgings with you. I will bring my books and your letters with me. Mr. Boyle has divers times offered to communicate and correspond with me in these [chemical] matters but I ever declined it because of his [freedom] and conversing with all sorts of people and being in my opinion too open and too desirous of fame. Pray let me know by a line or two whether you can have lodgings for us both in the same house at present or whether you would have me take some other lodgings for a time till [we can be together]. I am
>
> Yours most affectionately to serve you
Is. Newton

Similar letters follow. In 1692 Newton "cannot express how much I was affected" to hear of Fatio's being ill: "if you want any money I will supply you." Soon after, Fatio went to pay Newton a visit in Cambridge, and Newton would have liked to keep him there in the college room next to his own in order to strengthen his health. In 1693 he sent Fatio money for books and other things left in Cambridge and once more offered to make him an allowance to join Newton in residence there. This Fatio refused, though he continued to treat Newton as his financial and medical adviser until the correspondence was (apparently) broken off when Fatio went on a visit to his home. It was in the autumn following that Newton, fallen into a state of severe depression, wrote strange letters to his friends and had a spell when he "had not slept an hour a night for a fortnight together and for five nights together not a wink." Only in the late autumn of 1693 did he regain his normal health and behavior. News of this severe illness reached Huygens (and so Leibniz) from Fatio.

Was he its cause? Did he arouse in Newton, as Frank Manuel has maintained, a powerful homosexual passion? It is certainly true that Newton's letters to Fatio contain a warmth unique in his correspondence, but more than that is empty speculation. Strong friendships between men attracted fewer suspicions then than now, and that between Newton and Fatio was formally disrupted not by psychological forces but by Fatio's temporary departure to Switzerland in pursuit of an inheritance. All that need be remarked here is that at this point the history of Newton the mathematician becomes firmly enmeshed with personal feelings. For a time Fatio delighted to identify himself with Newton: absurdly, he assures Huygens that Newton will certainly take his remarks in good part or again writes (in a notably cold and upstage letter to a third party intended for Leibniz's eyes) of what "Mr. Newton and I" think on the subject of gravitation. Newton was so fond of him (or so Fatio reported) as to endorse some of Fatio's farfetched scientific speculations. Despite the apparent break in 1693, Fatio continued all his life (or at least so long as he retained the full use of his reason) to look up to Newton as an almost superhuman being.[24]

Thus, it is hardly surprising that in the 1690s Fatio's feeling for himself and Newton as injured parties fused together in detestation of Leibniz, in whom, unfortunately (perhaps because he never met him), Fatio could not detect those real qualities obvious to Huygens, to Newton (for much of his life), and to so many more.

In June 1696 Johann Bernoulli issued another of those futile and strife-provoking mathematical challenges: to determine the curve linking any two points, not in the same vertical line, along which a body would most quickly descend from the higher to the lower point. It was addressed to "the shrewdest mathematicians in the world," and copies were posted to Newton and to Wallis. Leibniz had solved the problem on the day it was privately received by him from Bernoulli, in the sense of giving a complete analysis of it, but he failed to recognize that the curve was a cycloid until Bernoulli informed him. Newton (according to his own story) also solved it in a single evening, and one of the solutions that arrived anonymously from England was recognized as his by Johann Bernoulli (it is a very elegant definition of the necessary cycloid).[25] Later, reviewing the whole competition, Leibniz noted that only five mathematicians had succeeded; besides himself and the author of the problem, there were Newton, Jakob Bernoulli, and L'Hospital (who received help from Johann, however). "It is surely worthy of remark," he commented, "that they only solved the problem whom I had guessed would be capable of solving it, as being those alone who had penetrated sufficiently deeply into the mystery of our differential calculus." The words may have been carelessly written, for Leibniz commonly worked under pressure, and perhaps he did not mean to imply that *Newton* was indebted to the differential calculus for his solution. He also says that Huygens would surely have solved the problem if he had lived, and he would not have claimed Huygens as his pupil in mathematical matters. But certainly the most obvious meaning of Leibniz's words was that Newton had learned the calculus from Leibniz and Fatio resented both this and the insult to himself as being a mathematician who could not solve the problem.[26]

It has been argued, and perhaps Newton himself imagined, that this brachistochrone problem was designed specifically to test Newton's powers. As Newton's last traditional biographer put it:

> Newton suspected that the problem had been devised and sent to him to prove, *by his inability to solve it*, that his fluxions was [sic] not the general and powerful method he claimed it to be. And by exerting his matchless powers, he routed his adversaries and vindicated himself.[27]

Newton, we may suppose, was ignorant of the private wars

waged by Johann Bernoulli with his brother Jakob and others, or of Leibniz with the Dutch philosopher Nieuwentijt, who assailed the title of the calculus to be called a branch of geometry, but there is no evidence that he thought the problem particularly directed against himself, though it obviously increased his prestige to solve it (he published the solution at once in the *Philosophical Transactions*). Nor is there any hint in the Bernoulli-Leibniz letters of a desire to score off Newton; on the contrary (as we have just seen), Leibniz specifically predicted that the problem was within Newton's powers, which the *Principia* had already proved to be great. It is quite illogical to charge Leibniz with the offenses of claiming both that Newton was a master of his own calculus *and* would fail to solve the problem. Leibniz could consistently have said only one of these things at a time. Equally, how would Bernoulli have guessed (without any air of surprise) that the English solution came from Newton's hand if he had previously believed that Newton was incapable of solving the problem? In fact, Bernoulli's intention seems to have been merely to "sort out the men from the boys," and to show the ascendancy of the "Germans" (including Swiss!) over the English and French, who had long disputed the leadership in mathematics; thus he observed with pleasure Varignon's admission that the brachistochrone problem was quite beyond him, even though Varignon was his friend and pupil, because he was outside the immediate charmed circle of Leibnizian initiates.

Reading Leibniz's words in the most naive way and reacting with great emotion, Fatio inserted into a lengthy and long overdue analysis of the brachistochrone a vehement and irrelevant (but true) assertion of his own mathematical originality. He had, he claimed, worked out his own mathematical processes during his association with Huygens from 1687 onward, and the proof lay in the letters they had exchanged. (That these statements might be true does not, of course, justify Fatio's complacent thoughts that he was as able a mathematician as Leibniz or the Bernoulli brothers – but Fatio's vanity may be partially excused by his ignorance). Then he went on

> Yet I recognize that Newton was the first and by many years the most senior inventor of the calculus, being driven thereto by the factual evidence on this point; as to whether Leibniz, its second inventor, borrowed anything from him, I prefer to let those judge who have seen Newton's letters and other

manuscript papers, not myself. Neither the silence of the more modest Newton nor the eager zeal of Leibniz in ubiquitously attributing the invention of this calculus to himself will impose on any who have perused those documents which I myself have examined.[28]

The unspoken cry here, "I am Newton's best friend," is almost as loud as the direct charge against Leibniz, and the former is productive of the latter. Probably no one ever, certainly no one living at that moment, had explored Newton's hoard of mathematics as Fatio was permitted to do, and one may well believe that he possessed a seeing eye as he went through those treasures that were already becoming historic. What is obscure is Newton's state of mind. The implication of Fatio's words might seem to be that Newton had himself made out a case for Fatio's benefit against Leibniz, pointing out this and that passage in his old papers and letters, but this is not a strong implication. It is certainly possible that Newton allowed or even encouraged Fatio's flattering interest in his own past without any such *arrière-pensée*, and that it was in Fatio's not Newton's mind that the formidable case against Leibniz began to assemble. Whether or not the case against Leibniz arose spontaneously or was thrust into Fatio's mind at the time of his reading the documents, he was certainly convinced at a relatively early stage of his friendship with Newton that Leibniz's was the second and inferior discovery of the calculus, for at the end of December 1691 (not long after Fatio's return from Holland on a visit to Huygens) he not only assured his Dutch friend that he understood the whole of the Leibnizian calculus very well but also that

> It appears to me by all that I have as yet been able to see (in which I include papers written many years ago) that Mr Newton is without difficulty the first inventor of the differential calculus and that he knew it as well as Mr. Leibniz does now, or even more perfectly, before the latter had even the least idea of it, which as it appears did not come to him before the occasion when Mr. Newton wrote to him on this subject. See if you please page 153 of Mr. Newton's book. Therefore I cannot cease to wonder that Mr. Leibniz takes no notice of this in the Leipzig *Acta*.

This is a less forthright declaration than Fatio was to publish eight years later, but it already goes beyond the claim made by Newton himself in the *Principia* scholium to which Fatio referred Huygens.

In response, Huygens was calm. He was not surprised, evidently, to learn that Newton knew as much as Leibniz and more, and merely expressed the hope that Newton would publish his knowledge; as for the disagreeable inference in Fatio's letter, he was content to throw salt on it: Newton in the scholium (he thought) allowed that Leibniz had hit upon the calculus about the same time that he himself had. Fatio, obviously, would never accept this tactful view of the situation. He at once strengthened his attack: Not only were the original letters of 1676 more positive in their message than the recent *Principia*, but they were such as would undoubtedly cause Leibniz pain should they ever come to be published, for they would underline his lack of generosity in failing to acknowledge Newton's achievement. Moreover, Leibniz's performance has been such that, in comparing it with Newton's, Fatio could not but perceive, he says, the difference between a spoiled and imperfect copy and the perfect original. Newton knew as much as Leibniz and all that Fatio himself knew and Leibniz did not; he had mastered not fluxions alone but the fluxions of fluxions of fluxions (a claim that Huygens did not understand, though it would have been clear to Leibniz). After this fresh outburst, however, Fatio let the subject drop and, within a few months, the closest period of his association with Newton ("a guide beyond comparison more enlightened and more generous" than Leibniz) came to an end. Perhaps the most interesting point in Fatio's last apologia for Newton is the criticism that Leibnizian differentials were only a "spoiled copy" of fluxions, for Newton made the same criticism later. Had Fatio already derived it from him?[29]

Officially, at any rate, all was smooth. In March 1693 Leibniz wrote in his own hand to Newton, of whom he received continuing news from Huygens and from at least one German visitor to England. Probably his interest was stirred by the news (rather garbled in transmission) of Newton's work being published in the new edition of Wallis's *Algebra* (which, in March 1693, neither Huygens nor Leibniz had yet seen). Leibniz's letter is frank and open. He praises Newton's achievement with series and adds that the *Principia* has proved also that "even what is not subject to the received analysis is an open book to you." Such compliments, be it emphasized, are equally found in Leibniz's letters to his friends.

He alludes modestly to his own endeavor, by employing a convenient symbolism, to extend the range of analytical geometry and then continues with an exhortation to Newton himself:

> But it is from you that I still await some triumph, to provide the finishing touches, both as to the best way of reducing to quadratures those problems which seek to determine curves from a given property of the tangents and of reducing the quadratures themselves (as I greatly desire to do) to the rectifications of curves, in all cases simpler than the measurement of surfaces or volumes.
>
> And above all things I desire you, who are a perfect geometer, to continue as you have begun to treat Nature mathematically, in which kind of investigation you have certainly, along with a very few others, accomplished something very worthwhile.[30]

If these phrases, generous but challenging, evoke any recollection in the reader's mind of any letter formerly addressed to Newton, it may be of that letter which Robert Hooke had written to him on 17 January 1680 containing the plea that Newton investigate "by his excellent method" the curve described by a body moving under an attractive force varying inversely as the square of the distance; this letter had (in a manner of speaking) provoked the *Principia*. Leibniz's letter, however, inspired in Newton no similar fury of creative work.

6

THE OUTBREAK: 1693–1700

IN MARCH 1693 Fatio de Duillier had been invited by Newton to rejoin him in Cambridge at Newton's expense, and on 11 April (apparently in reply to this invitation) Fatio wrote:

> I could wish Sir to live all my life, or the greatest part of it, with you, if it was possible, and shall allways be glad of any such methods to bring that to pass as shall not be chargeable to You and a burthen to Your estate or family.

Thereafter, the intimate and frequent correspondence between the two men ceases; the following summer was that of Newton's mental illness. We have no evidence as to what passed when Newton admitted his friend, at Cambridge, to the privacy of his manuscripts, nor subsequently do we have any record of how he reacted to Fatio's dramatic displays, first in private and finally in public, of his admiration for Newton and his conviction that Leibniz had stolen the calculus from Newton. If letters were exchanged between the two men, or if (as is unlikely enough, in fact) Newton disclosed his personal judgment of Fatio to others, the documents have failed to survive. The frustration of Fatio's hopes so far as Newton was concerned – Newton's decision not to entrust a second edition of the *Principia* to Fatio's hands (or at any rate, his implicit letting the plan slide) and Newton's ultimate impatience with Fatio's pet idea of an etherial hypothesis of universal gravitation – seems to indicate controlled revulsion against him and his enthusiasm. If Newton covertly provided Fatio at any time with ammunition against Leibniz, and there is no evidence that he ever did so, it must (in terms of the biographical relationship between the two men) have been before 1693; for it is inconceivable to me, at any rate, that the degree of intimacy such a communication would imply was ever reestablished between Newton and Fatio after the former's breakdown in health. And this is confirmed by Fatio's evident belief (as early as February 1692) – not yet amounting to a direct accusation – that Leibniz's

procedure in the calculus was so very different from and far inferior to that of Newton, that the latter might be recognized as the
finished original and the former perceived as a spoiled and imperfect copy. So impressive was Newton's draft "On quadrature" in
Fatio's eyes that he declared that Newton had gone far beyond
Leibniz and himself, whether with respect to quadratures or the
method of tangents. Of all this Newton had spoken positively
fifteen or sixteen years before, and it was only much later that
Leibniz had produced his differential calculus. In writing such
phrases to Huygens, Fatio may have been discreet; perhaps he already thought much worse of Leibniz than this letter admits, but
at least it is clear that it is but a step from the phrase about "original" and "copy" to the downright accusation that the "copy" was
the fruit of plagiary.[1]

On the other hand, there is little evidence to indicate that any
inkling of impropriety on Leibniz's part had entered Newton's
mind before 1693. It is true that in a draft letter to David Gregory
of November 1691, in speaking of his description of his method
of series in the *Second Letter* to Leibniz of 1676, Newton writes
that this depended upon a "certain process of analysis which I
there touched on rather obscurely," meaning of course the method of fluxions; Newton also wrote in drafting (but then struck
out) the words "which Leibniz is developing" (*quamque Leibnitius
excolit*). Read one way, the expression that Newton rejected on
second thought would indicate Leibniz's development of *Newton's*
method; read the other way, however, it means "which Leibniz is
independently developing though I had it first," a statement both
true and inoffensive. Newton could not possibly, at least at this
time, have supposed Leibniz to be developing anything from the
unintelligible anagrams in the *Second Letter*. In contrast to this ambiguous (but rejected) phrase, the triangular exchanges of letters
between Wallis, Leibniz, and Newton consistently suggest esteem
and amity. When Newton replied in October 1693 – belatedly, no
doubt because of his illness, though he gave another reason – to
Leibniz's conciliatory letter described at the end of the last chapter,
he emphasized the value he attached to the friendship of one
whom he had "for many years past regarded as one of the chief
geometers of this century, as I have made known on every occasion that presented itself." And he hoped that Leibniz's long silence had not sprung from a lessening of friendship. Newton went
on to speak of the printing of parts of his 1676 letters to Leibniz in

the projected reissue of Wallis's *Algebra,* writing that

[Wallis had] asked me to disclose a certain double method
which I there concealed in transposed letters. On which account I was compelled to explain my method of fluxions as
briefly as I could, which I had concealed in this sentence: "To
find the fluxions in any given equation involving any fluent
quantities whatever, and vice-versa." However, I hope I have
written nothing displeasing to you; if you judge anything
there worthy of reproof, do let me know by letter since I
value my friends more than mathematical discoveries.

He then embarked on a number of technical points raised by Leibniz. Nothing could be more cordial, even deferential, and obviously Newton felt embarrassed because he had not privately explained the full meaning of his letter to Leibniz before doing so in
public.[2]

Two years later, it is true, a letter from Wallis to Newton cries
alarm and regret because Newton's work was still unknown to
the world, whereas the fame of Leibniz had increased. Why did
not Newton at once print his *Opticks* in English and let foreigners
learn our language to read it? Further, Wallis had heard from Holland that Newton's mathematical discoveries were being anticipated in publication:

because your Notions (of Fluxions) pass there with great applause, by the name of *Leibniz's Calculus Differentialis . . .*
You are not so kind to your Reputation (& that of the Nation)
as you might be, when you let things of worth ly by you so
long, till others carry away the Reputation that is due to you.

And Wallis explained that he had endeavored to do justice to the
analogy between fluxions and calculus in the Preface to the first
volume of his works (printed *second,* in this year 1695), alluding to
the account of fluxions already given in the new Latin edition of
Wallis's *Algebra* (printed in the 1693 volume of his works). I have
discussed these passages already.

Certainly it is not wholly clear what Wallis meant to say to
Newton in this letter. Was he hinting at bad faith on Leibniz's part,
in the words "your Notions pass . . . "? Not necessarily; Wallis
might simply mean that Leibniz's ideas that were gaining circulation were the same as Newton's, as when in the printed Preface
referred to he writes of the "very similar nature" (*consimilis naturae*)
of the two methods. Nothing indicates that Newton took Wallis's
words in a bad sense. What one does see in Wallis – in his

1695 Preface explicitly and by implication in his letter to Newton – is a serious confusion between what Newton (in later explanation to Wallis himself) said he was thinking about when he
wrote those letters and what he actually expressed in them. As
already noted, the 1693 version of *Algebra* contains a great deal
more about fluxions than anyone could have guessed from the
bare text of the 1676 letters; yet, in his 1695 Preface Wallis writes
of Newton's "explaining this method [of fluxions]" to Leibniz,
which of course he had never done. It is likely that Wallis's historical vagueness in grasping exactly what Newton had and had
not imparted to Leibniz in 1676 sprang initially from a lack of
clarity in Newton's own mind, and that Wallis's conviction that
Newton had shared something of his great discovery with Leibniz
had its effect upon the picture that Newton formed of what had
happened in that year; particularly, of course, when Leibniz's access to the manuscript copies in Collins's hands was to crop up.[3]

As on many other occasions during his long and far from tran-
quil life, in correspondence with Henry Oldenburg and others
too, Wallis, ever conscious of Englishmen's readiness to hide their
light under a bushel (though foreigners said they were apt to claim
every innovation for some obscure member of their own nation),
repeatedly urged Newton to publish his mathematical papers,
without definitely maintaining now (any more than he had done
on former occasions) that anything worse was at stake than prior
discovery unacknowledged and later discovery proclaimed and
famed. Nor was Wallis alone in urging Newton into print – Leibniz himself did so, more than once. When the douce exchanges of
1693 led to nothing – exceptionally so as Leibniz was usually so
diligent a correspondent – he wrote a letter in October 1694 urging the Royal Society to put pressure on Newton to publish an
improved second edition of the *Principia*: Rumor was widespread
that Newton had such a work in hand. And he was no less anxious
to see Newton's other discoveries in mathematics and physics. As
this request also was without fruit, Leibniz made a new effort in
1697, this time writing through Wallis (presumably Leibniz would
not in any event have known how to address Newton since his
departure from Cambridge). Wallis quoted Leibniz's words to
Newton *verbatim* (in Latin):

> I would be so bold as to ask this favour of you [Leibniz had
> written in May 1697]; should the occasion arise, perhaps
> through some friend, offer my most humble greetings to Mr

Newton, who is a man of the highest talents, and beg him on my behalf not to allow himself to be diverted from publishing his most excellent reflections. Further, I did not only observe, after the publication of his book [the *Principia*], that the most profound Newton's method of fluxions was cognate to my differential calculus, but have also proclaimed the same [belief] in the *Acta Eruditorum* and advised others of it. I judged this [avowal] to conform to my honest nature no less than to his merits. And so I generally call it by the common name of infinitesimal analysis which is [a name] broader in scope than the quadrature method. Meanwhile, just as the methods of Viète and Descartes both came under the name of specious [i.e., algebraic] analysis, yet certain distinctions remain, so perhaps Newton's method and mine differ in several particulars.

Wallis himself reinforced Leibniz's plea, informing Newton that he meant now to give the full text of the 1676 letters in the final volume of his works (which came out in 1699) and urging him to publish his *Opticks* at once.[4] There is certainly no hint here of anyone's thinking of anything but parallelism between the two forms of calculus, nor (what is, perhaps, most significant) did anyone imagine that publication of the 1676 letters counted *against* Leibniz, though they certainly counted *for* Newton. And as Leibniz's final sentence hints, no one had as yet looked closely, technically, at the likenesses and differences between the two methods.

Few signs of turbulence in this calm and amity that prevailed through the decade 1689–99 are to be found either in the correspondence of Leibniz and his friends, where politeness to others did not always prevail. Newton is always referred to with respect, as an intellectual figure of considerable importance (though voicing strange ideas), and his great mathematical abilities were always appreciated. It was known that Huygens had respected Newton highly – though Huygens was far from sharing his physical theories and found much to amend in Newton's publications – and roughly speaking the Leibnizians continued to think of Newton with essentially the same sort of qualified esteem. But Newton and his writings figure far less in their correspondence than many other topics, not only of technical mathematics but of philosophy (such as the nature of infinity) and physics (the problem of vis viva, for example) and even trivialities like a set of elegant verses on Mme Scudéry's clever parrot. If, from Johann

Bernoulli, there are tart remarks about English mathematicians other than Newton, he did not spare the French either, whom he found insufficiently grateful for what they had learned from the German-speaking school. He found it very shameful that a French mathematician of the old school should be corrected in print by one of the "weaker sex" – Mme de L'Hospital. When Leibniz and the Bernoullis did refer to Newton, it was almost always in relation to the *Principia* – what else indeed was available for their appraisal? Although (like Huygens) they found Newton's concept of gravitational attraction unintelligible, they readily admitted its pragmatic usefulness and the sweep and power of the mathematical physics Newton had constructed with its aid. Thus Johann Bernoulli found good sense in Newton's idea that every particle in the universe has a field of influence (*sphaeram activitatis*) embracing every other particle, though he was puzzled by Newton's introduction of a discontinuity in gravitational action between space outside a large mass (where the force it exerts is inversely proportional to the square of the distance from its center) and *inside* the large mass, where Newton makes the force from the center proportional simply to the distance. (To us it may seem strange that Bernoulli had not thought attentively enough about Newton's propositions to see that the statements for the different cases are consistent and that the inverse-square law of force between the constituent *particles* of large masses always holds.) Leibniz shared Bernoulli's opinions, and noted his own surprise that Newton had not expressed the a priori argument (deriving, in fact, from Kepler) in favor of the inverse-square law, that a central force could be considered as acting on the surface of a sphere, like the intensity of light radiation, whose "density" per unit area is therefore inversely proportional to the square of the radius of the sphere.[5] But such comments are rare enough. Tschirnhaus, De Volder, Malebranche, Nieuwentijt, Hermann, and others are mentioned more often than Newton.

In fact, the contrast in this respect between Leibniz's later correspondence with the Bernoullis, when the calculus was moving forward very rapidly, and the earlier letters exchanged with his former mentor Huygens is marked, for during the period 1690–1 (when Leibniz's first reading of the *Principia* was fresh in his mind) Newton's mathematics was frequently mentioned. For example, Newton's flat assertion that "there exists no oval whose area cut off by appointed straight lines" can be represented by a finite equation was debated at length between Leibniz and Huygens, both of whom thought they could offer counter-examples, though neither was a true oval. Here too Fatio figured (in *his* correspondence with Huygens) as the self-appointed champion of Newton's generalization, rather ineptly as Leibniz was able to produce a general quadrature in another particular case under discussion that Fatio had declared to be impossible. Leibniz thought (as it proved, rightly) that there was some error in Newton's demonstration of the ovals theorem, quoting Horace's remark that an author may well nod in the course of a long work, but his admiration for Newton was undiminished.[6]

Inevitably, a more telling opportunity for the Leibnizians to assess Newton's contributions to the recent development of mathematics was presented by the appearance, as already mentioned, of something more than a bare account of fluxions in the second (first published) volume of Wallis's *Mathematical Works* in 1693. In fact, long before copies of the book arrived on the Continent (which, it seems, they did not speedily do – it is a large folio), Leibniz received from Huygens, who had them from David Gregory in 1694, copies taken from the relevant pages; Leibniz expressed in return his satisfaction at seeing the meaning of the enigmas in the *Second Letter* of long before and a strong preference for his own algorithm and procedures, but also some disappointment that Newton had not achieved greater perfection in the method of quadrature or integration. When, after two more years, an account of Wallis's volume appeared from Leibniz's pen in the Leipzig *Acta*, he chose to pass over the material that Wallis had received from Newton. As he had formerly remarked to Huygens after first reading it, "Newton's calculus agrees with mine," so now he quoted in the review Wallis's phrase that Newton's method of fluxions was analogous to the calculus of Leibniz, that Barrow's method was older than both, and that all were based on Wallis's own *Arithmetic of Infinites* without making any comment on Wallis's historical interpretation, which (as we know) was wide of the mark.[7]

Before he had seen Leibniz's review, Johann Bernoulli mentioned the publication of the first two volumes of Wallis's *Works* in a letter to Leibniz of July 1696, noting that much in them concerned Leibniz and that his calculus had not been praised as much as it ought. In reply, Leibniz commented on the "Newtoniana" in it (as formerly to Huygens), noting that he had hoped for more material bearing on integration; nevertheless, he added, "it must be admitted that the man is outstanding." Leibniz's reluctance to be much excited seems to have upset Bernoulli, ever ready to see the worst side of everything and everybody (not least his own elder brother), because in his next letter he broke out into a long paragraph asserting that the method of fluxions differed in no way from the differential calculus, save by change of name: Leibniz's differential was for Newton a fluxion, the former's sum [integral] the latter called a fluent, and so on. The actual operation was the same in both cases (putting $\dot{x}$ for dx, etc.); "so that," he went on, "I do not know whether or not Newton contrived his own method after having seen your calculus, especially as I see that you imparted your calculus to him, before he had published his method" (in the *Principia*). Bernoulli was equally cross with his brother Jakob for putting the idea into Wallis's head that Leibniz's calculus was founded on Barrow's method of tangents – "You see how much more honest a judge of your invention I am than my brother is."[8]

Here was an explosive suggestion. It was the first hint, in the story of the calculus so far, that someone had cheated – a hint made in private, it is true, and not to creep into print for another seventeen years, by which time very bad things had been said about Leibniz. There is nothing to lead one to believe that Bernoulli's suggestion did not come as a complete and unwelcome surprise to Leibniz, who seems to have lived calmly with the conviction that Newton had possessed a method similar to his own, but presumably less perfect, at the time of writing the two great letters in 1676. That Newton might have fashioned or refashioned his fluxions only after Leibniz's reply, which outlined his own algorithm, or even after 1684 had surely not occurred to him – after all, as he now told Bernoulli, he had not read Wallis's pages very carefully. Again, we should dissociate Leibniz from the technical error committed by Bernoulli of saying that fluxions and differentials were identical; as Newton could easily prove in due course (for the benefit of any impartial reader), they were equivalent but not identical in conception. (The fact that Newton himself later, and in retaliation, was to claim that Leibnizian differentials were merely fluxions spoiled does not exonerate Bernoulli from the fault of basing so serious a charge as plagiarism on so ill-examined an identity.)

Allowing that it was true that he had been the first to write openly about the new procedures, Leibniz felt (as he answered Bernoulli) unable to judge whether Newton had profited from his disclosure and would certainly not have dared to say so: "Moreover I could easily believe," he went on, "that he possessed some very remarkable knowledge at that time which, in his usual way, he has greatly polished up in the subsequent period." Leibniz was clearly anxious to drop the uncomfortable subject; Bernoulli, still raging (at great length) against his brother Jakob, took the hint, while the discovery (at the sale of books following Huygens's death) of an intriguing list of "errors" in Newton's *Principia* – nearly all of which, as we now know but the Leibnizians did not, originated with Newton himself and had from him passed through Fatio's hands – provided another focus of scandal-mongering excitement.[9]

Looking at the most private records of the history of mathematics during the decade 1689–99, the historian seems to discern a curious symmetry: Among the Newtonians, as among the Leibnizians, there was a natural ignorance of the genuine and highly personal origins of the calculus in the mind of the "other" claimant to the invention; partly for this reason, there was a tendency among the followers of either side (far less noticeable in the principals) to assume that there was a unique true calculus, which the "other" inventor had not discovered; on either side, the principal was supported by a chief lieutenant, who claimed to understand the master and his calculus better than any other mathematician; and before the end of the decade this lieutenant had in each case formulated the *possible* (not, by any means, *certain*) interpretation that the "other" claimant to the calculus was a fraud, a plagiarist, or at best a highly imitative second inventor. It was Fatio's tragedy that he first voiced this possibility, in anger and scorn.

Just as Leibniz certainly sought from Bernoulli no accusation of Newton, so Newton (to the best of our knowledge) has left no indication of an earlier awareness of Fatio's *Investigation* nor of involvement in its nasty aftermath. Nothing in the behavior of the two principals during the last decade of the seventeenth century spoils the picture of a European community of scholars engaged in the improvement of mathematics, within which the "British" (Wallis, Newton, Gregory, Craige, and Fatio) were neither ignored nor despised by their Continental colleagues. In public, rivalries were keen; in private, criticism of incompetence and pretense might be sharp, but there is even now little indication of a

possible buildup of tension, unless it is in the self-conscious suffi-
ciency of exclusive excellence that Johann Bernoulli attributed to
his (unacknowledged) epistolary partnership with Leibniz. He,
the master's chief and most faithful disciple, was also (as he knew)
in his own right a mathematician of great power and invention.
Jakob, L'Hospital, Varignon – all such were in a measure outsiders
and bunglers, and the "British" (in Bernoulli's eyes) really had
no share in the modern movement at all. Wallis was an antique,
xenophobic fuddy-duddy, the rest (except Newton) were mere
second-raters whose silly pride continually exposed their inade-
quacies, and Newton seemed to succeed by inexplicable flukes.
Bernoulli had a brilliant influence on the progress and diffusion of
the calculus, but he saw his work as that of a general organizing
subordinates in accordance with a strategy he (in privileged con-
sultation with Leibniz) should determine. Those outside the elite,
those who criticized its works, became increasingly his personal
enemies.

The "British," too, were restive, as one may most clearly see in
Wallis, because Newton had failed to put himself forward as a
national leader. They could not but feel excluded from the renaissance of mathematics taking place in Germany, which Craige,
rather feebly, had tried to bring to England. Everything points to
Newton's essential passivity, after renouncing academic life in favor of the King's business; it is utterly out of character and probability that he would set Fatio on to create a scandal. Hence it is not
merely possible but highly probable that the evil construction of
Leibniz's actions and publications first voiced by Fatio de Duillier
was devised by himself, not received readymade from Newton.
Of course, he received the facts about Newton's early work on
fluxions, about his correspondence with Collins and Leibniz, and
about Leibniz's remarkable ignorance of advanced analysis (as
Newton saw it) in 1676 from Newton's lips and papers; also Newton may well have indicated that he thought Leibniz's manner of
bringing the calculus to the world's attention less than handsome.
But it is difficult to believe that he could have disclosed to Fatio,
and to no one else, a hint of Leibniz's criminality; whereas Fatio is
just the person to have seized on this explanation of the events for
himself.

It may seem unfair to lay upon Fatio's shoulders the responsibility not only for beginning the calculus squabble (which is beyond doubt) but for inventing it in the first place. There is, how-

ever, an inherent probability that he would not have published the charge of plagiary against Leibniz in 1699 if he had received the idea from Newton, who would surely have pledged him to silence. Perhaps Fatio was made the more rash in his anger against Leibniz because he doubted that Newton himself shared his belief in Leibniz's rascality. The psychological complexities here are indeed fascinating, but as they must remain purely speculative, unprofitable.

In Fatio's defense, however, one can see that he had jumped for the only way out of the dilemma in which Newton found himself after the successful emergence of the Leibnizian calculus that was capable of salving Newton's own pride. Having failed to publish his early mathematical treatises, having failed again to lay his cards on the table after staking his claim in the *Principia* in 1687, after allowing a partial (in both senses of the word), if not false, picture of his relations with Leibniz to appear in the first two volumes of Wallis's *Works*, after failing to issue "On the quadrature of curves" in the 1690s, it was too late (in the face of the Continental achievement) for Newton simply to put himself up as a rival, or even as antecessor, of Leibniz. For fluxions to be announced as an alternative, superior system to Leibniz's calculus ten or fifteen years after the former was first published would have been almost absurd – and if Newton might have been content to leave the ultimate recognition of his mathematical genius to posterity like Thomas Harriot, that was not Fatio's way. The only tactic by which a "Newtonian" could (if he wished) at this late hour recover the ground lost to Leibniz was by showing that Leibniz's fame was spurious: that historically the great discovery had been made by Newton, and that in the highest tradition of technical mathematics, Newton's formulation of the calculus concepts had been superior. The last claim is, of course, very significant, as it was the only claim that raised the dispute above the purely personal level. If, as I myself think likely, it was Fatio who saw, not that in order to get Newton back into the fame it was essential to denigrate Leibniz, but rather the other way round, that by denigrating Leibniz (which was his object) he would open the way to secure for Newton the reward for his true merits, we must allow him the compliment of looking at the situation logically. The only other course open to Newton now was to maintain a complete silence, and this I think he would have preferred.

For the moment, Fatio's *démarche* seemed a damp squib. The

calm that had lasted through so many years was to outward appearances barely fluttered. The first news of Fatio's *Investigation*
reached Johann Bernoulli from Varignon in Paris, and he at once
passed on Varignon's letter to Leibniz. Fatio was, he remarked, a
man of a dark and misanthropic disposition; possibly he might
have received some illicit mathematical help from papers formerly
studied by his brother Jean Christophe when he was Johann's pupil in Basel. (One accusation of misdealing easily breeds another
in rebuttal.) Whatever Fatio might boast about the Newtonian
calculus while depreciating that of Leibniz, Bernoulli hoped they
would be able to show him something into which he would not
easily penetrate. However, Leibniz had already received a copy of
the *Investigation* from L'Hospital, which he, in turn, soon shared
with his friend; Leibniz reacted a good deal less strongly, remark-
ing that he meant only to make a gentle rebuke by way of reply
rather than to treat Fatio as he deserved, "for it would be a ridiculous spectacle, I think, if learned men who profess higher standards than others should exchange insults like fishwives." Nor did
he take the trouble to reply extensively to a very long letter from
Bernoulli written in August 1699 when he finally had the *Investigation* in his hands, which is entirely concerned with Fatio's mathematical merits (or otherwise) as shown by his handling of the
brachistochrone problem and the problem of the solid of least resistance (as defined and solved by Newton in the *Principia* though
without demonstration of the solution). The latter, Bernoulli said,
he had himself solved by a more natural and easy approach while
lying in bed without pen or paper – so much for Fatio's cumbersome arguments.

Although Bernoulli was mistrustful of Wallis's partiality toward
the English nation – which Leibniz was inclined to regard as no
more than laudable patriotism – Leibniz was by the autumn satisfied that Fatio's accusations only reflected a personal aberration.
He had received from Wallis the assurance (as courteous as could
be, he described it) given to Wallis by the secretary of the Royal
Society, Hans Sloane, that Fatio had only received the society's
imprimatur for his *Investigation* by means of trickery and felt confident – on what authority we do not know – that Newton himself
was annoyed by Fatio's accusatory conduct. It was obvious, Leibniz thought, that Newton would not wish to be involved by Fatio
in a dispute with Leibniz and Bernoulli, nor have the theory of
gravitation "improved" by untried notions of Fatio's, nor lastly

would he have wished to be covered with praise by Fatio (however well deserved such praise might be) as a cloak for Fatio's own unworthy activities. In all these exchanges there is no sign that either Bernoulli or Leibniz thought that the latter had actually been calumniated by Fatio, and they seem to have treated the issue as one of Fatio making a fool of himself.[10]

Of the three mathematicians who made public comment on Fatio's rare and trivial little publication (whose main ostensible topic was "Fruit-Walls Improved" – by sloping them to receive perpendicularly the oblique rays of the sun!), the Marquis de l'Hospital, noting that the method of Fatio's solution of the problem of the solid of least resistance did not seem the same as Newton's, went on to publish his own investigation of this problem. Leibniz, writing as an anonymous reviewer in the *Acta Eruditorum*, gave a very subdued account of it, entirely omitting those passages that were rather excitedly directed against Bernoulli and himself and merely indicating that no one had intended to insult Fatio by leaving him out of a list of Europe's most gifted mathematicians; the review states Fatio's personal grievances fairly, and if it allows him no praise, the worst fault of which he is explicitly accused is that of prolixity in his mathematical proofs. Finally, following immediately after this review, Leibniz also published extracts (in slightly revised form) from the letter about Fatio's *Investigation* that Johann Bernoulli had written him in August 1699. (Afterward, Bernoulli was surprised that his somewhat caustic expressions had gone into print unsoftened, but he had indeed given Leibniz the freedom to publish the letter as it stood.) Bernoulli also made no allusion here to Fatio's claim that Newton was the first inventor of the calculus, confining himself to a defense of his own conduct in proposing the brachistochrone problem and not failing to make some scathing remarks about Fatio's competence as a mathematician for which, perhaps, he had never had so much respect as Leibniz had shown. Fatio got a fairly rough handling, though in polite not abusive terms, as for example for his claim that the solid of least resistance problem (solved by Newton and himself) was more difficult than that of the brachistochrone proposed by Bernoulli as a test, a claim that Bernoulli denounced as absurd, but nothing of this touched Newton or the relative precedence (or merits) of fluxions and calculus.[11]

Leibniz's own rejoinder to Fatio had by then been completed and approved by his friend Bernoulli, though it was to be published only after a long delay in May 1700. The reason for this delay is not obvious – it may have some connection with Leibniz's engagement with the affairs of the newly founded Berlin Academy of which he was appointed president, a further manifestation not only of Leibniz's personal intimacy with the members of the ruling houses of northern Germany but of his now assured position as the most distinguished scholar and philosopher of the whole Germanic world. Leibniz had previously been able to congratulate Johann Bernoulli on his election as rector magnificus of the University of Groningen for a year's term; now Bernoulli could not only compliment his honored friend in return, but (as he said) congratulate the whole republic of letters on receiving such support from so powerful a prince as the Elector of Brandenburg, for the foundation of the Berlin Academy would bring glory to the Elector's name and to all Germany, "our common fatherland"; this new society, he hoped, would soon exalt itself over those of neighboring countries as the cypress above the viburnum – even in lighter mood, Bernoulli could not avoid a strident note. While all these solemn matters were pending, perhaps Leibniz had little time to finish his business with Fatio, though he had indeed got it out of the way before announcing the great new events to his friends and correspondents.

In his *Reply* to Fatio de Duillier, tardily published in the *Acta Eruditorum* in May 1700,[12] Leibniz, after explaining why the reply was necessary, first defended the issuing of challenge problems, then dealt with Fatio's indignation at not being listed among those judged capable of solving the brachistochrone problem. Had he not, Leibniz wrote, also omitted from the list to which Fatio objected the names of Wallis, Hooke, Halley, and Craige, not to mention Tschirnhaus, De La Hire, and Varignon? He had never said that only those few mathematicians he had then named could tackle the problem profitably, but only those who had mastered his calculus, or something like it. If Fatio had achieved something worthwhile when he at last came to publish, Leibniz would have cried his praise; but, in fact, his claim to have solved the problem of the solid of least resistance by a method more simple than Newton's was false, because Newton's method led to a construction by hyperbola areas, that is by logarithms, whereas Fatio's involved second differentials.

What was the reason for Fatio's hatred of himself, Leibniz asks? It was, that he had innocently preempted Fatio's mathematical dis-

coveries of 1687 by his own publication, ignored by Fatio, of three
years before; in the ancient saying, perish all those who spoke our
words before we did. Such is the weakness of human nature that
it would be surprising if a young man ambitious of distinction and
fame should not have fallen for the temptations to which Fatio
had yielded. Continuing in this moral vein (which surely must
have seemed insufferably pompous to its victim), Leibniz wrote
some unwittingly prophetic phrases:

> Few attain to such a state of virtue, that they can love that
> virtue the more dearly which is damaging to themselves; so
> much the less if (as Fatio does of me) they fabricate suspicions
> that another person has won fame not by the straight road
> but by devious practices, for surely mistrust is a feeling of
> hostility. We can readily conceal under a zeal for justice sen-
> timents which, plainly acknowledged, would disgust us.

Leibniz, when provocation really stung him, was to prove as human as Fatio or as Newton, and when he told Fatio again in the
next smug sentence, "In truth, the more I understand the defects
of the mind, the less I grow angry at any aspect of human behav-
iour," he showed excessive confidence in the solidity of his own
character under severe pressure.

Next Leibniz took up Fatio's appointing himself as the advocate
of Newton, with whom he had no quarrel. For it was certain that
Newton had several times conversed with Leibniz's friends in such
a way as to reveal an unfailingly high opinion of him, with never
a sign of any complaint; and in public too Newton's attitude toward Leibniz had been such that to resent anything would be unjust. As for himself, Leibniz went on, he had praised Newton's
great merits at every opportunity:

> and he himself knows best of all, and has sufficiently indi-
> cated to the public when he published his *Principia* in 1687
> that certain new geometrical discoveries which he possessed
> in common with myself, were owed by each of us to his own
> reflections without either receiving any enlightenment from
> the other, and that they had been disclosed by me ten years
> before. When I published the elements of my calculus in
> 1684, there was assuredly nothing known to me of Newton's
> discoveries in this area, beyond what he had formerly signi-
> fied to me by letter, that he could determine tangents without
> first removing irrational quantities (which Huygens informed

124

me, later, he too could do – although still wholly ignorant of
that calculus) but as soon as I saw Newton's *Principia* I perceived that Newton had gone much further. However, I did
not know until recently that he practised a calculus so similar
to the differential calculus . . .

that is, when the first two volumes of Wallis's complete *Works*
appeared, and Huygens promptly sent Leibniz a copy of the passages relating to Newton.

So Leibniz professed to take Newton's declarations at their face
value – save that he seemed to pass silently over Newton's claim
to have been master of the calculus of fluxions since 1666 – and to
accept gracefully the facts of independence and parallelism in their
mathematical development. Superficially, at least, all was perfect
harmony, and Leibniz was still secure (and on this point he was
always to be secure) that he had *published* the first outline of the
calculus *in print* in 1684. Behind this, as he had previously reminded Fatio, was his personal knowledge of keeping his discovery to himself for the Horatian epoch of nine years. What of Newton? For how long had *he* been master of the fluxional calculus?
For eleven years before 1687, or twenty-one – or perhaps only
three? Leibniz did not say, though implicitly in his words runs the
admission that Newton must have had at least the beginnings of
his own method at the time of the correspondence of 1676. There
was nothing here to which Newton could make a reasonable objection, nor could he well resent Leibniz's courteous request that
Newton should at last publish the work he had so long kept private nor Leibniz's explanation that the brachistochrone problem
had not been a challenge directed personally at Newton.

This was not yet the end of Leibniz's paper; he was not yet done
with Fatio's follies and meant also to add something to show that
he was by no means outworn as a mathematician. In the course of
this continuation, Leibniz invited Fatio to consider that what had
been at stake in the brachistochrone problem was not any particular line of descent but the method (of great importance and generality) by which problems of maxima and minima could be
handled. "No geometer that I know of," he wrote, "before Mr
Newton and myself, had that method; just as no one before that
geometer of great fame had proved by any public example that he
possessed it; and before Mr Bernoulli and myself no one communicated it. And yet it is obvious that this is the higher part of
the method of maxima and minima, extremely valuable in the

application of geometry to mechanics and physics, since by its means the most apposite shape for accomplishing something may be chosen from all possible shapes."[13] If the language here is a trifle compressed, the sense is obvious. Only within the calculus could the higher problems of maxima and minima be tackled; Newton and Leibniz first mastered this use of the calculus, while Newton had first published the solution to a problem obtained by so using the calculus – the solution in the *Principia* to the problem of the solid of least resistance. And finally Bernoulli and Leibniz first described this use of the calculus in print. The admission, and its tribute to Newton, are handsome, and Newton later was to quote it as such for his own ends. Though again conceding nothing to Newton by way of chronological *priority* in the general conception of the calculus, it certainly sets him and Leibniz on a level of equality and *does* allow Newton the priority of the first published solution (even if it was without analysis or proof). Implicitly, as any mathematician of the time would perceive, Leibniz's words speak a high – but by no means too high – esteem for Newton's mathematical prowess in 1687. Indeed, no mathematician then alive, unless it were Leibniz, could have tackled the solid of least resistance problem as Newton had successfully done.

Newton was later to argue, on the basis of Leibniz's *Reply* to Fatio, his earlier letters to Wallis and Leibniz, and other documents already discussed, that Leibniz before 1700 never made any claim to have preceded Newton in the discovery of the calculus, and suggested further that it was Wallis's death in 1703 that removed a restraint from Leibniz's claims.[14] At least the first part of Newton's assertion seems to be true, though not perhaps fully in the sense Newton intended. Leibniz always remembered, correctly, that he had laid the foundations of the calculus late in 1675; from his point of view, the fact that Newton afterward *claimed* to have been in possession of the method of fluxions in the following year, a claim that seemed to be sufficiently borne out by supporting evidence and therefore could be accepted by Leibniz himself, indicated that the origins of Newton's method must at least be roughly contemporary with his own. If Newton further maintained that his first work in the calculus was actually older than 1675, this need not disturb Leibniz either, so long as Newton correspondingly accepted Leibniz's claim (which was, and is, also supported by the evidence) that he had known nothing of this antecedent discovery. For so far as the exchanges between Newton and Leibniz of 1676 are concerned, it obviously made no difference, in respect to the state of *Newton's* knowledge of fluxions in 1676, whether that knowledge was a year old or ten years old. Looking at the issue still from Leibniz's point of view at this stage, the threefold conjunction of his independent discovery of the calculus in 1675, of his first publication of the calculus in 1684, and of the development of the calculus by his school in the last decade entitled him to an unassailably high position in the eyes of the world. At this stage, therefore, the fact that Newton's unpublished papers, of which he and the public had been equally ignorant, anteceded his own was irrelevant. If he himself were what Newton called the "second discoverer" of the new method, he nevertheless (and justly, one might say) would stand before the world in the first rank of discovery.

We may well attribute to Leibniz another reason for psychological assurance, which Newton did not yet understand nor perhaps did he ever do so fully. For Leibniz knew that he possessed the beginnings of a great system of mathematics, whose enormous potential was only starting to appear. It may be doubted whether he supposed Newton to be in an equally commanding position. That Newton had a "method" (just as Descartes, Fermat, Sluse, and others had formerly devised "methods" of tangents) that was capable of handling problems of tangency, maxima and minima, and so forth generally, just as he had another "method" for dealing with problems of quadrature by means of infinite series, Leibniz could well believe. Others, like Huygens, before they learned the calculus, had possessed similar "methods" of great usefulness and scope, which nevertheless lacked the great simplicity and generality of the calculus and did not for all their ingenuity in application to particular problems constitute a new system of mathematics. All these recent "methods" were, in fact, extensions, refinements, developments of earlier mathematical ideas – the ideas of Cavalieri, Descartes, Wallis, Fermat, and others – not the opening of a door to a completely new realm of mathematical invention, such as Leibniz believed his discovery of the calculus to be. Newton could well (as Leibniz saw it) have made enormous progress with such new methods without having hit upon his own discovery in its full beauty. Leibniz seems to have been really hurt by the voices of Newton's friends only when he at last realized that it was *this* claim, that is a claim for the full majesty of his own discovery and not for the discovery of some inferior method or methods that was being thrust against him. For a long time this assertion of Newton's total equality with Leibniz was merely impudent and foolish in the eyes of Leibniz and his friends, to be at first ignored, then firmly rebutted.

7

OPEN WARFARE:
1700–1710

EWTON'S claim that Wallis's death, by removing from the
scene the last of the older mathematicians, permitted
Leibniz to paint an exaggerated picture of his priority in
the development of the calculus does not seem plausible. However
Newton might view Wallis, it is perfectly evident to us that in his
correspondence with Leibniz, Wallis was far from displaying
skepticism of Leibniz's rights to the calculus. Moreover, it would
be evident to anyone having no more intimate source of information than Wallis's own *Mathematical Works* that Wallis had
known nothing of Newton's mathematical development before
1676, nor of the Newton-Leibniz letters of that year, until long
afterward. Wallis might indeed have proved, as an Anglophile, an
ardent defender of Newton, but not on the basis of independent
personal knowledge or (one might add without disrespect to one
who had been a considerable mathematician in his own day) an
independent personal capacity to judge the mathematical subtleties involved in the methods of differential calculus and fluxions.
In actuality Wallis's own role in the slow warming up of the calculus dispute had been to act as an uncritical mouthpiece for Newton. In his surreptitiously composed and anonymously published
Account of the *Commercium Epistolicum* (*Correspondence*) Newton
was later to quote with emphasis Wallis's declaration, in the Preface to the first volume of his *Works* (published in 1695), that in the
letters of 1676 "*methodum hanc Leibnitio exponit*" (He explains this
method to Leibniz), but, as already remarked, in any ordinary
sense of the word "explains," Wallis was mistaken, for Newton
did *not* there explain the method of fluxions to Leibniz or even so
much as say in what it consisted.[1] And in any case, Wallis was
there writing – as so many Newtonians were to write – under
Newton's instructions. The truth is, that just as there were no
witnesses to Leibniz's development of the differential calculus be-

tween October 1675 and 1684, so equally there was no one living who could speak from personal knowledge of Newton's mathematical discoveries between 1666 and 1685. The one man who might have done so was John Collins, long dead.

It seems far more likely that just as the next incident in the developing quarrel between Newton and Leibniz was caused by Newton's publication of two mathematical treatises in 1704, so it was the content of one of these treatises in particular that altered Leibniz's view of Newton's evolution as a mathematician. Not yet does he make any outrageous charge against Newton, such as Johann Bernoulli had privately proposed to him (chapter 6); the words Leibniz was to print were of subtle meaning and were possibly defensible as not amounting to a charge of plagiarism. They were words that Newton could and perhaps did silently tolerate for a number of years. But they were words that definitely claimed chronological priority for the Leibnizian calculus.

It is as though Leibniz at last discovered from Newton's treatise "On the Quadrature of Curves" just what the method of fluxions was – that it was far too close to the differential calculus for comfort; that it was equivalent to his calculus, was indeed identical to it in all but symbolism. Newton had not, after all, put together a series of ad hoc mathematical devices by improving on the tools that Sluse, James Gregory, and others had already used; he had adopted (in a modified form) the basic idea of the calculus itself. For Newton a varying quantity x increased or decreased ("flowed") through a moment of time denoted by o, the rate of flow being written $\dot{x}$. Then the product, $o\dot{x}$, or "moment of x," is the change in the value of x occurring in the time o. With o vanishingly small, this is equivalent to Leibniz's difference between two successive values of the variable x, so we may (as Newton recognized) write $o\dot{x} = dx$. But whereas dx is a "moment" of the same nature therefore as Newton's o, Newton's fluxion $\dot{x}$ is a rate (or ratio), though as Newton put it (referring to himself in the third person):

> when he is not demonstrating but only investigating a proposition, for making dispatch he supposes the moment o to be infinitely little, and forbears to write it down . . .

That is, operationally $\dot{x}$ might stand for $o\dot{x} = dx$, though conceptually the fluxion cannot be put equal to Leibniz's difference.

Leibniz, though a hurried reader, was a sharp thinker. He had been acquainted with Newton now for nearly thirty years. It was already fifteen years since he had seen what unconventional and difficult mathematical problems Newton could solve in the *Principia*, without giving overt evidence of an unconventional method of analysis. He had read nearly ten years before the narrative of fluxions published by Wallis. Now at last, for the first time, he had in his hands a pair of mathematical treatises by Newton himself, treatises that he had himself through the years begged Newton to complete and publish, treatises of whose immediate history and appearance to the public in this form, however, he knew nothing. Clearly, Leibniz, for all his great position in the world of scholarship and diplomacy, would not regard Newton's belated emergence in print as a trivial event, all the more because it came soon after an earlier shock to Leibnizian complacency.

This was the appearance, in the summer of 1703, of a book by George Cheyne (1671–1743), a Scot who had settled in London during the previous year. Cheyne was one of a group of ardent Newtonians eager to extend quantitative hydraulic analysis to the circulation of the blood in animals and man, and in his first book, *A New Theory of Fevers*, he had attempted to supply a "Newtonian" explanation of the feverish heating of the blood. His second book, however, was a treatise on the integral calculus based on what Newton and David Gregory had already achieved in calculus procedures. *On the Inverse Method of Fluxions* is not an important book, nor did Cheyne make any claim to display great powers of invention in it: His purpose was rather to expound and clarify. Whether even in this lesser object he was successful may be doubted, because both English and Germanic mathematicians detected most grievous errors in the book, involving Cheyne in a series of disputes. Yet such was the thin state of the mathematical literature at this time that Johann Bernoulli could write of it, in his initial comment to Leibniz, that it was "a most remarkable little book, stuffed with very clever discoveries; I know of no one in Britain since Newton who has penetrated so far into these deeper levels of geometry," though he also remarks on the many mistakes it contained.[2]

Equally obvious to Bernoulli was Cheyne's xenophobia. He used the Newtonian method, and although he exploited the Continental discoveries, he made no acknowledgment of them. Everything was attributed to the British, nothing was left to the credit of foreigners; his praising of Newton to the skies (though merited) placed all others in the shadow. This, Bernoulli told Leibniz, is how Cheyne finishes his book:

When I turn over in my mind all these discoveries of the great Newton, I cannot prevent myself declaring that all which has been published by others during the last twenty four years, roughly speaking, relating to these methods or to other not dissimilar methods, is only a repetition or an easy corollary of what Newton long ago communicated to his friends or the public.

Twenty-four years, back to 1678, before the time when Cheyne penned these words in 1702, would give Newton an unchallengeable priority in the calculus, if taken seriously; and the words "*vel publico*" could all too easily be seen as carrying an evil meaning. Not surprisingly, Bernoulli complained that Cheyne would make all the rest of us "Newton's apes, uselessly retracing his steps of long before," and he drew Leibniz's astonished attention to a sentence wherein Cheyne asserted that Newton's "method coincided with that published by Leibniz in 1693, that is at least 17 years after it was discovered by Newton." Bernoulli was at this moment much perplexed to determine the course of his future career, or he might have expatiated longer on this curious reading of history in Newton's still unpublished manuscripts, to the complete neglect of all that Leibniz and his followers had set before the eyes of the world. Well might Leibniz regret later that the English mathematicians were reluctant to show their powers by tackling openly posed problems.[3]

On this occasion Leibniz assumed a notably more Anglophobe position than Bernoulli; he took Cheyne (whose book he too had received) to be no more than a beginner with little understanding of the nature of series: "Whoever has once understood our work," he wrote, "can easily put together such a book; he furnishes no new series nor an elegant theorem." Contempt for Cheyne induced Leibniz to make for the first time in his correspondence a comment that definitely belittles Newton's mathematical achievements:

He tries ineptly to claim for Newton the method of series employing assumed arbitrary coefficients, determined by comparison of the terms, for I published that [in 1693], when it was not apparent to me or to any one else (at least, in the public domain) that Newton too possessed such a thing. Nor did he attribute it to himself, rather than to me. Which of us two had it first, I have not declared. I already displayed it in

my ancient treatise on the arithmetic circle-quadrature, which
Huygens and Tschirnhaus read in Paris [in 1675].

And again Leibniz approved of Bernoulli's rebuking Cheyne for
depreciating the discoveries of the Germanic school:

> . . . it may be the case that just as Mr Newton discovered
> some things before I did, so I discovered others before him.
> Certainly I have encountered no indication that the differen-
> tial calculus *or an equivalent* to it was known to him before it
> was known to me. [Italics added][4]

These last words, taken literally, contradict Leibniz's earlier, more
complacent admission that, in 1676, when writing his own first
outline of the calculus process to Newton in reply to the *Second
Letter*, he was satisfied that Newton too was master of something
analogous. Leibniz was already, early in 1704, through his annoyance with Cheyne, tending to dismiss from his mind the indirect
evidence of Newton's equality with himself as a discoverer of the
calculus, which after all had always rested upon his conviction
that Newton was honest and truthful. If Newton allowed his disciples to put about highly unjust and damaging claims for his own
benefit, Leibniz may have reasoned – perhaps not quite con-
sciously – was it possible any longer to regard him as honest and
truthful? Was not the master brushed with the tar of his pupils'
incompetence and partiality?

That Cheyne meant to be partial to Newton seems quite obvious. Early in 1702 he wrote a letter about the *Inverse Method of
Fluxions* to his mentor, David Gregory, explaining how he had
composed it for the benefit of the physician Archibald Pitcairne,
who obtained from Newton himself a grudging admission that it
might be worth printing (it would not have been in Newton's
character to view such a cobbled-up piece of work with enthusiasm). Now, Cheyne ingenuously confessed, "Necessity which be-
gets so many bad authors has forced me to let it go and I am about
printing it." In this letter he admits indebtedness only to Newton
and Gregory himself, but we are warned that Cheyne had studied
the Continental mathematical tradition in the *Acta Eruditorum* and
elsewhere – as indeed what shrewd student would not? – by the
inclusion of such names as:

> I proceed to demonstrate Mr Newton's second method of
> finding the fluent [integral] of an equation which may express
> the nature of a curve . . . I show how John Bernoulli found
> out his universal canon for such equations.

Because the sources of Cheyne's *Inverse Method* have not yet been determined, nor the extent of his indirect access to Newton's manuscripts, perhaps via David Gregory and other copyists, it is impossible to determine whether he was (as he maintained) wholly educated in the British school of mathematicians, or whether (perhaps by sheer carelessness) he incorporated in his book material that he had in fact derived from his reading of the Continental mathematicians. To Leibniz and Bernoulli, it seemed obvious that he had thus conflated the British and the Continental work entirely to the credit of the former, and later Cheyne was forced to acknowledge that he had not studied the British alone. In any case, even if it could be proved that all that Cheyne claimed for the British had been accomplished by them, it was wholly unfair to pass over the published achievement of Leibniz and his followers in the terms that Cheyne now repeated to David Gregory:

> all these are but a few examples of Mr Newton's (excepting yours) Methods, and . . . all found out within these 20 years by these or not unlike Methods are but either repetitions of, or easie corollaries from these things which he [Newton] has either imparted to his friends or the publick . . .

This, Cheyne continued, was "the part which Mr Newton would have altered but I am not for doing it unless you think otherwise."[5] As on later occasions, Newton was to show himself far wiser and more emollient than his self-appointed champions.

It is curious that just as on the Continent the center of gravity in mathematical interest shifted in the late seventeenth century from France to Germany, a similar and contemporary northward shift removed it within the British Isles into Scotland. Nearly all the mathematicians of this time and a little later, nearly all the ardent Newtonians, were Scots: David Gregory, Craige, Pitcairne, Cheyne, the Keill brothers, James Stirling, Mathew Stewart, Colin Maclaurin; a fact that no doubt provides ground for a comment on the characteristics of education in the English and Scottish universities. The Scots are a people who have prided themselves upon their courage, their dogged pugnacity, and their outspoken reluctance to compromise, all of which traits were in full measure manifested by the Newtonians, whereas the one British mathematician of this time who kept up ample and cordial relations with his Continental colleagues, Brook Taylor, was a Sassenach. One has heard of those who were "*plus royalistes que le roi*"; the same remark might be made of these Scots advocates of

Newton's cause. Nor perhaps is it wholly creditable to the Scottish training in logic that Cheyne should so plainly single out Gregory's own method ("excepting yours"), when he must or should have known that this particular way of using series had been first employed by Newton and was only rediscovered by Gregory (chapter 3), and was, therefore, not "exceptional" at all, being as much (or as little) to be credited to Newton as all else in calculus that Cheyne was unwilling to allow to Leibniz. Gregory's method was in no different position, as regards priority, from methods of Leibniz and Bernoulli. An instance of partiality could hardly be more conspicuous.

Though Newton had been obliging to Dr. Pitcairne's friend, he was not pleased to be thus published despite himself by a hand inferior to his own, an experience however that was to recur several times in these eventful later years of his life. Presumably as a consequence, David Gregory noted in his journal for November 1702 that Newton was now talking of a major program of publication: a new edition of the *Principia*, his long-dormant *Opticks*, "On the Quadrature of Curves," and "The Enumeration of Lines of the Third Order." After Cheyne's *Inverse Method* had actually come out, Gregory noted again:

> Mr Newton was provoked by Dr Cheyne's book to publish his *Quadratures* [*of Curves*], and with it his Light & Colours, &c.

There was also talk that he would publish his *Algebra* (which has nothing to do with the calculus), but he did not.[6] Newton's reaction to Cheyne's book seems too typical to be put down as mere gossip.

It had always been, as it was always to be, in Newton's nature to vacillate over the production of his writings to the point that whether, or when, they were ever printed becomes almost a question of chance or whim. The *Principia* itself – the only book ever written by Newton under the shadow of the printing press – was not exempt from doubt and change as his pen moved on. Its second edition slowly took shape over twenty years, through many rumors of its imminence, till Cotes at last brought it out in 1713. *Opticks*, the treatise on light and colors long famed by repute among Newton's acquaintances, was released only after the death of Robert Hooke on 3 March 1703, for Newton had adhered to his resolve to publish nothing more that Hooke could resent and belittle. The tacking on to this (unfinished) book the two hastily revised mathematical treatises in the first edition of 1704 – a conjunction not to be repeated in later issues – was a pure piece of opportunism on Newton's part. At last his dislike of printer's ink had been overcome when, to his long-standing awareness of the need to do himself justice in mathematics, there was added his particular dissatisfaction with Cheyne's "prepublication" of Newtonian fluxions.

The result is curious. The published *On Quadrature* is an abbreviated version of the incomplete treatise begun by Newton in 1691 when he had been stimulated by Gregory's work on the quadrature of curves by series; Gregory's was not a work on the calculus, whereas Newton's *On Quadrature*, like Cheyne's *Inverse Method*, certainly was as it contained what Newton always claimed as the fundamental enunciation of the method of fluxions. With the later published *On Analysis*, it constituted the principal exposition of this method to come from Newton's hands. To the mathematician, it was, and is, a magnificent and fascinating piece of work, manifesting (in its editor's words) all the qualities of "originality, insight, and penetration" possessed by Newton's acute and fertile mind. But *On Quadrature* and its exposition of fluxions was twenty years out of date – far from the world waiting twenty torpid years for Newton's enlightenment, as Cheyne would have his reader believe, the world, especially the German-speaking world, had got on remarkably well in mathematics without Newton's assistance. The quality of mind that would surely have amazed the world in 1680 or 1690 was of course still evident, but the ideas and procedures of the book were no longer new. To quote Whiteside again:

> Newton's historical importance as author of the "De Quadratura Curvarum" is the minimal one of a lone genius who was able, somewhat uselessly in the long view, to duplicate the combined expertise and output of his contemporaries in the field of calculus.

In other words, Newton by at latest 1690 – and indeed we may say, if I follow Whiteside's assessments correctly, well before 1690 – had reached roughly the point in the development of the calculus that Leibniz, the two Bernoullis, L'Hospital, Hermann, and others had by joint efforts reached *in print* by the early 1700s. They had nothing to teach him – but equally he had nothing to teach them, or even their qualified readers, and they did not fail to see that it was so. Far from being "Newton's apes," they had independently duplicated his hidden, single-handed achievement. From some points of view this is the saddest moment in a long tragic story, that Newton now, too late, brought to light the intrinsically superb work, which was only to draw fresh trouble about his ears.[7]

Perhaps, indeed, it was tactless of Newton to add now at the beginning of *On Quadrature*:

> Accordingly, considering that quantities increasing through equal times and generated by the increase do become greater or less in proportion to the velocity with which they grow and are generated, I cast about for a method of determining the quantities from the velocities of the motions or increments by which they are generated; and by naming these velocities of the motions or increments *fluxions*, and by naming the generated quantities *fluents*, I gradually in the years 1665 and 1666 hit upon the method of fluxions, which I have here employed in the quadrature of curves.

No such autobiographical assertion – the writing of which on a number of occasions caused Newton much trouble – is to be found in the early manuscript version of the treatise. Therefore Newton intended by adding these phrases to make a precise priority claim. But it was not a *new* claim: Wallis had already taken the origins of Newton's method back to the 1660s. Is it perhaps possible that until he saw the method for which this antiquity was claimed in *On Quadrature*, Leibniz did not appreciate the full significance of the claim? For certainly in the tract Newton had presented a very considerable account of the employment of the calculus to obtain quadratures.

As for its companion piece, the *Enumeration of Lines of the Third Order* (that is, the cubic curves), which is a piece of mathematical taxonomy involving very considerable insight and ingenuity, written in 1695, it attracted little interest until James Stirling took up the subject in 1717. It shows the "enduring quality and undulled acuity of Newton's mathematical mind in the middle of his fifty-third year" – well after the supposed mental collapse, which in the eyes of some permanently blunted his powers – but it held no fascination for either British or Germanic mathematicians. The *Enumeration* was the subject of a long but unperceptive summary by Leibniz in the *Acta Eruditorum*; it was left unnoticed in the rather moribund *Philosophical Transactions*; and it trapped the generally accurate Roger Cotes into a notable blunder. Whiteside, in a slightly mixed culinary metaphor, has called the tract "a boiled-downed puree of Newton's lifetime of discovery in the pure and analytical geometry of curves," which fills the reader with admiration "for the range and deftness of its techniques" but proves over-rich in its "drily bottled intricacies . . . its intricacies [that] were not to be absorbed into the common store of mathematical knowledge for many years."[8]

At Groningen in Holland, Bernoulli received the *Opticks* in December 1704; Leibniz, writing of the book from Berlin a few weeks later, gave his first view of it as "profound." But of the mathematical essays he had little good to say: The *Enumeration* was, he thought, correct and to be accepted as no mean improvement of geometry, whereas the *Quadrature* he dismissed as containing nothing new nor (for them) difficult. Newton himself (the phrase − "*ipse Newtonus*" is unconsciously revealing of Leibniz's respect!) − had relinquished the task of pushing integration processes beyond the limits he and Bernoulli had reached, perhaps for lack of fresh tricks. And Bernoulli agreed.[9]

Yet, because Leibniz devoted five full pages (anonymously) in the *Acta Eruditorum* to the two 1704 treatises, he cannot have judged them negligible, and if (despite many flattering references to Newton as "highly celebrated for his great merits" and so forth) Leibniz was to employ in the review a phrase that was to giye Newton mortal offense, I am still not certain that the damage it did was not accidental, rather than intended. The treatises are, Leibniz writes, most distinguished ("*praeclaros*"); he then plunges into an uninspired summary of the *Enumeration*, detailing Newton's technical nomenclature. Turning to the *Quadrature*, he explains how differences are generated by the momentary flow of a point tracing a line, which idea, with its inverse, the "calculus of summation," is the foundation of the differential calculus "discussed by its inventor Mr G. W. Leibniz in these *Acta*" and since developed by him and others.

> Accordingly [Leibniz goes on] instead of the Leibnizian differences Mr Newton employs, and has always employed, *fluxions, which are almost the same as the increments of the fluents generated in the least equal portions of time.* He has made elegant use of these both in his *Principia Mathematica* and in other publications since, just as Honoré Fabri in his *Synopsis Geometrica* substituted the advance of movements for the method of Cavalieri. [Italics in original]

Leibniz then gives more detail about the operations of differentiation and integration (using his usual symbols dx, dy, $\int$), pointing out that, because the inverse process to differentiation, that is integration, "from differences to quantities and from quantities to sums, or fluxions to fluents, cannot always be effected algebraically," it is necessary for the mathematician to know the cases in which an algebraic integration can be carried through as well as the auxiliary processes to employ where it cannot. Mr. Newton (Leibniz continues) has laboured very usefully at both of these tasks but (and the "but" is not without its sting) for further details Leibniz refers the reader to the recent treatises of Cheyne and Craige.[10]

The general tone of this review is one of modest approval. Newton cannot have relished the implication that there was nothing in *On Quadrature* that was not more readily available in the books of the two Scots, but the puzzling passage is that in which the names of the two Continental mathematicians are rather pointedly introduced by Leibniz. To say that Newton "employs, and has always employed," fluxions instead of differences is a true and harmless comment. It certainly does not signify that fluxions were younger than differentials and modeled on them. (The statement: "The Chinese employ, and have always employed, chopsticks instead of forks," it is obvious, does not tell us that forks are older than chopsticks; "instead of," like the Latin *pro*, says nothing about priorities.) For a number of years the passage was indeed left without comment, and Newton himself declared that it had remained unknown to him until John Keill brought it to his attention in 1711. Then, however, he interpreted it as a claim "that Mr Leibniz was the first Inventor of the Method, and that Mr Newton had substituted Fluxions for Differences. And this Accusation gave a Beginning to this present Controversy." Newton protested against Leibniz's explanation of the notions of calculus in his review in terms of differentials rather than fluxions and declared (of the particular passage now in question):

> The sense of the words is that Newton substituted fluxions for the differences of Leibniz, just as Honoré Fabri substituted the advance of movements for the method of Cavalieri. That is, that Leibniz was the first author of this method and Newton had it from Leibniz, substituting fluxions for differences.[11]

Is it possible that Newton, or rather in the first place Keill, was mistaken and that Leibniz (if tactless) did not mean to inflict a silent and deadly wound to Newton's pride? At first reading, surely, Leibniz's phrases seem innocent enough, and he could well have regarded it as appropriate to explain fluxions in terms of differences, as the differential calculus had so often appeared in the *Acta Eruditorum* before. The general (if tepid) friendliness of the review, the absence of any denunciation of Newton, make any injury done seem inadvertent. The sharp thrust is in Leibniz's comparison of Cavalieri and himself, of Fabri and Newton, which only those who knew something of the history of mathematics in the seventeenth century could appreciate. For Bonaventura Cavalieri, who reintroduced the method of infinitesimals to Western geometry, must be reckoned as one of the most conceptually inventive of mathematicians; Fabri was a competent second-rater who, through "the functional reinterpretation of Cavalieri's concept of indivisibles [infinitesimals] by means of a dynamically formulated concept of *fluxus* [flow] approached [toward] similar ideas put forth by Newton."[12] All this was surely very well known to Leibniz. He must have been aware of the *conceptual* analogy indivisible/differential, fluxus/fluxion. Nor could Newton have reasonably objected to this indication of a duality in mathematical thinking about changing quantity: on the one hand change by discontinuous, minute steps, on the other hand continuous change. And perhaps this was what Leibniz chiefly had in mind. But surely now in his mind also was the thought that Newton, like Fabri, was not the first inventor of the concept. Perhaps he did not mean to stress the matter of priority (or relative magnitude of inventiveness). But, if only in the unconscious manner of a Freudian slip, he had made an assertion, plain for all the well informed to see, that he was the originator, Newton the adaptor.

I do not really think that Leibniz was expressing in this most sly and secretive fashion a hot and nourished resentment against Newton, as the latter came to suppose; Leibniz worked in great haste. Indeed, he had by now a jealousy of the Newtonians (rather than of Newton); he had (stimulated by Bernoulli) a suspicion, which he was hardly yet willing to admit or recognize, that Newton had spent the years greatly improving his method of fluxions, with the example of the differential calculus to follow as model. But clearly in this review he did not mean to bare his inward doubts and grievances; they slipped out, with fatal results, to commit him to a position he had never meant to assume and from which at first (if it could be consistent with his dignity) he tried to extricate himself. Was it not Leibniz who (near this same time) wished for more humanity among the learned than among other men: "For if our knowledge does not make us better, we shall have done more for others than for ourselves"?[13] He was not naturally quarrelsome or vindictive. The parallelism Leibniz/Cavalieri, Newton/Fabri held far more, more emotional meaning than he had meant to express or would have expressed had he weighed his words at leisure and deliberately. As wit wounds when laughter is intended, so did Leibniz's too-clever historical analogy.

Despite ill omens and talebearers, the uneasy truce lasted for a further five years. Leibniz paid no further attention to Cheyne, and as for Fatio's insinuations, he had become so wretched a figure in London (condemned to stand in the pillory among other things) that Johann Bernoulli admitted that even his pity was aroused when he saw his suspicions confirmed that Fatio rather deserved treatment with hellebore (for his unsoundness of mind) than a serious refutation. The Leibnizian mathematicians were, in fact, in possession of the field outside the British Isles, and as yet no tough, clever, and persistent Newtonian champion had arisen to dispute their rights. On his side, Newton himself either did not read the *Acta Eruditorum* or failed to discover any malevolence in Leibniz's anonymous review of the two treatises. His own reputation and dignity steadily increased; he became master of the mint, president of the Royal Society, and a knight. He lived in good style and was familiar with the leading British statesmen. His authority as a master of mathematics and physical science was at home unquestioned, and even the Dutch and the French were beginning to court his favor, though, of course, Newton's full victory over foreign opinion only began to appear certain long after Leibniz was dead. Another mathematical book was on the way: This was his famous book on algebra, called *Arithmetica Universalis*, a formalized transcript of the university lectures that Newton had prepared, and at least sometimes delivered, over a number of years and which he had deposited in the Cambridge University Library in the manner required by statute as a record of his performance of his duties. There the manuscript lay for more than twenty years, consulted so far as anyone knows only by Roger Cotes until Newton's deputy and successor, William Whiston, decided to publish it in 1707. It was only with misgivings that Newton accepted Whiston's plan, because he was well aware of imperfections in his treatment, and he did not acknowledge the book by placing his name upon it. Some years later Newton issued an amended version of his own. The *Arithmetica Universalis* is indeed the least inspired of Newton's major compositions, and Leibniz at once correctly guessed the reason for the anonymity of a work thirty years old. "It is one of the ironies of [Newton's] posthumous reputation," Whiteside has written, "that the printed version of his *Arithmetica* was to become his most often read and republished mathematical work, while his more advanced papers on calculus and analytical geometry were relatively little studied or even ignored by the world at large till the present day. Such is fame."

Leibniz, after a hasty reading, made to Bernoulli the rather cool comment that the book was not to be despised, especially because of its examples, and sent Bernoulli a transcript of one passage in particular for his consideration. Bernoulli in turn found the passage interesting and turned it over to his young nephew Nikolaus, who as a result prepared a paper that Johann thought worthy of publication. Leibniz too found one procedure of Newton's so puzzling that he sought further help from Hermann. Not bad, one might say, for the intellectual content of lectures intended for students a whole generation ago! And to be fair to Leibniz, he gave the *Arithmetica Universalis* a good showing in the *Acta Eruditorum* for the benefit of non-English readers. Whiston, he declared, had been right to publish Newton's lectures, for the reader would find in that one small book things that he would seek in vain in vast tomes devoted to algebraic analysis. It was as yet far from true that Leibniz had formed a settled dislike of Newton, or that he was blind to Newton's importance as a mathematician. Moreover, he shared Newton's preference, in terms of mathematical rigor, for the classical geometrical analysis as compared with its modern algebraic equivalent.[14]

We may well imagine indeed that Leibniz would have been quite content to let the present ambiguous state of affairs last indefinitely, and Newton by himself would have been unlikely to disturb his tranquility. Leibniz had little if any reputation to gain by forcing (if he could) the British to moderate their esteem of Newton; as the progress of the future controversy makes plain, he had little desire to institute a definitive historical inquiry into the *origins* of the calculus (even though he believed it would clearly vindicate his own priority) because the *progress* of the calculus in the public domain since 1684 had made the preeminence of himself and his pupils obvious. Similarly, in the scientific and philosophical questions wherein Leibniz and Newton differed, the Newtonians were an insular, idiosyncratic minority. All sound thinkers in Europe were of the same opinion, broadly speaking, as Leibniz and his friends; if Leibniz differed in certain ways from Malebranche, the great power now in France for whom he felt no great respect, they were nevertheless both neo-Cartesians, and it was not until 1712 that Malebranche for the first time endorsed Newton's experimental study of light. Virtually no one on the Continent could as yet view universal gravitation as more than a perverse enigma or a reprehensible revival of occult powers in inanimate matter. The image of Newton prevailing about 1710 was almost the opposite of what prevails today: A nonspecialist will probably now think of Newton as one of the world's great scientists who happens also to have been a brilliant mathematician; in 1710 Newton appeared rather as a mathematician of great ability who was trying, mistakenly, to promulgate ill-conceived ideas about the physical world based (according to rumor) on a harebrained metaphysics. Leibniz had no reason to fear that Newton's physical theories would add to his reputation. He had only (it seemed) to sit tight to watch that reputation sink, as Fatio's had sunk.

Accordingly, a breach of the peace was intrinsically far more likely to begin with the Newtonians, revolutionaries fighting for recognition in a Cartesian universe. The creator of it was, in fact, a man who had already launched attacks on the mechanical philosophy of Descartes, charged with being conducive to atheism and with intellectual arrogance. John Keill, like his younger brother James, a physician, was an enthusiastic Newtonian and one, too, of a dangerous type, because he was a fanatic for a particular philosophical and indeed theological interpretation of Newtonian science. Moreover, he was a man with a reputation to build who had chosen deliberately to make himself known by polemics, in his attacks on the cosmogonic theories of Thomas Burnet and William Whiston. He was to prove a very different champion of Newton from any either Newton or Leibniz had encountered hitherto, almost equally fearsome as a friend or an enemy.

John Keill (1671–1721), born in Edinburgh and a pupil of David Gregory's in the university of that city, had followed his teacher to Oxford when the latter became Savilian professor of astronomy in 1694. He first attracted notice at Oxford by giving experimental demonstrations of modern physical science, probably for the first time in any British university, though such teaching with demonstrations had been started twenty years earlier in the Netherlands. Keill's plan was later practiced with profit in London by such professional teachers as Hauksbee and Desaguliers. He was made a deputy to the professor of natural philosophy, Thomas Millington (1699), and the two small books of lectures he published early in the new century, on the *True Physics* and the *True Astronomy* – by which of course he meant what Newton had established – went through several editions. He was elected a Fellow of the Royal Society in 1700 and Savilian professor of astronomy (in succession to John Caswell, who had followed Gregory) in 1712. He had proved himself an active and forceful propagandist, but not shown any preeminent scientific talent; Halley, it is said, greatly contributed to the improvement of the later editions of the *True Astronomy*. Keill's first paper of an advanced mathematical nature in the *Philosophical Transactions* was *On the Laws of Centripetal Force* (1708), which actually saw the light in 1710.

How did Keill come to play so large a part in Newton's life and affairs? It seems a plausible guess – and no more than guesses are possible – that he had been introduced to Newton by his former teacher David Gregory, who remained on good terms with Keill and consulted him on mathematical matters from time to time. Newton possessed, apparently, only one of Keill's books, the *True Physics* of 1702, which does not suggest any great degree of intimacy between the two men; reading the correspondence of Newton with Keill and others in subsequent years prompts the feeling that Newton did not wholly trust Keill and did not find his powerful advocacy always welcome or appropriate. It is not likely to be true, as has been suggested, that the *Commercium Epistolicum* of 1712 was *edited* by Keill, though Keill certainly drew Newton's attention to passages in print that Newton ultimately used in it. The evidence is rather that Keill made himself Newton's servant (as it were) than that Newton besought him to act in this way. One can only guess, then, that Newton had a slight acquaintance with Keill and his writings before 1709 (when Keill briefly left Oxford and England) and it is, again, more probable than not that Keill, like Fatio de Duillier, wrote his first defense of Newton without consulting Newton himself.

Keill's paper is based on a general theorem concerning central forces, which he had from the Huguenot mathematician Abraham de Moivre, but which was also (he points out) known to Newton (see chapter 9). About two-thirds the way through, Keill explains that, the law of centripetal force being given, the inverse method of tangents (that is, an integration process) must be used to determine the curve traced by a body moving under the action of this force. "All these things follow," Keill then went on quite unnecessarily after stating some results of the integration process, "from the nowadays highly celebrated arithmetic of fluxions, which Mr Newton beyond any shadow of doubt first discovered, as any one reading his letters published by Wallis will readily ascertain, and yet the same arithmetic was afterwards published by Mr Leibniz in the *Acta Eruditorum* having changed the name and the symbolism."[15] The assertion of Newton's priority was here unnecessary because de Moivre and other mathematicians had for some years published papers in the *Philosophical Transactions* employing fluxions, without sounding Newton's trumpet; it was offensive because Keill declared that Leibniz's differential calculus was not merely different in name and manner of notation from Newtonian fluxions but was *changed* (*mutatis*), as though to disguise its origin. Keill's words are virtually an open accusation of plagiarism from Newton's 1676 letters, and it is not possible to understand them as less than a deliberate provocation.

Why did Keill thus attack Leibniz? Many motives may be imagined: Keill may have wished to impress Newton and gain merit as Newton's champion, or he may simply have felt that Newton had been done an intolerable injustice to which he must at last and forcibly call attention. But I think it is more than likely that Keill had a stronger personal motive. Remembering that his article appeared in the *Philosophical Transactions* for 1708, which only came out in 1710, it is possible that the offending phrases were a late insertion, added to the original text; whether this is so or not, Keill would certainly have been aware, when writing them, of the constant and increasingly biting criticism maintained in the *Acta Eruditorum* against forces of attraction, which Keill and others took to form the base of Newtonian physics. And of this criticism Keill himself had been a principal target. Thus, by 1710, the English felt that they had two grievances against the Germans. I shall explore this matter further in the next chapter.

8

THE PHILOSOPHICAL DEBATE

L EIBNIZ'S idea of the basic structure of the universe was, within the context of his era, far more conventional than that which Newton developed and regarded as alone consistent with a mathematical science of mechanics. Newton conceived upon the foundations laid by the Greek atomists one of the grandest generalizations of modern science: the idea that all the matter in the universe consists of particles, of which the smallest are atoms, which are impelled or retained by the forces mutually acting between them into a myriad of different configurations and an endless variety of motions, from which by successive stages all the observed manifestations of nature result. The idea of particles, or atoms, was by no means new; the novelty lay in the idea of fundamental forces, forces of attraction and repulsion, operating directly between the atoms, or particles. The prevailing theory of Leibniz and Newton's time, originating with Descartes, was that what we may ordinarily call a "force," like magnetism or gravity, was only apparent, a kind of optical illusion; the reality lay in the movement of invisible, indetectable particles whose pressures on bodies cause the movements we attribute to forces. To this Newton's thoughts were completely opposed; forces, he thought, were real and prior, though he recognized that there might be still deeper explanations of the way the force worked. The first thing was to find out the nature of the force itself, the laws it obeyed, not to imagine hypotheses about streams of invisible particles. To Descartes, the totality of effects in nature was to be traced back to the ever-changing distribution of an invariant quantity of motion in the universe; nineteenth-century physicists were to conclude that it is the quantity of force (or more strictly, energy) that is invariant. Gravity was the force of which Newton's knowledge was most complete, and for which his corresponding mathematical theory was most perfect; but it was only one of the fundamental forces, virtues, or powers that he believed to be active in nature. "Would that we were able," he wrote in his Preface to the *Principia*, "to derive the rest of the phenomena of nature [that is, the phenomena not connected with the universal force of gravity] from mechanical principles by the same kind of reasoning. For many considerations cause me to suspect that they may all depend upon certain forces by which the particles of bodies are, by processes not yet known to us, either impelled mutually together so that they cohere into regular shapes, or repelled from each other so that they become separated; because these forces are unknown, natural philosophers have hitherto investigated Nature in vain."

To Leibniz, the idea of appealing to unknown mysterious forces was appalling, flying in the face of reason. To him, it seemed that Newton was trying to explain nature by inventing processes more inexplicable than the phenomena themselves. He, on the contrary, believed it right to rationalize phenomena by making use only of those effects that were known and clearly understood: when a piece of matter at rest is observed to begin to move, it must be for the reason that another piece of matter has struck it. In its simplest essentials, we may call this billiard-ball physics, though it was then known as the mechanical philosophy; if we watch any given ball on the table and see it move, we know that ball must have been struck by the cue or another ball; unless, more strangely, it was struck by some chance projectile or shaken by an earthquake. It is possible to complicate billiard-ball physics considerably by inventing a special kind of subtle and invisible matter, constituting an ether, whose particles are able to strike those of solid matter and sweep them along. But the basic idea is the same.

In his *Principles of Philosophy* (1644) Descartes had provided the first all-embracing, compelling model of a billiard-ball universe in which effects are brought about by etherial mechanisms of considerable complexity, and forces are made redundant. Newton learned much from Descartes, but even as a young man he reacted violently against Descartes's naïve idea of mechanism. His principles, Newton thought, were not only far too imaginary: They made the universe too simple and too atheistical. Leibniz agreed with Newton in finding much to criticize in Descartes's application of his principles themselves. He was satisfied, however, that the idea of nature underlying Cartesian physics was sound, only its detailed working out was erroneous, and so he became (in somewhat the same manner as Christiaan Huygens) a heretical exponent of the orthodox philosophical creed, whereas Newton,

147

by introducing quite a different model of scientific explanation, dependent not on eternal motion and incessant impact but on forces, began to establish a new philosophy of nature. Thus, for example, Newton contemplated (as a straight induction from his observations of capillary attraction, that is, the roundness of fluid drops and so forth) the existence of a force of cohesion in matter, distinct from the force of gravity, whereas Leibniz insisted (even after he had read the *Principia*) that there was no rashness or incongruity in imagining a kind of etherial pressure, which would explain both kinds of 'attraction' at once – a pressure making fluid drops round and also making them heavy. If he could imagine such a thing, why should it not be so?[1]

As a young man, shortly before he went to Paris, Leibniz published (with a dedication to the Royal Society of London) a *New Physical Hypothesis*, which is a Cartesian picture of the universe rendered (in its author's eyes) more logical and consistent, though no less freely speculative than the original. For example, near the beginning, rejecting Descartes's idea of a universal matter organized into three kinds of particles, Leibniz declares – almost like some moralists – that everything is made of bubbles (*bullae sunt semina rerum*); water is a mass of bubbles, and air nothing but a more subtle water. As for earth, it is not to be doubted (he says) that it too consists of bubbles, because the basis of earth is glass, glass in a thick bubble. Air, water, and earth are highly porous and are freely penetrated by a very subtle ether, which swirls around the globe in a vortex turning from east to west while the globe of Earth itself rotates from west to east. This ether rotation is the cause of gravity, for matter raised above the surface of the Earth causes turbulence in the ether, increasing as the body is higher up; and the vortex expels any cause of turbulence toward its center.[2] There can be no doubt that the wide variety of imaginary mechanical hypotheses of this kind, which the late-seventeenth-century reader found not only in Descartes himself and his direct followers but among the large group of neo-Cartesians (Huygens, Leibniz, Malebranche, Varignon, Villemont, etc.), was what Newton meant to condemn in the famous phrase "*Hypotheses non fingo*," and more at length in his *Opticks*: 'Hypotheses are not to be regarded in experimental Philosophy."

It need hardly be said that we now know, from his notebooks and correspondence, that Newton himself was willing to entertain, in his developing years, some equally strange mechanical hypotheses (such as the hypothetical mechanical effect of imaginary light-corpuscles) and ascribe many effects in nature to the omnipresence of ether. And in some of his later scientific speculations he turned again (almost, as one may suppose, in desperation) to the idea of an extremely rare, highly elastic Newtonian ether. But what brought his celestial mechanics into existence – after many gropings and after consideration of many tempting follies – was his realization of an extraordinary mechanical truth: If a body moves around some center of force toward which it is impelled (or attracted, either formulation will do) *by any law of force*, it will move around that center in the manner described by Kepler's second law of planetary motion. Now this was the law defining the planet's speed of motion in its orbit, an observational law that Newton found – but not before about 1680 – to be thus explicable on a most simple mechanical assumption. The proof of it was to stand as the first proposition in his *Principia*. The science of celestial mechanics starts with Newton as a mathematical science, not as a fabric of imaginary models.

Leibniz, who was the most intelligently perceptive of men, could not but be struck enormously by the contrast in style and content between Newton's *Principia* and his own little essay of sixteen years before. He appreciated at once the power of such a mathematical philosophy as Newton's – and the sharpness of the mathematical tools Newton was bringing into existence. Yet for all that he would not abandon his philosophy for Newton's. Instead, he tried to prove that his own philosophy could be no less mathematical.

Leibniz first had knowledge of Newton's book through a long summary of it – twelve solid pages – published in the *Acta Eruditorum* for June 1688. It is a factual, neutral review: Newton is praised as "a distinguished mathematician of our time," his hope for the future understanding of the universal role of forces in nature is quoted; his mathematical refutation of the Cartesian kind of celestial vortex is plainly noted; and his theory of the inverse-square law of universal gravitation is described without overemphasis or adverse comment. There is in the review no criticism of Newton as a proponent of mysterious "attractions." Leibniz, then pursuing historical researches in Italy on behalf of the House of Hanover, wrote to his friend Mencke, who was the chief editor of the *Acta*, that he had read the account of the *Principia* "eagerly and with much enjoyment" and added that Newton was a remarkable man, "one of the few who have advanced the frontiers of the sciences." His more detailed comments are rather mathematical than physical, as he went on to list a number of points where he thought he himself had got as far as Newton. As for the "physical cause of the celestial motions" (which was not, or at least the cause of gravity was not, Newton's concern), some meditations of his own had all the appearances of truth, he thought, and so he had now decided to bring them before the public.

They appeared in a strange paper called *An Essay on the causes of the heavenly motions*.[3] It is a paper that has remained almost incomprehensible from that day to this; Keill qualified it as such, and Huygens twenty years earlier had taken the same view. It was certain that an inverse-square law of centripetal acceleration, taken with the law of centrifugal force, which Huygens himself had first enunciated, and assuming an inertial motion in the planet, would produce Kepler's planetary ellipses; how then, by substituting his harmonic circulation in a vortex, Leibniz could deduce the same elliptical orbits was something Huygens protested he could not understand, "not comprehending how you can find room for some sort of deferent vortex in Descartes's manner, which you wish to retain." To Huygens, a deferent vortex was redundant, and indeed Leibniz himself admitted that the planet (he imagined) swam freely in the vortex; it was not driven around by it. We need to have at least an outline idea of this speculative physical model that Leibniz set up against Newton's mathematical propositions founded on the concept of a universal gravitational force. First, Leibniz realized that Newton had killed the Cartesian celestial vortex stone-dead; accordingly, he supposed his own to consist of numberless layers concentric with the sun, each layer's speed of rotation being inversely proportional to its distance from the sun. (This is equivalent to Kepler's second law; but Leibniz did not explain how these speeds could be so nicely arranged nor how a discontinuous vortex could fit in with continuous variation of speed in the planet.) To produce the elliptical shape of the orbit and also the observed changes in the planet's speed – fastest at perihelion, slowest at aphelion – Leibniz endowed it with a reciprocating movement to and from the sun along the line connecting the planet and the sun, much as Islamic astronomers had supposed several centuries before. This reciprocating motion was the product of a play between two forces, the one being the planet's weight of gravitation toward the sun, the second the centrifugal force of its revolution about the sun. We may presume that Leibniz had silently borrowed the notion of universal gravitation from Newton via the *Acta* review, and (like Newton) he was willing to let it be accounted for in any way possible; his own inclination was toward an explanation in terms of another ether (not the matter of the harmonic vortex) pushing the planets toward the sun.

The precise statement of Leibniz's ideas is complex, and the mathematical analysis of them that he provided was very involved (and contained some actual mistakes, besides many confusing misprints). Leibniz was to publish a revised version of it in 1706. He never abandoned the ideas expressed in his *Essay*, which throughout his life he always regarded as marking as great an achievement in celestial dynamics as the *Principia* itself, and far preferable from a philosophical point of view to Newton's work, because nothing in it was contrary to common sense and mechanistic principles. Yet in a long correspondence during 1690–2 Leibniz was quite unable to persuade his fellow neo-Cartesian, Huygens, that his theory made good sense or that it furnished an important contribution to the understanding of planetary motion, even though Huygens himself rejected what he took to be Newton's principle of attraction and thought it necessary to provide a mechanistic explanation of gravity. Huygens simply could not follow Leibniz in his arguments, his peculiar talk of "conspiring" motions, or his vortex of matter that bore the planet around without pressure and without resisting its radial oscillation. Leibniz, however, yielded nothing under Huygens's skepticism. He could not see why "attractions" should not have a mechanical cause. As between his own celestial mechanics and Newton's, he replied, it was necessary to investigate which account was better and if they could be reconciled. He repeated his conviction that his own harmonic vortex was as good a theory as Newton's of universal gravitation. Although Huygens agreed with Leibniz in finding the idea of "attraction" and its use by Newton to explain the tides of the oceans quite impossible, he still regarded Leibniz's vortex as redundant, and in the end declared that it was useless to carry on their discussion.

The most thorough recent student of Leibniz's *Essay* has indeed written of it that it is "regrettable" that Leibniz's vortex theory lacked influence, because it was "mathematically unexceptionable and outstanding among attempts to explain the planetary motions by the action of fluid vortices." However, this judgment passes over the serious mathematical oversight that the *Essay* contained (as the delighted Newtonians emphasized) before its revision in 1706 and the weaknesses and inconsistencies, which, in all fairness, Dr. Aiton has pointed out on earlier pages of his book, where he himself qualifies the harmonic vortex as (dynamically) "superfluous."[4] Perhaps the best point to be made in its favor, as Leibniz himself was fond of saying, was that it explained why all the planets and satellites revolve in the same sense in their orbits, which Newtonian mechanics cannot do, without the aid of extra hypotheses about the origin of the solar system. However, what matters here is not so much the conceptual or the mathematical validity of the ideas about the structure of the universe advanced by Leibniz in the *Essay*, though it is relevant that contemporaries qualified to judge, like Huygens and Varignon, were far from satisfied with its original version of 1689. What matters now is that Leibniz's notions were so very different in character from those of Newton, and that he remained faithful to them. In later years, before and after the revision of 1706, Leibniz always referred to his essay on celestial mechanics with great satisfaction, largely because it seemed to him to deduce from sound physical principles the inverse-square law of gravitation, which Newton had been compelled to leave unexplained as a brute fact of nature justified by induction from the phenomena. Thus the foundations of Leibniz's later open criticism of the *Principia* were always present in his mind.

In fact, Leibniz's personal reaction to the concept of "attraction," which he, like Huygens, found explicit in the *Principia*, was from the first clear and critical. As early as October 1690 he told Huygens that he could not imagine how Newton conceived of weight or gravity. "It seems that according to him," he wrote, "it [gravity] is nothing but an incorporeal and inexplicable virtue, whereas you have explained it very plausibly by the laws of mechanics." (The allusion was to Huygens' own neo-Cartesian *Treatise on Weight*, published in that year though first written before.) It surprised Leibniz that Newton omitted all physical cause for the inverse-square law of gravitation, and he thought it a mistake on Newton's part to have rejected vortices; yet he thought the *Principia* was one of the books that most deserved being rendered fully perfect.[5] An unpublished paper, presumably written when Leibniz was first thinking about Newton's force mechanics in the *Principia*, reveals his deep (and unchanging) philosophical criticism:

> The fundamental principle of reasoning is, *nothing is with-out cause* . . . This principle disposes of all inexplicable occult qualities and other figments. For whenever authors introduce some primary occult quality, they breach this principle. For example, if anyone posits in matter a certain primary attractive force that is not derivable from intelligible notions of body (namely, size, shape and motion), and he means that by this attractive force bodies tend without any impulse towards some body, just as some conceive gravity to signify the attraction of bodies by the bulk of the Earth, or their enticement towards it by a certain sympathy, so that they deny that the final cause of the effect can be derived from the nature of bodies, and that the process of attraction is beyond explanation: he is admitting that no cause underlies the truth that a stone falls towards the Earth. But if he posits that the effects [of gravity] depend not on an occult quality in bodies but on the will of God or a hidden divine law, thereby he provides us with a cause, but a supernatural or miraculous one.[6]

Such ideas were not new nor are they lacking in perpetual appeal. They lay at the roots of the materialist philosophy of Democritos and Epicuros and were to be given fresh currency in the nineteenth century, when briefly, as in Descartes's time, it appeared that all phenomena might be reduced to kinetics. Leibniz would certainly have approved of the suggestion of (among others) Sir William Grove in 1843 that all things might be

> effects of motion and matter. These two seem the most distinct, if not the only conceptions of the mind, with regard to natural phenomena, and when we try to comprehend or explain affections [i.e., properties] of matter, we hypothetically or theoretically reduce them to it [sic!]. The senses perceive the different effects of sound, light, heat, electricity etc., but the mind appears capable of distinctly conceiving them only as modes of motion.[7]

The *kinetic* view of nature is indeed radically distinct from the *force* view of nature to be associated with Newton, of which traces may already be found in the writings of Aristotle and which Grove's contemporaries Faraday and Maxwell were to transform into the *field* theory. Ultimately, the differences may be traced back to ideas of what is "natural" or "simple" or "self-evident," in other words to metaphysics, where exponents of either view may well feel, like Grove, that only one set of related apprehensions is possible to a normal human mind. So, to Leibniz, the nineteenth-century concept of "field" would have seemed as "unthinkable" as the combination of atoms and void or as the idea of action at a distance was to both Leibniz and Faraday. In the end, of course, the Newtonian idea of force is no less and no more a priori plausible than the Cartesian idea of the primacy of motion; the question "What is force?" is no more and no less unanswerable than the question "Where is motion?" But the Newtonian view has the pragmatic advantages of being more flexible and of not requiring the invention of imaginary bodies; for one cannot think of there being motion in a void space (containing nothing that moves), but one can certainly conceive of force in a void space.

The distinction of ideas is far from implying that Leibniz avoided the word "force" or avoided giving it a mathematical significance. Quite the contrary: He drew in his dynamics an important and original distinction between what he called "dead force" (our momentum, mv) and "living force" (almost the same as our kinetic energy, $\frac{1}{2}mv^2$), which he took to be the mathematical integral of the former. And Leibniz was even capable of writing "I find nothing so readily intelligible as force." Indeed, any differences between the mathematical treatment of force by Leibniz and Newton are rather accidental than fundamental. It is solely in Newton's refusal to admit a priori that forces must necessarily be resolved in particulate mechanisms that their serious difference lay. Leibniz made this clear in another unpublished fragment in which he wrote of the relapse into physical barbarism that occurred when this ultimate existence of mechanical causation of forces was denied:

> It pleases some to return to occult qualities or scholastic faculties, but because these have become unrespectable they call them *forces*, changing the name. But the true forces of bodies are of one sort only, namely those effected by an impressed impetus . . .
>
> He who would show that the laws of astronomy can be explained on the supposition of a mutual gravitation between the planets will have done something very worth while, even though giving no explanation of gravity. But if having reached this fine discovery he thinks he has provided an adequate cause so that nothing else remains to be found out, and that gravity is a thing essential to matter, he relapses into

154

physical barbarism and the occult qualities of the School-
men.[8]

Of course, one perceives a methodological difference also. For
Leibniz as for Descartes, there was a continuum between metaphysics and science, and though the truths of the former and the
latter might be found out by distinct processes, those of metaphysics were no less well founded than those of science. That any
particular mechanical hypothesis of gravity was true as well as
valid Leibniz would not have undertaken to prove, but he felt it
to be beyond doubt that *some* mechanical hypothesis must be true.
Newton did not believe this. Therefore he drew a distinction,
which Leibniz did not draw, between the scientific establishment
of the existence and laws of universal gravitation and the metaphysical discussion of possible hypotheses to account for the existence of such laws. He never denied that a cause of gravitation
might be discovered; he merely claimed that the search for this
cause did not seem to be, at least at that stage, within the province
of scientific inquiry. In other words, Leibniz claimed that a proposition about the mechanical cause of gravity was a perfectly proper
scientific proposition, because he believed *all* ultimate causes of
things to be mechanical; and this Newton denied, taking the view
that such propositions must be merely speculative. One can, perhaps, with utter anachronism, apply the Popperian test of falsifi-
ability to these two positions; it is clear that no argument or fact
could possibly ever falsify a belief in the reality of mechanical
causations such as Leibniz held.

Not that imaginary configurations of bodies or particulate
mechanisms in the manner of Descartes greatly preoccupied Leibniz in his maturer years. What he did observe was the spread of
attraction among the British. With whatever justice or error Leibniz (like Huygens) had found an occult concept of gravity in Newton's *Principia* – despite Newton's protestations that attraction
could well be interpreted as the result of impulse, that is, kineti-
cally – Leibniz now noted in the 1690s that gravitation was accounted for by the Newtonians not as an inherent or essential
property of matter (this view, on all the evidence, Newton always
rejected) but as a product of the divine will. The most eloquent
expositor of this interpretation of Newton's discovery was Richard Bentley, chaplain to the Bishop of Worcester and soon to be
master of Trinity College, Cambridge. Bentley was one of the
great classical scholars: His proof that the *Letters of Phalaris* were

forgeries was admired by Leibniz, who followed eagerly Bentley's subsequent controversy with Charles Boyle. In writing *A Confutation of Atheism from the Origin and Frame of the World* (published in English in 1693, in Latin at Berlin in 1696), Bentley was primed by Newton himself, from whom he took the arguments that gravity cannot be an essential attribute of matter and that action at a distance is unthinkable, including the celebrated phrases from Newton's own pen:

> 'Tis utterly unconceivable, that inanimate brute Matter (without the mediation of some Immaterial Being) should operate upon and affect other Matter without mutual Contact; that distant Bodies should act upon each other through a *Vacuum* without the intervention of something else by and through which the action may be conveyed from one to the other.

These refutations of opinions, which were attributed by Bentley to the mysterious and, in fact, virtually nonexistent "Atheists," would not surprise or disgust Leibniz, who could not, however, approve Bentley's equal rejection of mechanical explanations of gravity, nor his arguing that universal gravitation (which Newton, Bentley maintained, had proved to exist) was a "new and positive proof, that an immaterial living Mind doth inform and actuate the dead Matter, and support the Frame of the World." Mechanical explanations, according to Bentley, from their absurd complexity and multiplicity must prove

> repugnant to human reason; we have great reason to affirm, That Universal Gravitation, a thing certainly existent in Nature, is above all Mechanism and material Causes, and proceeds from a higher principle, a Divine Energy and impression.

Leibniz read these words as meaning that since the cause of gravity was "above all mechanism and material causes" it was above the realm of ordinary physical explanation, or as we might say the cause of gravity could not be understood scientifically or rationally by the human intellect and was therefore (as he put it) a perpetual miracle. He also, correctly, supposed that this "miraculous" idea of gravity originated with Newton himself. By 1699 Leibniz had learned that John Locke too was now of Newton and Bentley's opinion, revising the view that he had expressed in the first edition of his *Essay concerning Humane Understanding* (1690) "that bodies act by impulse alone . . . persuaded by the reading of the excellent Mr Newton's [*Principia*] that there is an attraction in matter even at any distance whatever. This [argument] is used by the very learned Mr Bentley too . . ."

As Leibniz made clear in discussing these questions with Thomas Burnet, he himself still adhered to the view he shared with Huygens that "both gravity and elasticity are only in matter by reason of the structure of the system [of the world] and can be accounted for mechanically or by impulse," and (as was ultimately to become clear) he always firmly rejected Newtonian attraction (taken as an attribute of the divine power) not only because he thought it followed from a false scientific metaphysics – one in which every difficult piece of explanation could be got over by saying "God wills it so" – but also because in Leibniz's eyes any attribution of activity to matter was bound to lead to a host of false conclusions in areas of thought having nothing to do with natural science. However, though Leibniz's concern from about 1699–1700 with this (to him) fresh development of Newtonian philosophy appears from time to time in his correspondence and private papers, he did not publicly criticize Newton for his views until he came to publish his *Essais de Théodicée* in 1710. These essays have nothing to do with mathematics or science, and Leibniz's main target in them was the skeptical philosopher Pierre Bayle; nevertheless, Leibniz went somewhat out of his way to insert a passage making his own opposition to even divinely inspired forces quite clear:

> Several philosophers have thought that in the order of Nature one body might act directly at a distance upon several other bodies all at once . . . but, for some time, modern philosophers have rejected the direct physical action of one body upon another distant one, and I confess that I am of their opinion. However, action at a distance has been recently revived in England by the excellent Mr Newton who maintains that it is in the nature of bodies to attract one another, in proportion to their masses and the rays of attraction they receive; upon which the famous Mr Locke has declared in response to Bishop Stillingfleet that having read Mr Newton's book he has retracted what he formerly wrote and recognises that God could put just such properties in matter that it could act at a distance.[9]

Was Leibniz deliberately malicious in thus misrepresenting Newtonian ideas? One can hardly suppose that so subtle an intellect as Leibniz failed to perceive the difference between what Bentley, for example, had written on the subject of God and gravity and his own statements in private to Burnet and in public, ten years later, in *Théodicée*. Again, Bentley and others were perfectly and deeply serious; they really believed that one could perceive God's own design in the law of gravity, just as other philosophers have seen it in the flower or (like Sir Charles Bell) in the human hand. Leibniz, surely, was deliberate in trivializing their almost reverent assurance by bantering words. He simply did not treat the Newtonians' metaphysical position as worthy of systematic criticism, at least at this stage, and therefore did not hesitate to distort it. Neither Newton nor any of his followers ever advocated action at a distance as a purely material interaction, in the manner foisted upon them by Leibniz; their view was that gravitational interaction is providential. Because they – and indeed, one supposes, Leibniz also – believed that God had providentially directed the evolution of the cosmos as the scene for man's creation, fall, and redemption, and had directed all human history since the birth of Adam, was it really so inconsistent in them also to believe that God endowed matter with properties that as mere matter it would not possess and sustained them by his perpetual intention? Whether one views the Newtonian point of view as consistent with a then normal Christian view of God's purpose or (with Leibniz) as merely naïve; whether one regards Leibniz's mechanical philosophy as marking a necessary independence of secular thought from theology or rather as a step toward excluding God from the universe altogether, it must seem probable that Leibniz's refusal to examine seriously the Newtonian foundations of scientific thought – the belief that to discourse of God does belong to natural philosophy – inevitably gave extreme offense.

The *Essais de Théodicée* were still, however, some years ahead, and meanwhile Leibniz continued to express the highest praise of Newton himself. The following passage from the undated letter (1699?) to Burnet in which Leibniz had commented upon Locke's change of opinion is remarkable in this respect:

> It seems that the object of all humanity should principally be knowledge and unfolding of God's wonders and that it is for this reason that God has given man the empire over nature. And Mr Newton being one of those men in the world who could contribute the most to this object, it would be almost a crime in him to allow himself to be diverted from it by

obstacles which are less than utterly insurmountable. The greater his talent, the greater his responsibility. For to my mind Archimedes, Galileo, Kepler, Descartes, Huygens and Mr Newton are more important in considering the chief goal of humanity than are great generals . . . Many can contribute to it by experiments which amass the raw materials but those, like Mr Newton, who can profit from them, in order to advance the grand fabric of Knowledge, and who can unravel the inner secrets are as it were members of God's privy council, and all the others work only for them. Therefore when one has such individuals (who are more uncommon than one might think) we must receive the best advantage from them that we can.

It is a majestic tribute; of himself Leibniz added that he saw his role only as that of an entrepreneur on behalf of the public good, urging others on to the task.[10]

From the beginning, then, Leibniz's persistent and consistent criticism of the metaphysical foundations of a system of scientific thought, such as Newton's, embracing the concept of attractive and repulsive forces that defy mechanical explanation, was balanced and indeed more than balanced by his admiration of Newton as a physicist and a mathematician. The *Principia* he could and did take very seriously; Newton's eccentricities about God and gravitation he could, for the moment, disregard as irrelevant absurdities. Accordingly, in public, he said nothing against Newton, or against Locke and Bentley, all of whom he esteemed, nor against John Keill, whose *Introduction to the True Physics* of 1702 was given six pages in the *Acta Eruditorum* in November of the following year.[11] Keill had not scrupled to denounce speculative mechanical philosophy in the manner of Descartes, writing in his Preface:

Although now-a-days the Mechanical Philosophy is in great Repute, and in this Age has met with many who cultivate it; yet in most of the Writings of the Philosophers, there is scarce anything Mechanical to be found besides the Name. Instead whereof, the Philosophers substitute the Figures, Ways, Pores, and Interstices of Corpuscles, which they never saw; the intestine Motion of Particles, the Colluctations and Conflicts of Acids and Alkalies, and the Events that thence arise, they relate so exactly, that there is nothing but a belief wanting in

the History of Nature, as often as they set forth the Miracles
of their subtile Matter . . .

and declaring that all the writing on mechanical philosophy from
time immemorial

does not amount to the tenth part of those Things, which Sir
Isaac Newton alone, through his vast Skill in Geometry, has
found out by his own sagacity.

All this a Leibnizian philosopher could assent to, and if he might
raise his eyebrows at Keill's declaration that magnetic attraction
"must still be reckoned among the occult Qualities" he would
nevertheless applaud Keill's arguments in favor of the infinite div-
isibility of matter. In any event, the *Acta Eruditorum* review, or
rather summary, was faithful but quite neutral.

The editors reacted similarly to a book of which Keill had written that it would last as long as the sun and moon endure: the
Elements of Astronomy of David Gregory, initially published in
Latin in 1702, and the first exposition of Newtonian principles in
a technical work on astronomy, or as Leibniz put it in a letter to
Varignon, a new system of astronomy founded upon attractions.
Unfortunately for the author, his futile attack upon the brachistochrone problem had by no means cast a flattering light upon his
mathematical powers, so that Leibniz and Johann Bernoulli wrote
his book down as no more than a piece of secondhand Newton,
from which little was to be expected. Such slight interest did the
book have for Leibniz that it was not until January 1705 that he
took in the fact that Gregory had included in it a refutation of his
own *Essay* of 1689 on celestial mechanics, though this fact had
been noted in the *Acta Eruditorum* review of October 1703. Gregory's handling of Leibniz is extremely delicate and respectful:
Leibniz is an acute philosopher and has a name celebrated among
geometers, no one could treat the vortex hypothesis better than
he, and so forth; but Gregory unerringly puts his finger on the
physical improbabilities of Leibniz's theory: some comets clearly
could not be carried around the sun in any material vortex, because their motion does not match that of the planets, and the
velocity law appropriate for the changes of speed observed in any
one planet does not corespond to that to be used in comparing the
speeds of different planets. Therefore there could not be what
Leibniz called a harmonic circulation throughout the solar system.
To this, in a letter to Bernoulli, Leibniz could only feebly rejoin
that if the existence of the deferent vortices in his system might be

160

denied, still the need for the gravitational vortex remained; nor could any argument upset his own basic assumption that nothing could be moved save by the contact of another moving body

> and that solar attraction without the motion of matter around the Sun, in which the planets swim, cannot be understood.

Bernoulli agreed that it was absurd to posit a centripetal attraction without the circulation of matter, but he did not offer an answer to the objections brought forward by Gregory. Nor indeed did the reviewer of the *Elements* in the *Acta Eruditorum*, who praised it as "a most excellent work" and "very elegant," and who seems persuaded of the merit of Gregory's choosing the law of gravitation as the primary law of nature:

> Accordingly, the shameful criticism commonly heaped in plenty upon mathematics will be brought to an end, that it builds nothing but fictions in the heavens, by which the human mind is vainly exhausted; for the application to Astronomy of the principles of Newton, a most profound investigator, has created a true physical astronomy.

And the reviewer goes on to desire a more detailed account of the calculations involved in developing this application, begging Gregory also to perfect the theory of the forces by which all the bodies in the world are acted upon. This review was certainly not written by Leibniz.[12]

A few years later, as Leibniz's irritation against the Scottish Newtonians (especially Cheyne and Craige) increased, reviews in the *Acta Eruditorum* show a very different temper; and as in the *Essais de Théodicée*, criticism of Newtonian philosophy becomes outspoken. The first to suffer was the English physician John Freind, whose Latin *Chymical Lectures* (1709) had been republished at Amsterdam. Freind treated chemistry as an offshoot of the natural law of attraction, that is, he supposed some bodies to be disrupted in chemical reactions because their attraction for each other was weak, whereas new substances formed because the attraction of their component parts was strong. For example, to explain why salts dissolve in water but not in spirits of wine (alcohol), he wrote that:

> Aqueous particles are more strongly attracted by the Saline Corpuscles [of the salt] than they are by one another: Whereas in Spirit of Wine, which is indeed much lighter than Water, but more impregnated with Saline Particles, they continue untouch'd . . .

Freind's attempt to create a chemical philosophy based on the idea of attractions of varying strengths between particles had been anticipated by John Keill, who had published a long paper on "The Laws of Attraction" in the *Philosophical Transactions* during the previous year (1708). Keill's treatment was somewhat more formal and mathematical than Freind's, and he had based it upon such general physicochemical laws as:

> the particles of the solute are more strongly attracted by those of the solvent than by each other; the solute has pores permeable by the solvent.

And both of these writers had been anticipated by Newton himself, who in *Quaestio* 23 of the Latin edition of *Opticks* in 1706 had provided a short monograph on chemical forces and their operation in chemical reactions, beginning with the words:

> Have not the small Particles of Bodies certain Powers, Virtues, or Forces, by which they *act at a distance*, not only upon the Rays of Light for reflecting, refracting and inflecting them, but also upon one another for producing a great Part of the Phaenomena of Nature? For it's well known, that Bodies act one upon another by the *Attractions* of Gravity, Magnetism, and Electricity; and these Instances shew the Tenor and Course of Nature, and make it probable that there may be more *attractive Powers* than these. [Italics added]

Here Newton had not scrupled to use the language that came most naturally to mind, giving his followers authority likewise to invoke action at a distance and forces operating at the microscopic level that were not observable as such at the macroscopic level; only Keill went beyond him in attempting to give Newton's ideas a mathematical dress.[13]

The *Acta Eruditorum* set itself firmly against this tendency of "Keill and his followers" to return to what it called "occult qualities, such as sympathy and antipathy were in the schools of philosophy." It was a grave error, the reviews maintained, to invoke "a certain attractive force which, if it be primary (as is his view) and suffices of its own essence to impel all matter towards all matter, surely cannot be explained by mechanical causes, and so must either be something absurd or must be reduced to a miracle or some extraordinary will of God, to which however it is agreed among those who understand the question one should not fly for refuge without need." Those who understand the question are, of course, the mechanical philosophers, and it is now the reviewer's expressed fear that the new attraction philosophy of Keill will, by returning to "a certain fantastical scholastic quality or even enthusiasm such as that of [Robert] Fludd," undo all the gains that the mechanical philosophy had won. "The great scholastic monstrosities," the dragons slain by the heroes of the scientific revolution, like Bacon, Galileo, Descartes, Pascal, and Boyle, would rise again in philosophy unless men like Keill were stopped short.

This blunt and forthright attack, amounting to the charge that the Newtonian philosophy of nature was "occult," reactionary, and an unfortunately mistaken innovation, was repeated in the *Acta* for the following month (October 1710) when George Cheyne's *Philosophical Principles of Natural Religion* was taken to task. Cheyne is rebuked for reiterating Gregory's refutation of Leibniz's vortex theory, and for holding the view that one particle of matter can attract another particle, "as though God should put laws in Nature, whose logic cannot be discovered from Nature itself," that is, as though God should legislate illogically. "To prove the existence of God," the reviewer states, harking back to the argument stated long before by Richard Bentley, "there is no need to have recourse to the occult qualities of the schools, such as is our author's [Cheyne's] gravitation." And in the following year (1711) the naturalist Martin Lister was praised for resisting the introduction into physiology of crude Newtonian speculations, which, it is argued, cannot be defended by appeal to the authority of Newton himself. For, the reviewer now maintains, when Newton wrote in his *Opticks* of attractive forces, he did not mean them to be *primary*, that is, occult qualities; he was employing the word "force" (as, indeed, Leibniz himself did) in a merely descriptive or phenomenalistic sense "in the same way as we admit of gravity, and magnetic or electric virtues." Here he quotes Newton's own caveat from *Quaestio* 23 in *Opticks*:

> How these Attractions may be perform'd, I do not here consider. What I call Attraction may be perform'd by impulse, or by some other means unknown to me.

Thus his object was to make it appear that Newtonians like Keill, Freind, and Cheyne were a good deal more foolishly incautious and extravagant than Newton himself.[14]

How much, if anything, of all this skepticism concerning Newtonian philosophy can have been known to Keill when he wrote the words that were to bring the calculus dispute to a crisis it is impossible to say. Nor can one guess whether he already knew that the *Acta Eruditorum* in its anonymous reviews of books had, in a sense, cast him in the role of chief, and most foolish, exponent of a "Newtonian" philosophy of attraction, which was perhaps not embraced by Newton himself. Seeing that Leibniz had maintained the same critical position for twenty years, although (so far as I know) not allowing it to appear in print before 1710, it may well be that it was known in Britain, not least, probably, to Newton himself, though again he never seems to have acknowledged the fact. Newton, like Gregory and Keill, was to work up a strong attack upon Leibniz's original *Essay* on celestial dynamics of 1689, but once more it cannot be established that he had even read this paper before 1710. What is certain is the fusion in that year of the philosophical difference between the two great protagonists and their mathematical dispute into a single quarrel. This is already clear from Keill's first surviving letter addressed to Newton (dated 3 April 1711). In this letter, to justify the harsh words applied to Leibniz in his *Philosophical Transactions* paper by a *tu quoque*, Keill directed Newton's attention to Leibniz's account of Newton's two mathematical treatises printed with *Opticks* (chapter 7) and also to the review of Freind's *Chymical Lectures*: "From thence you may gather how unfairly they deal with you," he wrote. Of course, the allusion to Freind's book does not necessarily imply that Keill had long been aware of Leibniz's hostility to the philosophy of attraction; but it certainly ensured that from this time forward the two issues were intermingled – to Newton's disgust, in fact, for Newton always insisted that the question of priority in the discovery of the calculus was a plain question of historical record, and he again and again blamed Leibniz for dragging in extraneous philosophical issues.

So far as my own examination of the *Acta Eruditorum* and of Leibniz's correspondence has extended – and the latter is so huge that no one could claim to know it thoroughly – I seem to see a change taking place after 1700 in Leibniz's attitude to Newton, no doubt produced by Fatio's declaration of 1699 that Newton was the first discoverer of the calculus and gradually hardening as Leibniz read more books from England written in the Newtonian spirit. Consider, for example, the opening sentence of a review by Leibniz of a now forgotten mathematical text, Charles Hayes's *Treatise of Fluxions* (1704), published in October 1705:

> The very noble author, thoroughly versed in the new infini-
> tesimal calculus, like several other writers calls the quantitites

which make up the infinitesimal part of a finite quantity, whether they be infinitesimal or unassignable differences, *fluxions*, in imitation of that accepted expression whereby we say a line is formed by the flowing of a point, though this does not truly signify anything other than that the line is described by the motion of the point.

The reviewer does not so much state as hint that the word "fluxion" is an improper substitute for the word "difference," but he certainly tries to indicate – perhaps rather ineptly – that the whole concept of the flow, rather than the motion, of a point is inappropriate. (Indeed, Newton himself had originally spoken of *motion* rather than *flow*.)

The anti-Newtonian reviewer, after 1706, was possibly Christiaan Wolff (1679–1754), one of Leibniz's disciples who took over from that date most of the responsibility for reviewing mathematical books for the *Acta Eruditorum*, though (according to his bibliographer Ravier) Leibniz continued to read and edit Wolff's reviews before they went into print. Therefore, as Leibniz continued to take an active interest in the running of the *Acta Eruditorum*, he must have known and presumably approved of its anti-Newtonian attitude. The last mathematical review Leibniz wrote was of *Phoronomia* (1716), a book by another disciple and correspondent, Jakob Hermann (1678–1733), professor of mathematics at the University of Padua. He owed much to Leibniz, but even during the years when the calculus dispute was at its height, he refused to deny that Newton possessed preeminent merits. Hermann's book is dedicated to Leibniz; but in a set of adulatory Latin verses also printed in it, a friend had written figuratively of Hermann's researches into mechanics and the heavenly motions:

> Newton, a dweller in the rich isle which yet contains nothing
> more golden than him, was the first to follow this path, and
> you perhaps have given the people nothing inferior.

Leibniz protested: Did not the poet's sentiment do too much injury to many others in favor of Newton? But Hermann defended his friend: What he had meant, he replied to Leibniz, was that Newton had first assembled into a system both the results of his own thinking about the subject matter in his *Principia* and the thoughts of other men, which again Newton had improved, without detracting from the repute of others (like Galileo and Kepler) who had earlier made particular discoveries subsumed in Newton's system. This apology failed to satisfy Leibniz who came back at Hermann – who thereafter, wisely, let the question drop without a further word about it – with the accusation that Hermann was too pro-English: Unlike the Greeks (as Tacitus saw them), the Germans were but too readily inclined to admire foreigners. The praise allowed by Hermann's poetic friend to Newton could not be justified, save by one who "believed the void to exist, and gravity to arise not from mechanical principles, but by means of an occult quality; which two hypotheses I regard as utterly false."

These words come from almost the last letter Hermann received· from Leibniz (the very last, in fact, contains some contemptuous comments from Leibniz upon the English mathematician Brook Taylor). They are, sadly, typical of the last years of Leibniz's life, deeply embittered by the calculus quarrel, when the great philosopher could no longer find one good word to say about any aspect of the work of the Englishman whom he had once praised so highly.

As in Leibniz's last letters to Hermann, so in the final extant letter to Thomas Burnet – though still in a more mocking spirit – Leibniz's revulsion of feeling against Newton becomes plain. It was many years since he had written anything to his British friend on subjects other than European politics, the war, and moral philosophy, but now, in a letter of August 1713, composed when he already knew that Newton and the Royal Society had declared openly against him, he went on from expressions of anxiety about the Tory reaction in English religion and politics to touch on more personal griefs:

> It is curious to see a more than papistical theology revive in England, and a thoroughly scholastic philosophy since Mr Newton and his followers have resuscitated the occult qualities of the Schools by their attractions. What you write to me, Sir, is rather witty "that it seems as if my opponents in the Royal Society have rather written against me lately as a Whig, than as a member of their Society." I had thought that Mr Newton was somewhat allied to the Whigs; and so I had not imagined that the spirit of faction would extend itself to the mathematical sciences.[15]

How little, after all, Leibniz understood Newton! And how vain his later hopes that the installation of his own Hanoverian ruling family upon the throne of England would bring about the collapse of Newton's party. Unlike Leibniz, Newton was never a courtier, and there is nothing to show that his own involvement in political life was anything other than unavoidable tedium to him. Failing to understand Newton, now his adversary, Leibniz made a series of mistakes, of which he adumbrates the gravest when, in this same letter to Burnet, he boasts a little of the Newtonians being "a little taken aback by a word of reply from a friend of mine." Failing to understand him, he had consistently trivialized – and ultimately before the public ridiculed – Newton's thoughts as a natural philosopher; whether or not Keill was brought to the point of combustion in his "Laws of Attraction" article by this incomprehension on Leibniz's part cannot be proved, but it is virtually certain that it determined Newton to give Keill his support. Newton, always a Whig, now believed that Leibniz, a dishonorable man, had attacked his own honor and competence.

9
THRUST AND PARRY: 1710–1713

EILL'S offensive remarks in his *Philosophical Transactions* paper of (officially) 1708 were too much for Leibniz's patience. He felt that the time had come to demand redress, and so raised the dispute to the level of international diplomacy by formally protesting, as a Fellow, to the Royal Society against Keill's conduct in a letter of 21 February 1711, in which he demanded that Keill should apologize for his libelous insinuations.[1] Newton himself, Leibniz alleged, had discountenanced ssuch "misplaced zeal of certain persons on behalf of your nation and himself" when Fatio de Duillier had first attacked Leibniz as a plagiarist; Fatio had then collapsed without support and clearly Leibniz expected that Keill would do the same, especially under pressure from the Royal Society, which would be conscious of Leibniz's dignity, distinction, and influence even if Keill himself were not. Leibniz's letter to Hans Sloane, the secretary of the Royal Society, rings with a genuine note of injured innocence; he had, he wrote, never heard "the name *calculus of fluxions* spoken nor seen with these eyes the symbolism that Mr Newton has employed before they appeared in Wallis's *Works*." How then could he have possibly plagiarized, more than ten years earlier, what was then unknown to him? To Leibniz, wholly ignorant of *On the Method of Series and Fluxions*, forgetful of his hurried inspection of *On Analysis* and of the exact content of Newton's 1676 letters to himself – where, indeed, the word *fluxions* was not to be found – the British accusations were manifestly absurd and contemptible.

The response to his demands in London must have surprised him, though he can hardly have underestimated Newton's power and prestige as president of the Royal Society. He probably overlooked the aspect of his own party (with the *Acta Eruditorum* as its mouthpiece) as seen from England, constituting a deliberately anti-British coalition. While Leibniz had a few friends in England, among them the journalist Edward Chamberlayne and students who had traveled abroad like Thomas Burnet, most Britons rallied patriotically to Newton's support against the criticisms voiced by Cartesians in France and Leibnizians in Germany. Attacks upon the Newtonian philosophy of attraction, whose origin was all too obviously in Leipzig and Berlin, had (one may suppose) not made smooth the way for Leibniz's complaint against Keill, who had no intention of submitting.

It seems that Sloane first sought Newton's advice on the letter, and Newton took it up personally with Keill, who quoted the *Acta Eruditorum* in justification. "I had not seen those passages before," Newton wrote, "but upon reading them I found that I have more reason to complain of the collectors [editors] of the mathematical papers in those *Acta* than Mr. Leibniz has to complain of Mr. Keill." For Newton agreed that he was there everywhere deprived of his discovery and (a particular insult) *On the Quadrature of Curves* was treated as though it were a mere compilation from earlier publications by Leibniz, Cheyne, and Craige. Thus swiftly did Keill win over the president of the Royal Society. When the society took up Leibniz's letter of complaint for the second time on 5 April 1711, Keill defended himself so successfully by reference to the unfair review in the *Acta* that far from the need to apologize being pressed upon him, he was required to "vindicate himself" by producing a written account of the dispute; and the following week this clear hint was reinforced by reference to the "President's right" in the matter, after Newton himself had drawn attention to his early correspondence with Collins.[2]

Preparation of his justification occupied Keill for two months; to what extent he consulted Newton personally on the development of his mathematical ideas, or was aided by Newton, is unknown. Newton certainly possessed a copy of the letter Keill wrote for Sloane to send on to Leibniz, and it is slightly altered in Newton's hand, but it may well be that Newton was glad to leave Keill independent in stating the argument on Newton's behalf. On the other hand, it might be argued that Keill displays so intimate a familiarity with Newton's mathematical history, and prefigures so exactly the lines of argument and selection of evidence to be followed later by Newton himself, that it is difficult to exonerate Newton from giving Keill assistance as his own champion. However, Keill refers to no document not already in the public domain with the exception of Newton's letter to Collins of 10 December 1672, which is quoted by Keill as the clearest piece of evidence for

Newton's mastery of the calculus before 1676. This letter, he says, had been found among the papers of John Collins, which long after his death had been acquired by the London teacher of mathematics William Jones, who earlier in this same year 1711 had published *Analysis by the Series, Fluxions and Differences of Quantities* as another contribution to the Newtonian documentation. Keill could, therefore, have obtained knowledge of it from Jones independently of Newton, and we have no reason to suppose he did not, nor that Jones would have been reluctant to reveal to Newton's friends the evidence that Collins's papers contained to his favor; these things must have been talked of among British mathematicians. Thus Newton's possible involvement at this stage in the advocacy of Keill before the society over which Newton himself presided must remain doubtful, especially as we know nothing of any earlier relationship between Keill and Newton (which would have encouraged Newton to embrace him almost as a fellow conspirator), whereas the one document we do have shows Keill actually trying to persuade Newton that a case against Leibniz did exist. It seems fairer to give Newton the benefit of the doubt, the more so because he hated controversies and restrained rather than incited Keill in subsequent years.

Although Keill's letter asserts formally that he means to press no "criminal" charge against Leibniz, it makes the factual claims that Newton "discovered the method of fluxions at least eighteen years before Leibniz had published anything on the differential calculus" and that "as some specimens of it were revealed to Leibniz it is not contrary to reason that these gave him an entrance to the differential calculus."[3] And still more emphatically Keill writes at the end of the letter "that the hints and examples of Newton" (contained in his 1676 letters to Leibniz via Oldenburg) "were sufficiently understood by Leibniz, [at least as to the first differences; for as to the second differences it seems that Leibniz was rather slow to comprehend the Newtonian method, as perhaps I will show more clearly in a little while]." The words I have enclosed in brackets were added to the manuscript by Newton himself, probably so that they might be printed in the text of the letter he gave to the world in the following year, 1712, where they appear without distinction, of course. In the face of these statements, and the whole burden of Keill's exposition, Keill's apology that he was not charging Leibniz with "borrowing" from Newton becomes a farce, made caustic by his sneer that "among Leibniz's other outstanding merits in mathematics" was his "unwilling[ness] to leave concealed . . . a discovery that was so noble and capable of extension to so many applications." It is a queer compliment to a thief to praise him for knowing the value of the thing stolen.

That part of Keill's case concerning Newton's absolute priority in the discovery of the calculus (in 1665–6) need not long detain us; it was valid, and more important, it was irrelevant. Leibniz could only be charged with plagiary on the basis of his understanding (even imperfectly!) Newton's work by 1676 (for again, no one supposed Leibniz could have gleaned anything material *after* 1676). Keill obviously did not mean to limit his defense to the assertion that Leibniz had been an honest, independent "second inventor," though this had been Newton's own, original view at least as expressed in public avowals. What Keill meant to show, and this is the heart of the matter, was that Leibniz was not an independent discoverer and therefore necessarily not an honest one, because of his unacknowledged obligation to Newton, though Keill left this part of his case as an implicit insinuation.

Of course, Keill wrote, he did not mean that Leibniz derived the differential calculus or indeed Newton's method of fluxions ready-made from the first discoverer; what he meant was that Newton had

> Given pretty plain indications to that man of most perceptive intelligence, whence Leibniz derived the principles of that calculus or at least could have derived them; but as that illustrious man did not need for his reasoning the form of speech and notation which Newton had used, he imposed his own.

These words were ill-judged. The escape clause – implying that despite the circumstantial evidence against him, Leibniz might not have followed Newton's deeper mathematical processes – was not pursued; it is pointless. And the suggestion that Newton possessed, let alone had communicated to Leibniz, a definite "form of speaking and notation" was a tactical error, because Keill was unable to quote anything of the sort. Indeed, whereas a new and serviceable notation was always a strong point on Leibniz's behalf, the inconsistency and imprecision of Newton's language of fluxions before 1691 was a weakness on his side, which he tried to conceal.

To strengthen the argument that "clear and obvious hints" about new mathematical processes would be quite sufficient to enable an intelligent mathematician to reconstruct those processes without the aid of a step-by-step explanation, Keill stressed the continuity of mathematical development: The fluxional calculus was an advance along an established path, not a revolutionary departure from it:

> if we substitute the $\dot{x}$ or $\dot{y}$ of Newton or the dx or dy of Leibniz for the letter o which represents an infinitely small quantity in James Gregory's *Pars Matheseos Universalis* or for the letters a or e which Barrow employs to designate the same, we meet with the formulae of the fluxional or differential calculus. . . .

To correct any impression that Newton himself had continued in the path thus struck out by Barrow and Gregory, however, Keill insists (correctly) that he had begun his own way before their books appeared; his point is that once they were available, it was the more easy for anyone to improve infinitesimal methods further. The obvious difficulty about this argument – and it still existed when the argument was used later by Newton – is that taken at one extreme it is of no force at all (for if Newton had really jumped a long way beyond Barrow and Gregory, the contents of their books can be supposed really irrelevant to the main story of the calculus, and this historians do now assume to be the case) or else, if Newton's jump is trivialized, what right had he to make a fuss? One could hold that the solid content of Barrow's and Gregory's improvements would be far more important and meritorious than any quantity of "hints" about the possibility of better things still. Keill was on firmer ground in writing about *On Analysis* (1669, and recently published by Jones) and its relation to *On the Quadrature of Curves* (published by Newton himself in 1704 and always regarded by the Newtonians as his major systematic exposition of calculus techniques). Nor is his assertion that by 1676 Newton had done more for the new infinitesimal calculus than any one else unfair, whereas in his claim (historically incorrect) that no evidence existed to show that Leibniz had formulated the differential calculus before that year – more specifically, before receiving Newton's two letters in that year – Keill was overlooking no fact then accessible to him. There *was* at the time no evidence of Leibniz's invention before his reply to Newton of 11 June 1677.

Thus for the central issue, Leibniz's indebtedness to Newton before June 1677, Keill made everything turn on Newton's *First* and *Second Letters.*[4] He nowhere suggested that any valuable information might have reached Leibniz by a more circuitous correspondence or through Collins's open-handed attitude to the mathematical material in his possession when Leibniz made his second visit to London. Therefore Keill had to state a case for believing that the two letters, though admittedly containing no explicit "theory of fluxions," or even the word itself, were rich in hints so potent and direct that no mathematician as acute as Leibniz, following in the path of Barrow and Gregory, could miss their meaning. Naturally, it never crossed Keill's mind that Leibniz might not be (and indeed was not) searching for such hints.

Keill was, certainly, adopting a strong debating position in his advocacy. What is a "hint"? How useful can a hint be? For example, the mere hint that a problem has actually been solved by *A* is certainly of value to *B* (assuming he knows in a precise way the nature and significance of the solved problem), for he can proceed on the principle that what one man has done, another can repeat. He is assured that his effort is not futile. In this vein Keill quoted Newton's allusion (in the *Second Letter*) to a general method of tangents that avoided the simplification of compound expressions, permitted the finding of maxima and minima, and so forth; a method that (Newton had opined) was not very difficult and yet would also assist in obtaining the quadrature of curves. Keill saw these claims for achievements already effected by Newton as constituting "a beacon guiding him to the perception of Newton's method," and in opposition Leibniz was powerless except in the affirmation, or proof, if he could produce one, that his own discovery had preceded Newton's letters. How could anyone possibly have proved that he was not inspired by a hint so vague as that quoted by Keill or vindicate his denial that he had received encouragement from it?

But clearly Keill could not be satisfied with this. He had to show that Newton had instructed Leibniz in more specific ways, and to achieve this end he unashamedly read history backward; that is, according to Keill (and afterward according to Newton himself), because we with hindsight see that Newton was able to discuss certain mathematical procedures and specific examples with Leibniz because he was master of an advanced infinitesimal calculus, any competent person studying these procedures and examples could reconstruct that same calculus from which they were descended. As a rough analogy, we trace in the appearance and mannerisms of children attributes and traits already familiar to us

in their parents; according to Keill's implicit principle of reversa-
bility – that it is as possible to deduce the general rule from particular cases, as the particular cases from the general rule – we
ought to be able, by studying children, to reconstruct the appearance, character, and mannerism of their two parents. Manifestly,
Keill's rule is not wholly false, but neither is it wholly valid: In
particular, just as one could assign the characteristics of children
to the mother or father quite wrongly, so it is nearly always possible for any finite number of particulars to be embraced under
more than one general rule. This is the nontrivial possibility that
(with respect to Leibniz) the Newtonians never contemplated.

Thus, not only was Leibniz historically innocent of the process
of "reconstruction" of Newton's method from hints given, in the
manner postulated by Keill, but it certainly makes no logical sense
to believe (as Keill and Newton alleged) that since Leibniz had
followed such a process, he would inevitably have come up with
a replica of Newton's method. This is the basic fallacy of the
whole Newtonian argument against Leibniz. It was the fallacy
that led them into declaring both that Leibniz was clever enough
to have profited from Newton's hints and that he had "imitated"
Newton's method. For, obviously, if Leibniz ever knew enough
of Newton's achievement to imitate it, he had no need of hints;
but if he only had hints, what was there for him to imitate and, as
Newton said, spoil? The two accusations against Leibniz are really
inconsistent, but they both had to be made because the weak hints
alone could be demonstrated, and therefore the equivalence between Leibniz's calculus and the method of fluxions had to be accounted for as an unspecific "imitation." This argument was, as
will be seen later, a double-edged one for the Newtonians to employ.

To illustrate Keill's conduct of his case more specifically: in the
First Letter of 1676, he declares, Newton

> taught the method by which quantities may be reduced to
> infinite series, that is by which the increments of flowing
> quantities [variables] may be displayed . . .

Keill then continues, as though still with reference to the *First
Letter*, to show how by means of the binomial expansion a sequence of successive differences (or fluxions) of a quantity may be
found. His statement is remarkably disingenuous: Newton did indeed explain his binomial theorem to Leibniz in the *First Letter*,
that is, truly enough, "the method by which quantities may be

reduced to infinite series," but nowhere did the *Letter* apply the binomial expansion to finding differences (or fluxions) or, indeed, to bringing flowing quantities into the discussion at all. Newton's whole emphasis was on series and excludes infinitesimals; so (later in the *Letter*,) he wrote:

> How the areas and lengths of curves, the volumes and sur-
> faces of solids . . . and their centres of gravity are determined
> from equations thus reduced to infinite series, and how all
> mechanical curves may be thus reduced to similar equations
> of infinite series . . . all this would take too long to describe.

This passage is then illustrated by several examples again wholly without reference to infinitesimals. And, in fact, Keill's own explanation of the relation of Newton's binomial expansion to the successive differentials of an expression is adapted not from the *First Letter* at all, as the nonalert reader might suppose, but from *On the Quadrature of Curves*. Keill was correct in claiming that the series and integrations laid before Leibniz by Newton in 1676 had been known to him since 1666 (or thereabouts), and that Leibniz did not at first follow Newton's processes, but the whole mathematical context of infinitesimal calculus introduced by Keill, which he supposed Leibniz capable of reconstructing so easily, had been absent from the *First Letter* that Leibniz received. But for the fact that Keill puts the whole matter forward in so calm and matter-of-fact a manner, one would really find the suggestion that mere inspection of the methods of binomial or Taylor expansion could serve by itself as a sufficient introduction to the differential and integral calculus quite ridiculous. Similarly, turning to the *Second Letter* of 1676, Keill maintains (referring to Newton's first example in the letter) that Newton explained how quadrature of curvilinear areas might be effected by proceeding from differentials to integrals. Keill then shows how integration is the converse of differentiation. But in the *Second Letter*, in fact, Newton himself had done nothing of the sort, nor (let me repeat) employed such words as "differential, integral" or "fluxion, fluent," at all or any equivalent or any such symbolism. He left the whole algorithm, here disclosed by Keill, completely in the dark.

One cannot but suppose Keill (and Newton later) blindly sincere. They were writing of things that they believed not merely possible but actual. It may now seem astonishing to imagine that the "hints" discussed by Keill in the 1676 letters could have furnished a reader with the means to reconstruct Newton's method of fluxions, not to say Leibniz's differential calculus. The Newtonians, however, found the matrix among Newton's mathematical papers, from which the material for these letters had been taken, so natural that they were wholly blind to the far different reaction of an unprepared reader whose mathematical evolution had been very different from Newton's. Surely, they might have argued, with so many pieces of the jigsaw puzzle in his hand, Leibniz must have glimpsed the whole picture? Forgetting the difference between the open and the predisposed mind, ignoring Leibniz's own interest in (as it seemed to him) another and distinct puzzle picture, it was the more easy for the Newtonians to make their mistake because seventeenth-century mathematics was indeed converging by several avenues upon the concepts of the calculus; they could not conceive that Leibniz, in fact, came to them by a different approach from Newton's.

Keill's letter of counterattack rather than apology was read at a meeting of the Royal Society on 24 May 1711, where the decision was taken to transmit it to Leibniz without the society's officially expressing any regret or disclaimer. Or so Sloane, as the society's secretary, wrote to Leibniz, prompted by Newton himself (it seems) if one may judge by the drafts among Newton's papers. Considering Leibniz's fame and eminence, Sloane's curt note is barely courteous, especially in view of the document to which it served as a cover, so it is not surprising that Leibniz took his time about acknowledging his receipt of it. His tone, in doing so in December 1711, was dignified: Why should such a man as himself be expected to apologize for his life, like a suitor before a court of law, in order to refute a person who had neither knowledge of the events in question nor the authority of him who was chiefly concerned, Newton? As for Keill's justifying himself by the plea that false charges against Newton had been made in the *Acta Eruditorum*, this was worthless; for in that periodical, Leibniz thought, everyone had received his due. He and his friends had always allowed Newton the credit of "arriving by his own efforts at basic principles similar to our own." His own right to the discovery was just as good, as Huygens had acknowledged, and as Leibniz had kept it to himself a full nine years (that is, since 1675), no one could claim to have forestalled him. Should not Newton, he appealed, be brought in to tell Keill, the upstart (*homo novus*), to back down?[5]

Here, perhaps, if matters had remained as they were for the

previous twenty years, the issue might have rested. Leibniz's declaration that he had not been forestalled *in publication* was unexceptionable; there was no difficulty about the letter as long as Leibniz was not asked to admit that he had been forestalled *in the idea*. But the situation was not as it had been. Keill was a man not to be suppressed, nor to admit Leibniz's vindication of his own independent discovery of the calculus. It must, in Keill's view, be universally recognized that Newton had been first in the field, way ahead of the Germans. And now Newton himself was beginning to believe that, his honor having been touched, nothing but a full disclosure of the cards in his hand could clear him. Moreover, with both men, the two different anti-Leibniz issues were now fused into one: It was not only necessary to prove that Newton's had been the first and original calculus (as Leibniz had since called it) but that, because the calculus was unique, Leibniz had stolen it from Newton. The security of Newton's fame demanded that Leibniz be made infamous: There could no longer be any question of sharing honors.

We have no sign that Newton did anything more until Leibniz's second letter to Sloane reached London in January 1712. With that, however, he embarked upon the manifestation of his own historical right to the first discovery of the calculus, which had been in his mind since April 1711 and for which he may already have made quiet preparations. It is likely that he addressed the Royal Society on the subject at its meeting on 14 February 1712; at any rate notes for a speech survive. The controversy, Newton reminded the Fellows, was not of his making, as the *Acta Eruditorum* articles had escaped his notice before "last summer." Leibniz's right to demand an apology from Keill was no stronger than his own right to receive an apology from the reviewer of the "book of Quadratures" in the *Acta* who had "taxed me with borrowing from other men." It was possible, Newton admitted, that he and Leibniz might both have independently discovered the same calculus, "because the same thing is often invented by several men"; his own creation of the infinitesimal method, however, had occurred before 1669, whereas he had not heard of Leibniz's before 1677. Thus far the speech is mild enough; Leibniz is not accused of misconduct. A related draft by Newton, in contrast, returns to his favorite emphasis on the unique right of the "first author" and his doubt that Leibniz was such a "first author" of the calculus. This draft recites with some indignation the offensive words about Newton's employing fluxions in place of differences and asserts that Leibniz learned from the 1676 letters that Newton had written "a treatise of the methods of converging series and fluxions six years before I heard of his differential method."[6]

Although Newton's public reactions to Leibniz's two letters to the Royal Society are not precisely known, it is yet clear enough that they were defensive rather than counteroffensive; he himself was not willing to seem to go nearly so far as Keill had done in his letter of self-justification. Newton justifiably claimed priority; he did not unjustifiably deny independent invention to Leibniz. Whether this was his own true belief, whether in his heart he did not already believe what Keill had plainly written, is another question; for Newton was now taking devious steps to improve his position. Having (presumably) heard his side of the story, as well as that of Leibniz, the Royal Society decided to appoint a committee to look into the questions in dispute and report back; in subsequent weeks the committee was enlarged by other members (including the King of Prussia's minister in London), and on 24 April – just fifty days after the first nomination of the committee – its report was ready.

One may wonder how such expedition was possible. Many letters and papers, collected together at the offices of the Royal Society, had to be studied, and there were no copies for circulation. Three of those who signed the report had only sat on the committee for a week before doing so – they can hardly have come impartially to their decision. And, of course, nothing from Leibniz's side was available to them at all, save some of his letters to Oldenburg. The fact is that the report was drafted by Newton himself – the draft, varying little from the subsequently printed version, still exists in his own hand – and the committee (or those of them competent to form any opinion) allowed themselves to be used as men of straw. Some of them must have known that the supposedly impartial report was their president's own work; one can only hope that they honestly believed all its findings to be true. Someone, presumably, must have acted particularly as Newton's stoolpigeon, but it is useless to try to guess his name; it was not Keill, for he was not of the committee. However it was stage-managed, the "Royal Society's" report came out overwhelmingly in Newton's favor. It dragged out the old story of Leibniz's unfortunate attempt to claim for himself "another difference method properly so called," which had already been printed by Mouton;

and declared that he had possessed no other differential calculus before June 1677

> which was a year after Mr Newton's letter of 10 December 1672 had been sent to Paris to be communicated to him, . . . In which Letter the Method of Fluxions was sufficiently described to any intelligent person.

The conclusion that Newton was "the first inventor" of the calculus was obvious, and Keill was absolved of doing any injury to Leibniz in saying that this was the case. No formal accusation of plagiarism against Leibniz features in the report. As Newton put it later (1718):

> whether Mr Leibnitz found the Method by himself or not is not the Question. The Committee of the Royal Society did not enter into this Question, but on the contrary said: *We take the proper Question to be, not who invented this or that Method but who was the first inventor of the method. And we believe that those who have reputed Mr Leibnitz the first Inventor, know little or nothing of his correspondence with Mr Collins & Mr Oldenburg long before, nor of Mr Newton's having the Method above 15 years before Mr Leibnitz began to publish it in the Acta eruditorum of Leipsic.* Here the Committee of the R. Society treat Mr Leibnitz as second Inventor . . . [Italics in original]

Not as a plagiarist pure and simple was Leibniz condemned, but as one guilty of concealing his knowledge of the prior, relevant achievements of others. This was the essence of Leibniz's misconduct and, if proved, it would defame him as effectively as the worse crime of open theft: that he had first silently ignored, and later explicitly denied, Newton's genuine right (of which Leibniz was necessarily aware) as "first inventor." As Newton, and hence the "committee," saw it, Leibniz's correspondence of the 1670s had not only proved Leibniz's moral turpitude but had been vitally important for his later publication of the calculus. "Second inventors have no right," Newton declared harshly.[7]

The society accepted the report as correct and fair, ordering its publication along with extracts from the relevant documents. The whole, along with a number of pointedly anti-Leibnizian footnotes and comments, constitutes the *Correspondence of John Collins and others about the development of Analysis*, invariably known by its shortened Latin title as the *Commercium Epistolicum*, which the Royal Society distributed free to suitable recipients in January–February 1712–13. This was to remain the permanent basis of Newton's case against Leibniz; selection of the documents and appropriately pointed annotation of them occupied Newton throughout the latter part of 1712.

As to the exact form of this case, the report exhibits some uncertainty in Newton's mind. Keill had relied chiefly on the content of the two *Letters* of 1676, treating these as sources of direct "hints" to Leibniz about calculus procedures. Newton in the report, however, ignores these *Letters* (though both were to be printed in the *Commercium Epistolicum*) and relies instead on his letter to Collins of December 1672. This letter had indeed been quoted by Keill, too, but only as evidence of Newton's early mastery of calculus techniques; he quotes from it Newton's statement of the tangent rule and his general promise that

> This [rule] is one particular [case] or rather a corollary of a general method which extends itself without any troublesome calculation not only to the drawing of tangents . . . but also to the resolving [of] other abstruser kinds of problems about the crookedness, areas, lengths, centres of gravity of curves.

It is a general affirmation very like others in the *First* and *Second Letter* of 1676. Keill did not argue that Leibniz had seen the Collins letter, as Newton does in the report, making now one of the great historical blunders of the whole dispute, which was only corrected in the nineteenth century.

The origin of this blunder was, one may suppose, Newton's discovery in the course of his documentary researches of the joint intention of Collins and Oldenburg, after the death of James Gregory in 1675, to send to Leibniz a collection of Gregory's mathematical letters to Collins to which some of Newton's were to be added. We may assume that Newton had previously known nothing of this plan, because the documents were not in his possession. In the *Commercium Epistolicum* Newton printed (accurately) Collins's opening words to Oldenburg at the head of his collection:

> Forasmuch as you have much pressed me yourself, being incited thereto by the earnest desires of Mr Leibniz and others of the Royal Academy at Paris, to give an account of the great pains and attainments of the late learned Mr James Gregory . . .

And also Collins's endorsement:

Extracts from Mr Gregory's Letters, to be lent Monsieur Leibniz to peruse; who is desired to return the same to you [Henry Oldenburg]

This collection, not of Gregory's original letters but of extracts copied from them, does indeed contain a full statement of Newton's tangent rule as he explained it in his letter to Collins of 10 December 1672. Therefore, when Newton saw this, he declared in the *Commercium* that here

Newton says he has a general method of drawing tangents, squaring curved lines, and effecting similar things, and explains the method by the example of drawing tangents; which method Mr Leibniz afterwards called "differential."

Newton naturally concluded that the collection (known as the "Historiola") had been sent to Leibniz early in 1676, Gregory having died in the previous October. This was an error, for the "Historiola," being both bulky and precious – for it was of interest to all mathematicians – was not sent to Paris at all. Instead, Leibniz received (but this was not ready before the end of July 1676) a short version called the "Abridgement," in which Newton's December 1672 letter to Collins was condensed to a few uninformative lines only. When this was first realized, more than a hundred years ago, it seemed obvious that Newton had maligned Leibniz fearfully, in a way that could have been avoided had he examined more scrupulously what Oldenburg actually sent to Leibniz in July 1676. Augustus De Morgan writing in 1852 of the "atrocious unfairness" of Newton's partisans and of the "gross suppression of facts" in the *Commercium Epistolicum* did not realize, however, that the *complete* text of the "Historiola" had been examined by Leibniz during his visit to London in October 1676. There he duly found Newton's letter of December 1672 and carefully noted down for himself the outline of Newton's method of tangents.

Thus, though the *Commercium Epistolicum* was wrong as to the facts of its statements about Newton's tangent method, these statements were correct in spirit. Leibniz *did* see the December 1672 letter, though many months later than Newton had supposed. It was far too late to have any effect on his own mathematical development, and what is even more to the point, Sluse's method of tangents, identical in all but language with Newton's, had been in print in the *Philosophical Transactions* since January 1673: There was no "secret" about it if we except the reciprocity of tangent finding and quadrature hinted at by Newton in his letter.

Taking the equivalence of the Sluse-Newton methods into account, the importance that Newton attached to the December 1672 letter in the report and the *Commercium Epistolicum* becomes hard to understand, and not least the assertion of the "judgement of the Committee" at the end of the book that in this "letter the method of fluxions was sufficiently described to any intelligent person." If this were so, why should not Sluse also be claimed as an inventor of fluxions? De Morgan called this claim "the most reckless assertion ever made on a mathematical subject," whereas Leibniz denoted it as merely "silly" – Newton's 1672 ideas were (he thought) worlds apart from the differential calculus. We can only account (in part) for Newton's paradoxical claim by supposing that in his own mind he confused the question of his own priority in mathematical discovery (as to this, the 1672 letter could fairly count as evidence) with different questions about the extent of Leibniz's infamy (in Newton's eyes). The 1672 letter gave some, if slight, measure of Newton's mathematical progress – Newton had far stronger documents at hand. To a demonstration of what Newton had accomplished before 1672, the question of Leibniz's knowledge of Newton's mathematics before 1677 was, however, irrelevant. If Newton had wished to prove that Leibniz knew that he had made great advances in the infinitesimal calculus *as well as the method of series* (which latter point Leibniz never contested) without being informed of the nature of these advances, and therefore (when he had trod the same path by his own independent efforts) ought to have admitted his awareness that Newton had been before him, then again Leibniz's demonstrated *early* knowledge of the 1672 letter and other materials would have been to the point. But to prove Leibniz guilty of this lesser fault, of not acknowledging that he must have come after Newton, was not by any means the same thing (as Keill and Newton often seem inclined to suppose) as proving that he was guilty of the graver charge of actual plagiarism of Newton's methods.

It must, to go further, also be observed that when Newton in the report descended to general insinuations that Leibniz must have known of his work because Collins had imparted news of it to his correspondents, he made disgraceful and irrefutable accusations, rendered all the more unworthy by the underhand method he adopted of giving currency to them.

Another mistake in the report, never confessed nor corrected by Newton, relates to the time lapse before Leibniz's reply to the *Second Letter*, which first conveyed news of the differential calculus to Newton. The latter always maintained that Leibniz had enjoyed a long interval of leisure in which to work up his own method as a variant of Newton's after Newton's method had been brought to his knowledge; here Newton makes that interval at least a year and perhaps four years. Once more, Newton's opinion was quite at variance with the facts (chapter 5), and though circumstances chanced to make the case against Leibniz in this respect look worse than it was, some of the evidence for believing that no such long period for "cooking" was available to Leibniz was actually in Newton's hands, if he had ever cared to look at the question from Leibniz's side.[8]

Curiously enough, just at the moment when Newton was moving toward an "impartial" presentation of the historical documents, telling in his own favor, the reviewer in the *Acta Eruditorum* of William Jones's *Analysis* (who was not Leibniz) was proposing the very same thing – the publication of John Collins's correspondence; to this suggestion he was prompted by Jones's moderate allusions to the evidence in that correspondence of Newton's mathematical precocity. Otherwise, the reviewer seized the opportunity to enlarge upon the merits of Leibniz's differential calculus, which could even serve as a touchstone for the detection of many mistakes in physical science, and to treat the Newtonian method as a poor derivative of or substitute for the Leibnizian algorithm while carefully denying that anything known to Newton had escaped Leibniz. Newton (perhaps prompted by Keill) did not fail to examine the review of Jones's book and drafted a letter of protest about it to the editors of the *Acta*. He resented the omission of any reference to the early date of *On Analysis* (1669), the continued ascription of the "first invention of the differential method to Mr Leibnitz," and the assertion of the identity between his method and Leibniz's; his own method, Newton claimed (rather speciously), avoided the uncertainty of infinitesimals and performed "the whole computation . . . in finite quantities by the geometry of Euclid." He was often to try to make out such a claim but never succeeded to the satisfaction of independent contemporary critics.[9]

While the *Commercium Epistolicum* was in preparation, the philosophical dispute between Leibniz and Newton was not altogether without fresh incident. In May 1712 a London periodical called *Memoirs of Literature* copied from the French reviews an open letter to the Dutch philosopher Nicolaus Hartsoeker written by Leibniz during the previous year, the two having been for some time in dispute over the nature of infinitesimals (on this score and others Hartsoeker was equally critical of Newton). Into this paper Leibniz gratuitously inserted a derogatory allusion to Newton, much as he had done not very long before in his *Essais de Théodicée*: Those who supposed God to have made it a law of nature that every lump of matter should attract every other were invoking a miracle in order to explain gravity, when what was necessary was to find a natural physical explanation for it, not to appeal to the supernatural. Leibniz's words did not escape Newton, for he prepared an undated draft of a letter of protest to the editor of the *Memoirs of Literature*, which is interesting because he clearly takes up a national cause. Because the English, having found out the universal law of gravity:

> do not explain gravity by a mechanical hypothesis, he [Leibniz] charges them with making it a supernatural thing, a miracle and a fiction invented to support an ill-grounded opinion . . .

But was the law less true for being unexplained, Newton asked, and what of such other properties of matter as hardness, inertia, duration, and mobility:

> yet no man ever attempted to explain these qualities mechanically, or took them for miracles or supernatural things or fictions or occult qualities. They are the natural real reasonable manifest qualities of all bodies seated in them by the will of God from the beginning of the creation . . .

The defect of Newton's argument is that it begs the vital question: Is gravity a "real reasonable manifest quality of all bodies"? Let us suppose that a logical analysis of our experience of the world convinces us that all matter is made of particles, and that these particles must necessarily be hard, extended, movable, and so forth. Let us agree that this is a common tenet of the mechanical philosophy of nature. Does it *also* follow that these particles possess active powers or forces? By no means: Mechanical philosophers before Newton had taken precisely the opposite view, and like all their predecessors back to remote antiquity, they had defined material substance as naturally inert. Just as Aristotle had found the cause of the motion of matter in psyche or pneuma, so the mechanical philosophers had reasoned that matter alone can never create, but only conserve motion. Thus Newton was exceptional, not ordinary, in seeking to include force within the ontology of matter – if this is what he meant by saying that force was no more the consequence of God's will than extension and hardness. The mechanical philosophers had been especially prudent to avoid the attribution of active powers to matter (which was Spinoza's sin) because to do so would open the path to atheism, for if mere matter were supposed active enough to organize itself into a universe, no divine creation was required. Nor do "active powers" in matter overcome the problem of action at a distance; if gravity is, like hardness, inherent in matter, how does its activity operate across empty space? It is obvious enough that in seeking to find arguments against the need for mechanical hypotheses to explain gravitation, such as Leibniz preferred, Newton was starting on a difficult task.

In the end he decided to leave it to others – to his young collaborator in preparing the second edition of the *Principia*, Roger Cotes, and to his old friend Samuel Clarke. Newton's own participation was virtually restricted to the celebrated "*Hypotheses non fingo*" of the new editions of his books (chapter 8). For his part Cotes, valiantly arguing the Newtonian case in the Preface to the new edition of the *Principia* (1713), did not hesitate to take the same line as Newton himself in the draft just quoted, maintaining that

> if the nature of things is not correctly explained by the gravity
> of bodies, it will not be rightly explained by their extension,
> mobility and impenetrability.

This was indeed taking the war into the enemy camp, challenging the Cartesians and Leibnizians to explain why, philosophically, Newtonian forces were any less respectable than the accepted "mechanical" properties of bodies.[10]

In fact, the early years of the second decade of the eighteenth century are marked by the Newtonian counterattack. With the exception of his *Opticks*, and its companion mathematical treatises, nothing of which was in the least controversial, Newton himself had not published at all after 1687, and the Newtonian literature since that day, the work of Wallis, Gregory, Keill, Craige, and others, had been truly independent and in large part nonpolemical. There had been little effective effort to rebut Continental criticism of Newton's influence either in mathematics or in physics or even to interpret Newtonianism properly. With Jones's publication of Newton's mathematical treatises in 1711, followed by the *Commercium Epistolicum* (1712–13), and the second edition of the *Principia* (1713), a decade of Newtonian publication opened, which also saw a pirated Continental edition of the *Principia* (Amsterdam, 1714), Joseph Raphson's *History of Fluxions* (1715), the Leibniz-Clarke *Correspondence* (1717), and in the same year the second English edition of *Opticks*, Keill's *Introduction to the True Astronomy* (1718), and finally two books that signalize the beginnings of Newton's Continental influence – the first French translation of *Opticks* (Amsterdam, 1720) and Willem 'sGravesande's *Mathematical elements of Physics, confirmed by Experiments* (Leyden, 1720), the first Newtonian text composed by a Continental author. The same years saw Keill's robust invasion of the Continental journals on Newton's behalf (aided by 'sGravesande in Holland) and, with the death of Leibniz (1716), the disappearance of the chief actor on the other side. This second decade of the century was that in which Newton's ideas no longer seemed absurd to most Europeans, in which defenders of his physics – if not of his fluxions – began to make themselves known here and there, and the slow retreat of the neo-Cartesians began, though that was to be far from final at the time of Newton's death.

As he seemingly had no notice of the proceedings at the Royal Society in the spring of 1712, Leibniz was taken aback by the appearance early in the next year of the *Commercium Epistolicum*, with what amounted to an official censure of himself as a plagiarist. It was some time before the copy sent to Hanover reached him in Vienna, where he was at that time, so that his first information came from friends who were no less amazed to find so dignified and esteemed a scholar as Leibniz treated as a criminal, and worse, an ignoramus. Pierre Varignon, for example, wrote of his distress at the unjust trial "to which Mr. Keill has just subjected you in England; we are the more astonished at it here because Mr. Newton himself has in the *Principia* recognised you also as a discoverer of the calculus in question, and because for almost thirty years you have peacefully enjoyed the fame which you have reciprocally allowed to each other, with a courtesy edifying to all decent people . . . "; yet these phrases already indicate a pro-Newtonian change of sentiment on Varignon's part; only a few weeks before, when he knew of the *Commercium Epistolicum* merely from Johann Bernoulli's report of it, he had been willing to agree with Bernoulli (and Leibniz himself, of course) that Newton's method of fluxions – or at least his *original*, pre-1684 method of fluxions – had not been an algorithm equivalent to Leibniz's differential calculus at all, but something of altogether lesser scope "as the ancients knew *loci*, but not their calculation in the way that algebra has furnished it to us." Presumably, actual examination of the *Commercium Epistolicum*, together with recollection of what Leibniz and the authors in the *Acta Eruditorum* had previously credited to Newton, made Varignon less inclined to take Bernoulli's word on the matter; and even Bernoulli was shaken by what he now saw of Newton's early letters and papers, if only momentarily.[11]

The *Commercium Epistolicum*, as first published in 1712–13, was a selection of texts – some extracts, some complete – with brief interlinking passages and annotations, ending with the report, which the Royal Society had approved on 24 April 1712. The documents are, within the ordinary limits of error, faithfully reproduced and nothing obviously favorable to Leibniz's side was omitted; Leibniz hoped to be able to demonstrate cheating in the presentation of the documents but never succeeded. The chief factual errors deliberately introduced by Newton relate to *On the Quadrature of Curves* and are not profoundly significant: He wished to make it appear both that this was an early treatise and that its dot notation for fluxions was one of those he had employed from the beginning. The basis of the fluxional calculus was, after all, transferred to this treatise from early writings, and the question of notation is not of the first importance. The collection opens with a quotation from a letter of Isaac Barrow to John Collins dated 20 July 1669, referring to a "friend" who

> brought me the other day some papers, wherein he has set down methods of calculating the dimensions of magnitudes like that of Mr Mercator for the hyperbola, but very general, as also of resolving equations . . .

This was Newton's *On Analysis*, which then follows complete, within a couple of pages of the start of the *Commercium Epistolicum*. It may at first seem strange that Newton chose to insert this essay as his substantive evidence of early achievement, rather than the slightly later *Method of Fluxions*, because *On Analysis* is not overtly concerned with fluxions at all but, as its title states, with the method of series. There were two reasons for his choice. First, because this essay had been communicated to Barrow and Collins – and, as Newton knew in 1712, copied by the latter – mentioned by Collins in his correspondence, and further made available by him to Leibniz, Newton could claim that in some sense *On Analysis* had become known to the inner mathematical world of the 1670s. Of course, he could make no analogous claim for the *Method of Fluxions*, which had not left his own hands. Second, in Newton's eyes, the method of series was itself an indispensable part of his new infinitesimal analysis. As a recent writer puts it:

> It is quite clear that Newton's use of infinite series belongs to analytical mathematics. These series had no meaning in traditional terms until Newton's development of the method of ratios [fluxions].

The calculus is, after all, differential and integral and it is the second element of the pair that contains far more difficulty for the mathematician, who has to discover appropriate methods of approximation where the obtaining of integrals by the simple procedure of reversing differentiation is impossible.

Newton had quickly realized that the methods of series expansion that he had already used for solving algebraic equations had a further utility in developing his method of fluxions; series expansion was, as it were, a mathematical tool of general usefulness extending beyond this new method and prior to it, but one which the new method required for its perfection. That is, the relation between the two areas of mathematics was not quite reciprocal because series could certainly be used without infinitesimals, whereas the perfected infinitesimal method required series expansions and, as Newton later maintained, the two constituted a whole together.

There seems to have been a real difference between Newton and Leibniz here. Taking their stand on Newton's centering his discussion in *On Analysis* upon series expansions and not upon the use of infinitesimals, both Leibniz and Johann Bernoulli maintained, as probably no modern historian of mathematics would do, that Newton's advancement there of the study of infinite series had nothing to do with a knowledge of the "real" infinitesimal calculus, even though (as we have seen in Chapter 2) *On Analysis* does contain explicit allusions to a new infinitesimal method. Their blindness, as it may seem to us, was not simply the product of a biased perversity but was historically conditioned by Leibniz's approach to the calculus through the study of the "method of differences," in contrast to Newton's initial interest in problems of interpolation and quadrature by means of series. Newton took a broad view embracing several lines of progress in mathematics of which the fluxional method was only one, whereas Leibniz tended to see every development in a role ancillary to the new concept of algebraic differences. Certainly even in August 1676, nearly a year after the idea of a differential calculus had first come to him, Leibniz was still far behind Newton in the mastery of series expansions and integration procedures, writing frankly to Oldenburg:

> I would very much like to know, for I cannot yet satisfy my-self in this, how the roots of equations can be made known by infinite series, and also how the tables of sines and loga-rithms may serve for the resolution of equations.

He could, he goes on (perhaps with excessive assurance), obtain any root approximately, but only in a very cumbersome way

> however, I do not doubt that to your colleagues, who have given so much effort to this business [of series], a certain brief and subtle elegance is accessible.

Although by mid-1676 Leibniz had himself obtained interesting results by the use of series – not least his circle quadrature – that method did not occupy for him the central position it had occu-pied in Newton's mathematics.[12]

As so often in collections of documents, the bite of the *Commercium Epistolicum* is in the "editorial" comment from Newton's pen. He lost no opportunity of ramming his message home to Leibniz's disadvantage. Thus, in *On Analysis*, picking out the place where he had referred to area moments (infinitesimal areas generated in a geometrical figure by an infinitesimal change in x producing a corresponding change in the area y), he adds

> N.B. There is here described the method by fluents and their moments. These moments were afterwards called differences by Mr Leibniz: hence the name differential method.

In a sense the passage does indeed contain one of those "hints" Keill insisted upon; but it certainly gives no "description" of a new "method" in the ordinary sense of words. Yet Newton was *not* here claiming that Leibniz had gleaned anything from a perusal of *On Analysis*, so the invidious comparisons are really irrelevant to his purpose of establishing priority. Newton's comments also insist that the *Second Letter* of 1676 had "explained the methods of infinite series and of fluxions at the same time" to Leibniz, that he had had no competence in geometry before the middle of 1674, and had possessed no knowledge of quadrature series until those of Gregory and Newton were imparted to him by Collins and

Oldenburg (which Leibniz had then passed on to his friends in France as if they were his own discovery), and that Leibniz had not formulated his differential method until after his return to Germany in the autumn of 1676 – that is, more than a year later than, in historical fact, we now know that Leibniz had his first notions of the differential calculus. It is surely needless to say that all this poison poured by Newton into his readers' ears was distilled from distortion and illusion. Take the last point, where Newton constructs (out of almost nothing) a beautiful imaginary chronology. Collins, he says, received from Gregory in 1671 his arithmetic quadrature of the circle (this was, of course, in the form of an infinite power series) which Oldenburg transmitted to Leibniz in 1675; this same series, Newton goes on, Leibniz then imparted to his friends as his own and began to prepare a short paper about it; in the next year, 1676, he polished the paper in order to send it to Oldenburg. In the third year, 1677, his return home allowed Leibniz, according to Newton, no leisure to get on with this task and he soon realized that his new analysis made a particular treatment of the circle series unnecessary. Accordingly (*igitur*), this analysis must have been discovered only after Leibniz had put his paper aside and become involved in public affairs, that is, after his second visit to London and return to Germany. Newton's "historical reconstruction" here does not closely correspond to facts accessible to him, and he simply made out of nothing the supposed dates of Leibniz's work on the circle-quadrature paper.

It should be needless to repeat that Leibniz discovered his arithmetic quadrature of the circle in 1673, whereas the British series to the same effect (though discovered much earlier, it is true) were imparted to him only on 12 April 1675 (not 1671!). Equally, Hofmann has shown how false was Newton's unshakable supposition that Leibniz had obtained the Newton-Sluse tangent rule, as it were sub rosa, from his own letter to Collins of December 1672, making this typical of Leibniz's whole procedure; Leibniz got the rule, in 1673, from Sluse's paper published in the *Philosophical Transactions*. The point is not, of course, to blame Newton for ignorance of Leibniz's development as a mathematician, as knowledge of it was necessarily inaccessible to him. What was reprehensible in Newton was that he arrogantly believed (on the basis of very few facts, distorted in his own mind) that he *could* follow that development, thus fabricating in his own interest a chronology for which he had no authority.[13]

On the other hand, one should not overlook the provocation Newton had received, not only on his own score. It was certainly true that Gregory had sent the circle-quadrature series to Collins in 1671, and if Leibniz had been completely candid he would not have claimed to be the *first* to discover it, though he was certainly an *independent* discoverer. Collins and Oldenburg had put him in the picture, but in his August 1676 letter he (conveniently?) forgot what they had told him, permanently it seems, for it came to Leibniz as a shock in reading the *Commercium Epistolicum* to discover the British claim for Gregory's priority, which he was at first inclined blusteringly to deny as a fabrication: Evidently Leibniz (who, like most of us, hated to be forestalled in anything) closed his mind to the truth of this matter. Moreover, Leibniz made mistakes, for surely Newton was correct in deducing that even in 1676, even after he had formed the idea of the differential calculus, Leibniz was an ill-trained mathematician (who usually wrote his letters in haste). The mistakes gave Newton grounds for asserting in the *Commercium Epistolicum* that the discovery of the calculus was beyond such a duffer. One of Leibniz's more notable misconceptions was his doubt that the method of infinite series would be capable of resolving certain intricate problems "such as neither depend upon equations nor upon quadratures, such as problems of the inverse method of tangents, which Descartes himself allowed were beyond his powers." How could the inventor of calculus have written these words in August 1676, Newton scornfully asked, when it is basically obvious in the calculus that quadrature, the inverse method of tangents, and integration of differential equations are all the same thing? The fact seems to be that Leibniz, temporarily entangled in his integrals, mistakenly thought they were not; which is as much as to say, what is true, that in 1676 Newton was a far finer and more accomplished mathematician than Leibniz without the comparison bearing on the independence of Leibniz's discovery.

The same point could be made concerning Leibniz's relative lack of development in the process of integration by means of series, again pointedly noted by Newton in his comment on Leibniz's letter of August 1676: "Leibniz had already received the method of arriving at these series from Newton at his own request . . . and merely changed the sign for the hyperbola; for the circle he subtracted [the series for] the versed sine received from Newton in order to obtain the sine of the complement." Leibniz seems not

only to have been very casual in perceiving the parallelism between his own results and those of the British mathematicians but
slow to perceive the fact that Newton was telling him a good deal
about the way in which the second part of his calculus – integration – could be effected; though again one must qualify by re-
marking that the instruction hardly descended to step-by-step
procedures. This comment, of course, fits in with Bernoulli's later
claim that having learned differentiation from Leibniz he himself
had virtually to construct the integral calculus independently: Johann Bernoulli had not had the convenience of studying Newton's
letters.[14]

For all its mistakes, weaknesses, and special pleading, the *Commercium Epistolicum* completely altered the state of the dispute between Newton and Leibniz. Before 1712 the case for Newton had
been untidy; apart from general assertions, and the letters to Leibniz printed by Wallis in 1699, only three of Newton's mathematical treatises were (relatively recently) in print, and on the face of
it their authenticity as genuine early works might be uncertain.
After 1712 it could not be doubted that Newton had been a most
fertile mathematician since before 1669, and the sequence of his
achievement (so far as this affected Leibniz) could be followed;
moreover, the book established effectively (and not, in my own
view, unreasonably) the junior status of Leibniz as compared with
Newton. Certainly the book improved the historicity of Newton's position, even if it did not bury Leibniz's reputation (as
Newton had hoped) under a weight of adverse testimony. In
Newton's mind no one who worked through the documentation
he presented could fail to do him justice: Was it not simply *obvious*
that he had been years ahead of Leibniz? Although we may doubt
whether reading the *Commercium Epistolicum* actually changed any
one's opinion, Newton always insisted that the root of the matter
lay in it, that, in other words, the matters in dispute between himself and Leibniz related uniquely to historical events of the 1670s
and could be reduced to two simple questions to be answered
from the published documents: (1) Had he not known the new
infinitesimal calculus before Leibniz, and (2) had he not communicated this knowledge to Leibniz? All else was irrelevant to these
issues of historical matter of fact. But, as we shall soon see, the
original dispute was about to be extended to other, nonhistorical
disagreements.

Partly this regrettable and largely pointless diversification arose from the natural eagerness in combatants to use any weapon, partly it sprang from the inability of Leibniz's party to counter Newton's display of historical scholarship. Both Johann Bernoulli and Christian Wolf urged Leibniz to issue a true historical narrative of the evolution of the genuine calculus, urging that "most people may deduce from silence that the English case is a good one," and Leibniz himself (who admitted the wisdom of this advice) also spoke often of printing a *Commercium Epistolicum* of his own, in refutation of Newton's. He did neither. The paper on *The History and Origin of the Differential Calculus*, which he began in 1714, remained forever a fragment. Assuredly, Leibniz had not Newton's patience in combing through his *juvenilia* (which it would have been difficult to interpret to the public); perhaps Leibniz felt that in bandying texts and dates with Newton he could easily be the loser, because it would in any case have been impossible for him to take his own knowledge of calculus more than one year before Newton's *Second Letter*, perhaps he felt that the historical argument as Newton had opened it was really irrelevant, that rather questions of method, understanding and scope were at stake. At any rate, Leibniz and his friends made no effort to refute the *Commercium Epistolicum* historically, merely denying (and in the most strict of literal senses this was true) that it contained anything of the differential calculus; though even Leibniz did not cease to admit, as from the beginning, that Newton had been in possession of a "precursor" of the calculus.[15]

In the broadening quarrel, two main lines of development may be detected. Along the first, Newton's competence as a mathematician – in his writings on pure mathematics, but above all in the *Principia* – was assailed and every effort was made to convict him of error and ignorance on the ground that so feeble a mathematician could not conceivably have devised the calculus. The leadership here was exercised by Johann Bernoulli, Leibniz's lieutenant and his junior by more than twenty years, who was most anxious to hide his attacks behind Leibniz's back. In the end, after the philosopher's death had removed the possibility of contradiction, he amiably but pusillanimously made an outright declaration that Leibniz had been quite mistaken in things he had said about Bernoulli's involvement in the quarrel, which we now know to be true. He found a number of specious, though incorrect, reasons for supposing that Newton had really made little progress with the calculus, and that little by borrowing from the German school, and he extended this low view of Newton to the whole group of English mathematicians, whom he proceeded to challenge in trials of expertise and insight. In this second decade of the century Johann Bernoulli was beyond doubt Newton's most effective and dangerous technical critic; if his criticisms of Newton had been advanced in a constructive rather than a destructive manner they would have been more useful to the advancement of science.

Along the second line, under the direction of Leibniz in person, his earlier objections to Newton's theory of gravitation were developed, and Newton and his followers were charged with promoting a dangerous counterrevolution. Just as they had held aloof from the promising new mathematical wave on the Continent, so (it was held) they wished to undo the success of the mechanical philosophy. They were obscurantists. Newton's whole effort in mathematical physics had been misconceived, it was argued, and the true route of advance lay with Leibniz's dynamics. And, of course, this view too had elements of truth in it, and again, if it had been possible for Newton to accept some of Leibniz's formulations, notably his conception of "active force" (vis viva), as constructive suggestions, it would have been for the general benefit of physics. But the quarrel in both mathematics and physics forbade, or rather postponed, what would have been a creative synthesis of different ways of looking at things.

Johann Bernoulli's debut as an informed and technically qualified critic of Newton actually took place while the *Commercium Epistolicum* was in preparation and had, directly, no relationship to the outbreak of the row between Keill and Leibniz in 1711. It developed from Bernoulli's continuing study of the more difficult problems in mechanics, notably problems of the movement of bodies in fluids resisting their passage, a topic opened up long before by Christiaan Huygens and at one time of interest to James Gregory as well as Leibniz. Newton had examined such problems in a series of propositions in the second book of his *Principia*, already debated many years before between Huygens and himself and between Huygens and Leibniz. To his surprise, Bernoulli discovered that his own investigation of a special case yielded a result not agreeing with a general one of Newton's (by a numerical factor of 3 to 2) and, on further investigation, that Newton's answer to the problem entailed an absurd consequence that Newton had failed to perceive but which made his error certain. He wrote an account of his discovery to Leibniz in 1710 and sent another to the French Academy in the following year, which, however, did not appear in print until 1714. Of this Newton learned through a visit to London made by Johann's nephew Nikolaus, a very promising young mathematician (as indeed was Johann's son, confusingly also Nikolaus), in September 1712. The story of their meeting is given by the Huguenot refugee mathematician Abraham de Moivre, who was closely associated with Newton during the last twenty years or so of Newton's life, though we know all too little of what passed between them; de Moivre served at this time as one of several links between Newton and Continental scholars (one must remember that Newton was now seventy and much occupied in public business):

> I have had the honour [de Moivre wrote to Johann Bernoulli about Nikolaus] of introducing him to Mr Newton and to Mr Halley . . . We have met Mr Newton three times and he was so amiable as to invite us twice to dine with him . . . your nephew told me that he [!] had an objection against a result in Mr Newton's book concerning the revolution of a body in a circle within a resisting medium, and when he had shown his objection to me I, on his behalf, showed it to Mr Newton . . . Two or three days later when I went to his house, he [Newton] told me that the objection was valid and that he had corrected the result, which now proved to agree with your nephew's calculation. Thereupon he added that he intended to see your nephew in order to thank him, and begged me to bring him to his house, which I did.[16]

The unseen subtleties behind this story are intriguing. In the first place, despite the reasonably amicable correspondence between his friend and Johann Bernoulli extending over some years, Newton well knew that the Swiss mathematician (now reestablished back home at Basel since the death of his elder brother Jakob) was a vehement partisan of Leibniz's, and therefore Newton would guess that Johann and his friends would have crowed a little over Newton's technical slip, which as will be seen in a moment was not really so very dreadful. He presumably did not learn from Nikolaus that Johann had already sent a paper off for publication, with some rather mean comments on Newton as a mathematician, but again he presumably foresaw the possibility of this. Moreover, Newton was left a sort of loophole by which he could redeem his credit, for although Johann had found Newton's *result* to be impossible, he had not been able to fault Newton's published *proof* of this false result; this, laboriously, Newton had to do for himself and then correction was relatively easy. So there was a sense in which Newton was not "put right" by Johann Bernoulli, merely warned that something must be amiss.[17] Fortunately it was still possible to correct this part of the second edition of the *Principia*, which Roger Cotes was producing at Cambridge; as the book was not yet complete, although the pages containing Book II, Proposition 10 (the seat' of the trouble) had been printed off, the whole of that particular section could be reprinted, and this was done as may be seen in all copies of this edition (1713). In the new matter that he wrote for the reprinted pages Newton was less than generous. He made no admission that neither he himself nor Cotes had found anything amiss when revising the text for the new edition, he made no graceful acknowledgment of the perspicuity of the Bernoullis, and he passed over the whole incident in silence, leaving any careful reader who might compare the old edition and the new to make of the difference what he could. He did put a grudging notice at the beginning of the new edition that errors in the first had been corrected, but again no thanks to the Bernoullis and no thanks to Roger Cotes (who had saved Newton from many more serious slips). All this was remarkably ungenerous of Newton, who certainly never enjoyed admitting errors any more than most people, but then he had had experience over the years of his mistakes being thrown back against him: hence his policy here of least said, soonest mended.

Perhaps, too, Newton's handling of the business seems a little less strangely arbitrary when one recalls that Bernoulli had treated Newton and his friends in an equally offhand way only a few years before. Writing to Bernoulli in July 1705, de Moivre had told him of his discovery of an important theorem expressing the centripetal force at any point in an orbit (which, however, he had learned in conversation with Newton and which was already noted in Newton's *Principia* papers); Bernoulli soon hit upon the proof of this theorem of de Moivre's and published it in the Paris *Mémoires* as his own, without acknowledgment to the British (chapter 7). Though elsewhere he did confess that he had had the theorem from de Moivre, this did not save him from a severe rebuke administered by John Keill in 1716. And if Newton feared already in the autumn of 1712 that Bernoulli would prove very troublesome to him in the future, his fear was justified not only by Bernoulli's subsequent meddling in the calculus dispute behind the scenes but also by Bernoulli's open criticisms of points of theoretical mechanics in the original *Principia*, which he published in the *Mémoires* of the Paris Academy of Sciences and the *Acta Eruditorum*, appearing in 1713 and 1714.[18]

In the present context the significance of Newton's mistake lay in the ingenious explanation of it furnished by Johann's nephew, Nikolaus, in a postscript to his uncle's paper on mechanics in the Paris *Mémoires* for 1711 which, mischievously, does Newton the maximum of discredit. It was his "method of changing indeterminate, variable quantities into converging series, and making the terms of these series serve as successive differentials, which has led Mr Newton into error," Nikolaus declared. He drew attention to a rule for obtaining the first, second, third, and so on increments (or differentials) of fluents, and the fluxions proportional to them, stated by Newton in a Scholium at the close of the 1706 reissue of *On the Quadrature of Curves*, where Newton had written that the successive fluxions would be *as* (or in proportion to) the terms of an infinite converging series – which was correct; but in the following example, by an oversight, Newton gave the unmodified terms as the derivatives. Putting o a very small increment of z, so that the fluent z^n becomes by flowing $(z + o)^n$, Newton expands this last quantity into the converging series

$$z^n + noz^{n-1} + \frac{n^2 - no^2 z^{n-2}}{2} + \frac{n^3 - 3n^2 + 2no^3 z^{n-3}}{6} + \dots \text{ etc.;}$$

the second term of this series, he says (without further explanation), is the first increment or difference of the fluent to which its first fluxion is proportional; the third term is the second-order increment or difference proportional to the second fluxion, and so on. His statement is true for the fluxions, but incorrect for the increments and differences.

Properly speaking, of course, the numerical denominator of each term should be removed. This done, the process would accord with the general rules for finding fluxions or differentials outlined correctly by Newton at the beginning of the book. If now, wrote Nikolaus, you modify the series reached by Newton in proving his result in Proposition 10, so that instead of successive incorrect differentials you insert the correct ones, i.e., multiply the successive terms as given by Newton by 1, 2, and 6 (the first term being correct, of course), and then proceed with the argument as Newton has it, you come out with the right answer as already found by Uncle Johann! Newton did not properly understand how to differentiate. Q.E.D.

Nikolaus was, in fact, mistaken, though that was hardly relevant to the force of his criticism. The terms in the series employed by Newton in Proposition 10 of Book II were not successive derivatives of some quantity, as the great French mathematician Lagrange pointed out in 1797, and (in Whiteside's words) the success of Nikolaus's "reconstruction" – which was no doubt fully as malicious as Newton's "reconstruction" of Leibniz's route to the circle-squaring series – was due to "the jesting hand of coincidence at its most playful." Newton told de Moivre that "this error is the simple consequence of [my] having considered a tangent at the wrong end," and there is no doubt that it was indeed geometrical and not in any way connected with the calculus technique. This retort, though true, was neither strong nor readily provable, unfortunately, and Keill's later excuse that the false statements about successive differentiations of z^n in *On the Quadrature of Curves* were due to the printer's carelessness in omitting *ut* ("in proportion to") three times from the text was neither plausible nor true. Newton – as elsewhere, as Johann Bernoulli delighted to emphasize – had taken for granted what was trivial and obvious to him in a careless formulation, not caring to dot *i*'s and cross *t*'s. Leibniz was guilty of a similar informality (and was charged with it too by Newton's friends).[19]

Nikolaus's discovery, inevitably adopted as his own by Johann, was the explosive charge in the next grenade hurled back to England from the Continent. Just as the Bernoullis' trenchant, technical criticisms of the *Principia* in the *Mémoires* of the Paris Academy of Sciences and the *Acta Eruditorum* were made known to Newton only in late 1713, so the *Commercium Epistolicum* was slow in making its main effect, though presumably copies reached Paris in the first days of 1713. Only in the spring, returning from Paris, was Nikolaus able to put a copy of the pamphlet into the hands of his uncle, who at once sent off a report about it to Leibniz in Vienna, not forgetting Newton's ignorance of differentiation as revealed by his nephew. Much was to turn on this letter. After summarizing the background of the whole affair, including frankly the British case that Leibniz had "stolen" the calculus from Newton and published it under disguise in 1684, Bernoulli gave it as his own opinion that Newton "did not so much as dream of his calculus of fluxions and fluents, or of its reduction to the general operations of analysis in order to serve as an algorithm or in the manner of the arithmetical and algebraic rules," until after instruction from Leibniz. Here Bernoulli put his finger on one of the weak spots in Newton's defense, the lack of systematic development and exploitation of calculus techniques, especially the explicit use of differential equations, in Newton's writings, which are really detailed and explicit only on integration and infinite series; even today capable historians have tended to misunderstand or underestimate Newton's handling of infinitesimal methods in the *Principia*. Then Bernoulli made the rather feeble point that dotted letters to denote fluxions appeared only in the last volume of Wallis's *Works* (1699) and elaborated the contention about higher differentials derived from Nikolaus. "At any rate," he assured Leibniz, "it is clear that the true way of differentiating differentials was not known to Newton until long after it was familiar to us." Despite all this confidence, at the end he begged Leibniz not to involve him in the dispute, nor to make him "appear ungrateful to Newton, who has heaped many testimonies of his goodwill upon me."[20]

Whatever one may think of that piece of diplomacy, Bernoulli was well served in like manner by his friend Leibniz, now eager to make a riposte to Newton but without giving himself too much labor. Bernoulli's letter filled him with fury against the vanity-swollen English, who made a habit of appropriating German work to themselves: Boyle's investigation of niter had come from Glauber, his famous air pump from von Guericke, and now they meant to deprive Mercator of Holstein of the glory of the first discovery of series, not to mention his own affair! Keill's letter about all that he had put aside unread in contempt for the man, for Newton's claim was quite absurd: "He knew fluxions, but not the calculus of fluxions which," he told Bernoulli, "he put together at a later stage after our own was already published." Nevertheless, with the Royal Society's name involved, and Newton himself obviously active behind the scenes, Leibniz meant to clear his reputation even though he knew so little of the basis of Newton's grievance against himself, which he never did carefully examine.

The device he adopted was a strange one, evincing a certain deviousness of character not inconsistent with those anonymous reviews in which he praised himself as another person might. He decided to distribute widely a leaflet (or "flying paper" – flysheet – as Newton called it, hence it is always known by the Latin label *Charta Volans*) depreciating the *Commercium Epistolicum* and including the supposedly impartial opinion of a "leading mathematician" on the book, taken of course from Bernoulli's recent letter, without revealing his identity. The printing and distribution of the *Charta Volans* was entrusted to Christian Wolf, and it was soon copied into various learned periodicals in several languages. Unfortunately, in editing Bernoulli's letter, Leibniz (or Wolf) made the "leading mathematician, highly skilled in these matters," refer to one of Johann Bernoulli's Paris *Mémoires* papers as being by "a very eminent mathematician" with the ridiculous result that when Bernoulli's identity as the "leading mathematician" who had taken Leibniz's part began to be guessed (it was pretty obvious), then the joke went around that he had praised himself as the "very eminent mathematician." But all this took a few years to become notorious.[21]

The *Charta Volans*, as it were in the words of a third person, traces the gradual revulsion against Newton in Leibniz's mind in the face of the excessive claims made on Newton's behalf by his English partisans, who not only meant to gain glory for Newton but to exclude Leibniz from any share in "the analytical discovery or differential calculus first discovered by Leibniz in numbers and then transferred (after having contrived the analysis of infinitesimals) to Geometry." This is one of the first indications in the controversial writings of a truth to which Newton paid no attention, that Leibniz, unlike himself, had come across his first independent notions in mathematics – which led along a clear line to the calculus – while studying the summation of numerical series.[22] The paper then makes for the first time the open and blunt accusation that Newton had stolen the calculus quite barefacedly from Leibniz – no mincing with "hints" here – "having undeservedly obtained a partial share in this [discovery], through the kindness of a stranger, he longed to be given credit for the whole – a sign of a mind neither fair nor honest." The claim that Newton's earlier work in mathematics, before he took advantage of Leibniz's disclosure, was *not* the calculus, but was concerned rather with "advancing geometry synthetically or directly by infinitely small quantities" was buttressed by the "independent" technical arguments kindly furnished by Bernoulli; and Leibniz did not scruple to bring in the hearsay that had reached him about the grievances of Hooke and Flamsteed against Newton to further bolster his case.

The appearance of the *Charta Volans* advertised the fact that neither party was willing to seek a further reconciliation; each had now claimed priority, each had accused the other of plain plagiarism. Newton had appealed to his private prehistory of the calculus to assert his primacy, Leibniz to the public history of its enormously successful development in association with himself to demonstrate his de facto possession of rights. In a sense, the future determination of the dispute was already presaged. For all the British could do, the Leibnizians were not to lose their strong practical grip on the notation, the language and the articulation of the calculus. Because Newton's work of the sixties and seventies of the last century was already archaic, and because even the able, inventive men of his own party, like Cotes and Brook Taylor, were unable to dispute Continental leadership in the growth of mathematics, the best that Newton could now hope for was a historical distinction. But naturally he did not see the situation in this light nor could he perceive that the practical balance of usage was already tilted against him. Newton, laboring to make the true record of the past redress the inequity of the present, was a Canute vainly trying to stem the rising tide. Fluxions were to be for no more than a century the insular English language of English mathematicians, as little understood elsewhere as their spoken tongue.

10

THE DOGS OF WAR:
1713–1715

MANY years later Newton recollected for the benefit of Pierre Varignon:

> In autumn 1713 I received from Mr Chamberlain (who then kept a correspondence with Mr Leibnitz) a flying paper in Latin dated 29 July 1713 . . .

John Chamberlayne certainly had some acquaintance with Newton and had exchanged letters about political affairs with Leibniz since 1710. He was a journalist, proprietor of an annual resembling *Whittaker's Almanac*, which had been begun by his father. However, it was only at the end of February 1714 that Chamberlayne wrote to Leibniz deploring the dispute between him and Newton, as though he had recently learned of it, and offering his services as a mediator "between two of the greatest Philosophers & Mathematicians of Europe." Chamberlayne was the first of several aspirant mediators; but would he have waited four months or so before reacting to the message of the *Charta Volans* that he had passed on to Newton – even if it were the case that Leibniz (who was of course not supposedly the writer of the "flying paper") had compromised his position by openly posting it to his friends? However this may be, Newton as yet gave no outward sign of its existence; he did not (so far as we can tell) acknowledge its existence before early April 1714, nor did he do so until he and John Keill and many others were aware that Leibniz's replies to the *Commercium Epistolicum* were appearing in the Continental literary periodicals.[1]

We need not be surprised at Newton's inertia; after his burst of personal activity in prosecuting the publication of that volume and the "committee's" censure of Leibniz, he returned to his former policy of leaving his defense as far as possible to others; and he may at first have felt that an anonymous "flying paper" deserved no attention. How Newton had expected Leibniz to respond to the *Commercium Epistolicum* we shall never know; it was and it remained (especially when the *Account* was added to it in 1722) the supreme statement of Newton's case. It was meant to establish finally Newton's right as first inventor of "the Method" (though not, of course, specifically of the differential and integral calculus) and in this, save among Newton's irreconcilable enemies, it ultimately succeeded. At the same time it left open to Leibniz the possibility of quietly accepting for himself the role of second inventor (not *explicitly* denied to him in the *Commercium*) of "the Method" and inventor and exponent of his own, highly successful calculus. Perhaps Newton expected Leibniz – whom he supposed, wrongly, to be conscious in his soul that this was his utmost due – to tolerate the ignominy of making no further retort against Newton and the historicism of the *Commercium Epistolicum*. Did he imagine, as the worst possible alternative, that Leibniz would, on the contrary, reject Newton's position absolutely and retort an open charge of plagiarism against himself? Possibly not, as (from Newton's point of view) this would have placed Leibniz's character in the worst possible light. At any rate, Newton had now to cope with the situation that his policy had failed in its immediate object of forcing Leibniz to keep silent; indeed, his policy had made the dispute worse by making it more public, by transferring it from the decent obscurity of a learned language and the memoirs of academies to the vernacular monthlies. Newton turned to Keill – whether he had deliberately done so before or whether Keill had presented himself as a self-chosen champion – but *this* time Newton took the initiative. Now that his own side of the story had been made so public by Leibniz, he told Keill, "I think it requires an answer." And he went on

> If you please . . . to consider of what Answer you think proper, I will within a Post or two send you my thoughts upon the Subject, that you may compare them with your own sentiments & then draw up such an Answer as you think proper.

How often and in what precise manner Newton had, both before and after 1714, inspired the pens of other men we do not exactly know: Here the facts are precise.[2]

In charging Leibniz with making the dispute public, Newton was being rather less than fair, because Keill had already published in the bimonthly *Journal Literaire de la Haye* for May–June 1713 a long article relating the history of the discovery of the new infinitesimal calculus from the Newtonian point of view, a sort of popularization (in French) of the *Commercium Epistolicum*. Keill had gone back to Fatio's claims for Newton, considered his own previous intervention with the letter of protest from Leibniz and his own reply to it, and, after describing the appointment of the *Commercium Epistolicum* Committee by the Royal Society, had printed its report in full. He also printed part of Newton's first important letter to Collins (10 December 1672) to buttress the documentation. Keill, who cannot have acted in this way without Newton's knowledge and approval, would no doubt have argued that his article contained nothing that was not already a matter of public record or contained in the *Commercium*, but it is hardly strange that Leibniz sought to set the record more in his own favor in the same journal.

The "Remarks on the Dispute," the reply to Keill's "Letter from London," was published by Leibniz at the end of 1713; it wasted little politeness on the Newtonians. Leibniz, according to the *Remarks*, had known nothing of Newton's pretended discovery of the differential calculus before the publication of the *Commercium Epistolicum* and had expected Newton to disavow what ill-informed partisans had put forward on his behalf. Leibniz had never submitted himself to the judgment of the Royal Society, which had not heard his side of the story. It was evident from Wallis's published correspondence that what Newton had invented (and had since tried to pass off as the calculus) he had concealed, whereas Leibniz had openly communicated his method; it was evident from the *Principia* that Newton had not known the differential calculus in 1687, because he had avoided its use in circumstances that demanded it. Further, as an illustrious mathematician (Johann Bernoulli) had pointed out, the mistakes in that work would not have been committed by a mathematician competent in the calculus. What Newton had first published as his method of fluxions in Wallis's *Works* was no more than Leibniz's own discovery dressed out in other names and symbols. Until now, Leibniz had been content to believe that Newton, as he asserted, had discovered something similar to the calculus, but the contrary now appeared. And why had Newton so long kept silent in the face of Leibniz's possession of the discovery if not to wait for the deaths of Huygens and Wallis, who would have spoken as knowledgeable and impartial judges?[3]

The *Remarks* were followed by a French version of the *Charta*

Volans, with its declaration by the illustrious mathematician to whom Leibniz had so often referred.

They point directly to the weaknesses in Newton's position – weaknesses, as we would now say, not relating to the brilliance and real extent of his original mathematical ideas (the subject then in dispute) but to Newton's development and exploitation of them during a long life. Neither the *Remarks* nor the *Charta Volans* addresses itself to the historic evidence of Newton's early progress in the calculus; and although it may be true that a thing (like the gold at Fort Knox) that is hidden and unused is useless, concealment and disuse are nevertheless not evidence for nonexistence. If Leibniz had been arguing that Newton had no moral right to the discovery of the calculus, because he had failed to make it useful to mathematicians, his arguments would have been valid; but as arguments against the historicity of that discovery, they are invalid. Leibniz might analogously have argued that Columbus did not discover the New World because he never ceased to believe he had reached Cathay. The later observer sees again in these exchanges, as in the past and in the future, the essential mismatch between the Leibnizian and Newtonian contentions: the latter always emphasizing the historical reality of the discovery of fluxions from Newton's early papers, and later on the distinction of this *original* discovery from Leibniz's subsequent differential calculus, the former steadfastly holding to the view that more recent events (following Leibniz's publication of 1684) proved that Newton could not have been master of a similar calculus before 1684; or (as Newton wrote himself):

> the Author of the Remarks has laid aside the Records of the first seven years which make for Mr Newton & begins his report with the years 1676 & 1677, & thereby confesses that he has no way to defend Mr Leibnitz but by laying aside the Records which make against him.

To which in turn the Leibnizians might have riposted that the pre–1676 records making for Leibniz were wholly ignored by the Newtonians. How could there by any rational resolution to a dispute so ill-defined, and which was soon to ramble into yet less relevant side issues?

Newton's characteristic approach to the debate is strongly manifest in a draft letter intended as a retort to the *Remarks* and the *Charta Volans*, as published in the *Journal Literaire de la Haye*, which is still extant in a French as well as an English version. It is not clear why Newton began these rather elaborate preparations, later supplanted by Keill's efforts. The draft letter is chiefly interesting because it represents Newton's first attempt to compose a narrative based on the documents printed in the *Commercium Epistolicum* and other sources. The effect of the comments of the "great mathematician" (Johann Bernoulli) was to make Newton draft a far fuller explanation than before of the history of his notation and the ideas underlying it. *His* method, he claimed, could be used with any symbolism, whereas

> Mr Leibnitz confines his Method to the symbols *dx* & *dy*, so that if you take away his symbols you take away the characteristick of his method

And in any case, Newton asserted, his own "letters with pricks [dots] must be allowed as old at least as the year 1676, & by consequence older than the differential Notes of Mr Leibnitz" – an assertion which (if it means anything) is untrue; but really Newton was right in saying that the question of the dotted fluxional notation was quite trivial. He also considered at some length the distinction between fluxions and differences, which were "not quantities of the same kind." Moments or differences, Newton argued, are infinitely little quantities: They are "small parts of things generated by fluxions in moments of time." Fluxions, on the other hand, are velocities and finite quantities. From this Newton could contend that the method of fluxions was geometrical, whereas Leibniz's calculus was not: "For Geometry admits not of approximations nor of lines & figures infinitely little."[4]

Newton's view that Leibniz "doth not understand the Method of the first & last ratios," by which the whole mathematical operation "is performed exactly in finite quantities by Euclides Geometry . . ." so that it is "throughout as evident exact & demonstrative as any thing in Geometry," was at least as old as his reaction to Leibniz's review of *On Analysis*, and Leibniz's claim there that fluxions – or the method of first and last ratios – differed from the Leibnizian calculus "only in the manner of speaking."[5] Thus before February 1713 – for the *Commercium Epistolicum* had not yet been published when this draft was written – Newton seems to have been concerned to develop the logical superiority of his method over that of Leibniz (whom Newton does not even yet, in these pre-*Charta Volans* days, accuse of plagiarism) rather than its temporal priority. His discussion is, in effect, filling out Fatio's jibe of long before that Leibniz had not invented the new infinitesimal calculus, but merely spoiled it. The consistency with which Newton adhered to this argument and the trouble he took to make it effective prove that it was important to him; yet since Newton recognized that a simple manipulation converts a fluxion into a differential, and that it was often convenient to work with fluxions as though they were infinitely little quantities rather than ratios, his efforts can hardly have been considered successful. Later commentators do not seem to have found that in their logical foundations the two methods differed; certainly Berkeley did not. The fact is that Newton's attempt, as it now becomes – for it was not so originally – to answer Bernoulli's contention about the absence of notation from the early records of Newton's mathematics with the charge that, irrespective of notation, the foundations of his own method were geometrical, whereas those of Leibniz's calculus were not, is just as irrelevant to the basic historical issue as the matter of notation. Considerations of mathematical rigor had been, after all, as secondary to Newton as they had been to Leibniz.

Newton was on firmer ground in his reply to Bernoulli's assertion of his ignorance of higher-order fluxions. He dismissed it by saying it was "all one as to say that he then understood not how to consider motion as a quantity increasing and decreasing" (for acceleration is the second derivative, or fluxion, of distance) while also adopting (like Keill) the low device of arguing that *ut* had been omitted in the printing of *On the Quadrature of Curves* (Chapter 9). The rest of Newton's first abortive reply to the *Charta Volans* and Leibniz's *Remarks* is pretty much a rehash of the documentary material already given in the *Commercium Epistolicum*, what Johann Bernoulli rudely called "twice cooked cabbage." But it now led Newton to the downright declaration that Leibniz's method was "not demonstration, without the method of series is not universal, nor has any advantages which are not to be found in the method of Fluxions, nor has Mr Leibniz added anything to it of his own besides a new name & a new notation."[6]

Before the end of April Keill was deep into his "Answer to the Authors of the Remarks" (at whose name he did not yet guess, though he thought that Christian Wolf was the "eminent mathematician" who had written the Latin letter), which was to appear in the *Journal Literaire* of The Hague in the summer of 1714. This was to bring up particularly the question of higher-order fluxions and the Bernoullis' alleged "mistake" made by Newton, where

Keill was very anxious to retort with a *tu quoque* argument, though again it is hard to see how blackening the Leibnizians' reputation really advanced Newton's. Keill maintained, and Newton found his "Demonstration good," that Newton had not and could not possibly in geometry have taken the third term in his (mistakenly expressed) series given in Proposition 10 of Book II of the original *Principia* to be a second-order fluxion. Indeed, only an ignoramus would think such a mistake possible. Anxious to improve the position further, however, Keill recalled that Newton had pointed out to him geometrical errors in Leibniz's "Tentamen de motuum coelestium causis" (An Essay on the causes of the celestial motions) of 1689, amended in 1706; and evidently he turned to Newton for assistance in furbishing a relation of this error on Leibniz's part, thus turning the tables.[7]

It may well be that until the calculus dispute became critical (from 1711 onward) and until Keill pressed upon Newton the evidence of Leibniz's long-standing bad faith toward himself, Newton had remained unconscious of Leibniz's "*Principia*" essays in the *Acta Eruditorum* for 1689. Perhaps it was in the autumn of 1712 that Newton drew up a paper about them, which he later sent to Keill. Passing through another, intermediate draft, Newton's comments were printed, much condensed, in the *Commercium Epistolicum* in a note on Leibniz's record of publication in the 1680s. After Leibniz had read an epitome of the *Principia* in the *Acta* for 1688, Newton reports, he put together a letter on optical lines, a sketch of the resistance of mediums and the motion of projectiles in them, and the essay on the causes of the celestial motions:

> and had them printed in the *Acta* of Leipsig at the beginning of the year 1689, as though he too had discovered the principal propositions of Newton concerning these topics, and that by a different method serving to open up new techniques in geometry; and yet he had not seen Newton's book.

And in a footnote to these lines Newton probed more deeply into the nature of Leibniz's misconduct:

> If this liberty [of following too closely upon another's heels] be permitted, any innovator may easily be robbed of his discovery. Leibniz *had* seen the epitome in the *Acta*. Through his correspondence with scholars everywhere he could learn the content of the propositions in that book [the *Principia*]. If he had not seen the book, he ought nevertheless to have seen it

before he published his own papers on the same subjects written on his travels.

It seems that Newton did not mean to urge any actual charge against Leibniz that he had written the three *"Principia"* essays with the book before him, but he did, and justly, accuse Leibniz of bad manners – of greedily trying to steal fame for himself by climbing a ladder that another man had prepared for him. Not once only but in three essays – for all of them, Leibniz had found his stimulus in the *Principia*. There are, of course, excuses and exonerations to be offered on Leibniz's side, not least the points that published work is open to every man's use and that Leibniz had genuinely published original thoughts of his own, but Newton might reasonably feel that Leibniz had singled him out for an attention he could easily have spared. Continuing, Newton adds (a little mysteriously):

> Some say that Propositions 11, 12 and 15 of the *Tentamen* are false, and that Mr Leibniz by his calculus had deduced Propositions 19 and 20 of the same essay from them. However, such a calculus could be adapted to propositions discovered previously, without denominating the discoverer.

The purport of these sentences seems to be not so much that Leibniz was incompetent in his own calculus as that he had really taken from Newton the physical ideas that (by fudging argument) he tried in the *Tentamen* to claim as original with himself.[8]

If the Bernoullis' attack upon Newton in the matter of second-order fluxions had been as malicious as ill-conceived, did Keill and Newton not make themselves look foolish by picking up the same weapon to use against Leibniz? That Leibniz had made errors in the 1689 *Tentamen* is certain, as is the fact that he only corrected them in 1706 after they had been pointed out by his friends. Keill and Newton were confident that the errors arose from Leibniz's poor handling of second-order derivatives; indeed, in Keill's view Leibniz's correction (in 1706) of his earlier mistake in the *Tentamen* was itself a botch, because Leibniz had always failed to perceive the true source of that first mistake. Whether there is any virtue at all in such criticism of the *Tentamen* – which has been vehemently doubted – its relevance to the main issue seems just as doubtful as that of the Bernoullis', already firmly rebutted by Keill in the same long paper. To suppose that the true inventor of the calculus made himself known by his miraculous inability to make mathematical mistakes is an idea roughly equivalent to the medieval submission of supposed wrongdoers to the ordeal by fire. At least Nikolaus Bernoulli had the justification of connecting Newton's supposed error in *Principia*, Book II, Proposition 10 with his careless general statement about fluxions and differentials in *On the Quadrature of Curves*; Newton and Keill could allege nothing similar against Leibniz.

In my own view, Newton could well have strong reservations about the physical speculations upon which the *Tentamen* was founded – Keill called it "the most incomprehensible piece of philosophy that had ever appeared" – for certainly nothing could be more remote from his own notions of scientific methodology or better exemplify the kind of extravagant, not to say redundant, hypothesis that the mathematical physicist, in Newton's view, ought to avoid. More personally, Newton could very reasonably feel that Leibniz had treated him less than honorably in the three *Acta* essays. True, in the first of these, "On Optical Lines," Leibniz had praised Newton's mathematical talents and admitted that it was the *Principia* that had inspired him ("*excitavit me*") to publish his own work – all earlier, as he claimed, than his reading of the epitome of the *Principia*. For example, the optical lines now discovered by Newton had been known to (but not published by) Huygens and to Leibniz himself. In the paper on resisted motion, there is no mention of Newton at all; Leibniz claimed to have presented his ideas to the Academy of Science when he was in Paris. Again, he makes it appear that his thoughts on celestial motion were mature ones, and in the *Tentamen* he wrote of the inverse-square law that "it was already known to the celebrated Isaac Newton, though how he got it I cannot tell." Considering that on the last two topics especially Newton had been first in the field with ample and masterly discussions, richly orchestrated and illustrated with empirical evidence, Newton might well feel that this strident cry of "Me too," accompanying rather unsatisfactory technical discussions of the issues, was indicative of an envious and greedy character. He was hardly likely to recall the evidence (in his own possession) that Leibniz's interest in "the matter of reducing all mechanics to pure geometry" was at least as old as 1676.[9]

On the other hand, it was the feeblest possible comment to make on the *Tentamen* that it showed Leibniz, in 1706, still ignorant of second differentials, for one matter surely beyond all doubt was the healthy Continental development of the differential and integral calculus since 1684. And it was certain that Leibniz had been throughout the guide and mentor of the Continental mathematicians. Only on the crassest ignorance of the mathematical publications of the thirty years before 1714 could any denial of these facts be based; what was in question – as Newton so often and properly insisted on other occasions – was the first origins of the calculus in the 1670s.

Although Newton's support and encouragement of Keill was kept wholly private, and although Newton's Huguenot friend and coadjutor Abraham de Moivre loyally indicated to Bernoulli that Newton was leaving his defense entirely in Keill's hands, as soon as Keill's "Answer" appeared, the conjecture was made that Newton himself had had a hand in it. Although he strove to maintain an air of Olympian superiority and an assurance that the glory of first discovering the calculus could not possibly be challenged, Newton was never able to detach himself completely from the battle nor confidently resign his generalship in it to others.[10]

Meanwhile, on the other side, matters were becoming embarrassing for Johann Bernoulli. He was not yet much concerned about the use Leibniz had made of his letter in the *Charta Volans*, for at this stage he seems to have combined a genuine respect for Newton's ability with the belief that Leibniz would easily overwhelm him on the priority issue. His criticism of the *Principia* (first edition) and proclamations of Newton's "errors," however, obviously placed him in an exposed situation now that a battle line had been drawn between British and Continental mathematicians. (As we shall soon see, not *all* Continental mathematicians were eager for the impending struggle.) It was Bernoulli who had found that something was amiss in Proposition 10, and drawn an unwelcome moral from it; it was he who had blamed Newton for failing to demonstrate central forces inversely (that is, not only to show that if the inverse-square law holds, the path of a body in the force field must be a conic section, such as an ellipse, but that if the elliptical orbit, for example, is given, it must necessarily be produced by an inverse-square central force, and no other). And it was Bernoullis' "impudence," of course, that had impelled him to supply Newton's "omission." In the spring of 1714 he began to be concerned that Newton had formed an ill opinion of him, for he had received no copy of Newton's revised *Principia*, and later he thought Newton had had him expelled from the Royal Society. Although these fears were baseless, as de Moivre tried to convince him, they were nevertheless, indicative of Johann Bernoulli's state of mind. He might in turn advise de Moivre that he was delighted not to be supposed to have taken Leibniz's part in the calculus dispute, pointing out that his own papers critical of Newton's mechanics must have been written months before the *Commercium Epistolicum* was sent abroad, and thus before he knew "what was at issue between Mr Newton and Mr Leibniz," whereas his suggestion about Newton's error in higher-order fluxions (really an accomplishment, Johann now explains, of his nephew!) had first been put forward as much as two years before, "long before anyone in England dreamed of putting Mr Leibniz on trial"; but he must have seen how flimsy a screen this chronology provided, because the English protests against Leibniz's claims went back fifteen years and his own known association with Leibniz's calculus for more than twenty.[11] Despite such protests, Johann Bernoulli in his private correspondence with Leibniz was urging his friend to examine scrupulously the documents in the *Commercium Epistolicum* and produce a counter-blast of his own. On a more careful reading, evidently, he found the evidence of British mathematical prowess as early as 1671 disturbing: Surely, therefore, some of the documents must have been "if not wholly fabricated, then at least altered and falsified." He could not believe, for example, that "Leibniz's" arithmetical quadrature of the circle had been so early anticipated by both James Gregory and Newton, especially as the former's nephew, David, had spoken of it as *published* by Leibniz without comment. Might it not have been fraudulently inserted in the *Commercium*? Leibniz complained through the years – not completely without just cause – that the historical record as set out in that book was erroneous in some particulars unfavorable to himself, and he threatened to produce a parallel documentary collection of his own concerning his relations with the British long before; he never matured this plan, however, nor did he ever indicate more than trifling discrepancies in the *Commercium Epistolicum*. In essence, and taking its brevity into account, Newton had done his work of selection fairly and the documents (though not his related comments upon them) have stood unassailed to this day.[12]

Here, for the moment, the debate reached a temporary stalemate. Leibniz made no retort to Keill's "Answer," and, equally, Bernoulli let Keill's "Observations" (*Philosophical Transactions,*

July–September 1714) upon his own treatment of the inverse problem of central forces go by without immediate comment. Keill's energetic firmness and the (at least) passive support of the Royal Society for its president seemed to have silenced Newton's opponents who – as it is now clear from their correspondence – relished the public row no more than Newton did, and had begun to fear Keill as a truculent polemicist. As regards the historical discovery and nature of the calculus there was now little new that could be said on either side, unless Leibniz were to tell his own inner story (which, indeed, in 1715 he began to prepare). Leibniz did permit a letter to be published in the autumn of 1714 in which he tried once again on the one hand to confine Newton's genuine achievement in mathematics to the method of series, and on the other to link Newton's use of infinitesimals with that earlier employment of them by Fermat and others from which (as the British said) his own calculus had sprung. Newton, Leibniz argued, had not taken the essential step of using "not zeros, nor [quantities] infinitely little in the strict sense, but quantities which are incomparably or indefinitely little, and greater than some given magnitude . . . " As with Newton's attempts to express the distinction between differentials and fluxions in verbal formulations, Leibniz's meaning is far from clear. In any case, as it went without saying that the relationship between fluxions and differentials was close, so close that either might be said to be imitated from the other by a mathematician inferior to the first inventor, it was almost impossible that some logical distinction of conception should exist to identify the first invention – and who could prove that the *second* invention might not be logically superior to the first, if less original, just as Brigg's logarithms have many advantages over Napier's? As for Chamberlayne's attempt to effect a reconciliation, it dragged on without making progress, for the Royal Society declined to reconsider the *Commercium Epistolicum* or to encourage Leibniz to state a case.[13]

In Germany the *Acta Eruditorum* was inevitably upon Leibniz's side; in the Netherlands the periodicals would print manifestos from either contestant (though one at least, the *Journal Literaire de la Haye*, inclined to Newton's party from the sympathy of one of its editors, Willem Jacob 'sGravesande, who was soon to emerge as one of the most effective Newtonians on the Continent);[14] and in France very little public attention was paid to the calculus dis-

pute, perhaps because of the great war. Germany, the Netherlands, and Britain were all allied against France. The French Academy of Sciences inclined toward Leibniz, to the extent at any rate
that it permitted Johann Bernoulli to criticize Newton's *Principia*
in its *Mémoires* without ever accepting a rejoinder from the other
side. Pierre Varignon, its leading mathematician, seemed safe
enough as a member of the Malebranchist group, which had sat
at Bernoulli's feet long before. Fontenelle, its secretary, who was
to be (through longevity) the last living Cartesian, was no natural
friend to Newtonian theories, although, in the preface to
L'Hospital's *Analyse des infiniment petits*, he had made a dissection
of reputations that Newton was prepared to accept without resentment (though, obviously, Newton would have taken the
words to be L'Hospital's own). Barrow, Fontenelle wrote, had
invented a proper but incomplete calculus for handling the problem of tangents, "whose deficiencies had been supplied by the calculus of the famous Mr Leibniz who had begun where Barrow
and others left off . . . " This was in the Newtonian spirit, even if
Fontenelle then went on to emphasize the enlargement of mathematics effected by Leibniz's calculus and the astonishment created
in other mathematicians by Leibniz's discoveries, whereas Newton had only discovered "something similar" as might be seen in
the *Principia*. Obviously, in 1696 and without inside knowledge,
the writer of the Preface could not have said much more on Newton's behalf.[15]

In later years the academy in general and Fontenelle in particular
began to take Newton (from 1699, like Leibniz, a foreign associate) a good deal more seriously. A notable event was the attention
given to Newton's *Opticks* (1704), of which a partial rendering
into French by Etienne-François Geoffroy was read at several
meetings of the French Academy of Sciences during the winter of
1706 –7. After the restoration of peace, scientific relations between
France and England warmed considerably: There was polite (if
not consequential) correspondence between members of the
Academy of Sciences and Newton's circle – facilitated, perhaps,
by the presence in it of such Huguenots as Abraham de Moivre
and Pierre Coste; some Englishmen, of whom the most distinguished was the Earl of Bolingbroke, went to live in France; and
others, including Fellows of the Royal Society, went on visits,
while even more French came to explore London and the extraordinary cultural life of England. Many of those who were distin-

guished by noble birth or intellect sought an introduction to Newton and his elegant niece, Catherine Barton, who in these years was (presumably) at one and the same time the mistress of Newton's household, the intimate and admired friend of Jonathan Swift, and a notable beauty in London society. Far from rejecting Newtonian ideas about the operation of physical forces to produce phenomena as philosophically absurd, these young French savants were eager to learn more about them, just as they were also anxious to witness the English experiments on the vacuum and on light and color of which they had read.[16] Whatever prestigious authority belonged – and rightly belonged – to Leibniz and Johann Bernoulli, formidable intellects both, Vienna and Basel lay at the extremities of civilized Europe, remote from the cities of Paris, London, and Amsterdam, cities whose geographical triangle constituted the economic, intellectual, and military core of European civilization. From Bernoulli's letters one can sense his frustration at being dependent on Parisian friends or chance travelers for books and news, and Leibniz could (for the most part) act only through the hands of Christian Wolf. Especially during the last two years of Leibniz's life these two were almost spectators, passionately interested in events that were occurring elsewhere (above all in London) but scarcely able to influence them.

Although, in France, one Malebranchist mathematician (Varignon) was almost overwhelmed by the honor of his election to the Royal Society and another (Reyneau) was captivated by the republished *Principia* and rebutted Bernoulli's insinuations, political events nevertheless suggested a possible change in Newton's rising fortunes. Or so Bernoulli thought, supposing that a German king, for whose mother Leibniz had been a philosophical confidant, would strongly tilt the balance against the "Tory" Newton and cause the Royal Society to change its tune. Even Keill speculated whether, as a consequence of the Hanoverian succession, Leibniz might not "have the impudence to show his face in England"; "if he does," Keill continued stoutly, "I am persuaded he will find but few friends." Though George's "Court" did play a minor part in subsequent events, there is no sign that any pressure was put on Newton to do more than answer Leibniz's counter-accusations; certainly there is no evidence of any official interest in dismissing or disgracing Newton. The change of regime, which brought some relief and revenge to Flamsteed for the humiliation of the *Historia coelestis* edited by Halley, did Leib-

niz's cause little good. Newton, a permanent civil servant, remained on as good terms with his new Whig masters as he had
been with the Earl of Oxford.[17]

Seeing this hopeful opportunity, however, Johann Bernoulli
again urged Leibniz to push on with his exposure of the *Commercium Epistolicum* in order to publicize the passages omitted by
"Newton's toadies, or craftily suppressed, because they judged
them possibly less favourable and pleasing to Newton." At the
same time he proposed a new expedient, the issuing of some challenge problems "where Newton would, as you know, find himself in difficulties." The process was otherwise to become known
as "feeling the pulse of the English." "Doubtless there are many
[topics] to hand," Bernoullli wrote, "which were once discussed
between us, and which are not easily dealt with by the ordinary
method of differences, such as those we considered concerning the
passage from one curve into another, problems which are dealt
with by employing a certain particular [method of] differentiation."[18] The kind of problems Bernoulli had in mind are those that
are treated by what is now called the calculus of variations, in
which branch too (as it happens) Newton had made some notable
progress.[19] Bernoulli had relied on his skill in dealing with problems of this kind in his wrangles with his brother Jakob, as far
back as 1697. What he proposed was certainly a real test of mathematical skill and ingenuity – and, not to mince matters, the British
did not in the next few years come out well from this contest –
not least because even when the right approach to the problem
had been found, serious difficulties of integration had to be
solved.

The details of this ill-regulated mathematical contest are not important, save possibly to the specialist historian of mathematics.
Its only effect was to enhance irritation and ill-feeling; it did nothing to clear up the points at issue.

Leibniz's problem of defining the normal to a family of curves
was expressed in its original form in a letter dated 25 November
1715 to the Abbé Antonio Conti, then visiting London, who put
Leibniz's letter into Newton's hands. Leibniz was clearly not very
enthusiastic about it and, in fact, used such terms in expressing it
to Conti that they seemed to limit the problem to the case where
the curves were hyperbolas, and it was not until the following
January that, after perceiving the ambiguity of the letter to Conti,
he made it known that a completely general solution was re-

quired. Clearly, the letter to Conti was written in response to pressure from Wolf and Bernoulli that Leibniz should demolish Keill's defenses of Newton; Leibniz had urged in reply his reluctance to dispute with a rude and ignoble bully (for Keill's opponents were circulating apparently quite fictitious stories about his morals) and his lack of leisure to search through old correspondence; in fact, he begged Bernoulli to provide him with a suitable problem from the mathematical wrangles of the late 1690s. And, seemingly, the problem was only written on a note added to the letter to Conti, which is mostly of interest for quite different reasons. Leibniz's mind, in fact, was not now on mathematics at all, still less on the matter of the origins of the new calculus: He wanted to attack Newton's philosophy.[20]

The Abbé Conti plays in the calculus quarrel the role of an adventitious and slightly foolish figure. He was a Paduan by birth, presumably of good family, who for the last two years had lived in France. He too was an Oratorian and a great admirer of Malebranche. It was his wish to win fame by an intimate acquaintance with distinguished philosophers, though his own talents were not remarkable. "He was devoured by a rather hollow intellectual fervour," it has been said, "his thought was tentative and woolly, as is proved by his philosophical writings and his uneasy pilgrimage amid ideas and men." He sought in vain for a system to satisfy his needs. From Malebranche he turned to Leibniz, to whom he wrote in April 1715, just before his departure for England; probably the introduction was facilitated by one of Leibniz's French correspondents of Malebranche's circle, Nicolas Rémond. In England, he wrote fulsomely, he would support Leibniz's cause, just as he had done at Paris. Leibniz was not at once able to reinforce Conti's loyalty while actually within the lion's den because the letter did not reach him for many months, and meanwhile Conti found himself extremely happy in London. He got on friendly terms with Leibniz's friend Caroline, Princess of Wales, but, what was far worse, found Newton very reasonable and cordial also. Conti flattered himself on his philosophical conversations and dinners with Newton – who, despite a legend of taciturnity and indeed boorishness, could be a generous and affable host and was, after all, thoroughly at home in the "great world" of politicians and diplomats. And Newton, the oracle in public, in private treated Conti as a distinguished visitor, setting before him his opinion of Descartes ("I was myself when young a Cartesian") and of Cartesian metaphysics ("a tissue of hypotheses"). He favored Conti with the story that the first book of mathematics he had ever looked at dealt with astrology, and that, failing to understand it, he had turned to trigonometry, of which also he understood little. Then he had taken up Oughtred and Descartes's *Geometry*. In short, Newton made a great impression on Conti, and Conti (at first) won Newton's trust as a sympathetic foreigner.

So much so that soon Leibniz in his letters to Caroline, Rémond, and other friends lamented Conti's faithlessness, which he hoped Conti would repent on returning to the healthy air of the Continent from the atoms and void of England, for he was a chameleon taking up the color of his background.[21]

It was no doubt from a desire to remind Conti of the principles to which he should adhere that Leibniz chose him as the recipient of a renewed attack on Newton's philosophy, together with the challenge problem. Moreover, as one of the Princess's friends, Conti was the more suitable as she was already interested in the same philosophical questions about which Leibniz was writing in more detail to Dr. Samuel Clarke.

Leibniz left Newton to a postscript following his long-delayed answer to Conti's letter. After a sober statement of his confidence that Newton had never been master of his own algorithm or notation of the calculus, and his belief that the *Commercium Epistolicum* might be the truth but not the whole truth, Leibniz turned to Newton's philosophy, which he thought strange. If every body is weighty, then gravity is an occult or miraculous quality. To say that God has created a law of nature does not make it natural if it is contrary to the nature of the creation. God is not the soul of the world, and he has no need (as Newton had seemed to suggest) of the material creation as an organ of sensation. As for universal gravitation, atoms, and the void, none of these beliefs was demonstrated by the Newtonians nor maintainable on the basis of experimental science. Newton's method of induction from the phenomena (Leibniz went on) was an excellent one, but when the *data* are defective it is permissible to frame hypotheses and to put provisional trust in them until better facts are accessible. Newton, however, when he abandoned experimental philosophy, had no rational case to offer in favor of his extraordinary ideas. What a pity too (Leibniz added, unkindly now) that Newton had no able disciples, whereas he himself had been so much more fortunate!

Some of this had been in Leibniz's mind for months and years, and indeed it was his *Essais de Théodicée* (1710, above) that indirectly gave rise to the correspondence between Leibniz and Newton's friend Samuel Clarke. The Princess of Wales had asked Clarke to translate the *Théodicée* into English, which he refused to do on the grounds that his own views were so very different. When the Princess showed Clarke a letter from Leibniz to herself expressing anxiety about the current Newtonian trends of thought in Britain, Clarke composed a reply seeking to prove (but not to the Princess's satisfaction, who said that he was merely "gilding the pill") that Newton's metaphysical conceptions were not what Leibniz supposed them to be. Leibniz in turn replied to this paper, and a correspondence ensued, amounting in all to five documents from either party and terminating a year after it began with Leibniz's death in November 1716. It was published (in English and French) the following year.

It is unfortunate that Newton's relations with Clarke are as obscure as they are. Samuel Clarke was well known in his own day as a classical scholar, philosopher, and theologian. As a clergyman he became notorious for the Arian (or "Unitarian") tendency of his writings, and no doubt this prevented his advancement to the highest posts in the Church (indeed, Voltaire reports the quip that Clarke would be the best man in the Kingdom for the See of Canterbury if only he were a Christian). Like Newton, but unlike William Whiston, another of Newton's circle, Clarke managed by a discreet formula to avoid expulsion from his offices. It is possible that Clarke sought out Newton at Cambridge after going up to the university in 1691, for certainly he soon made his support of Newtonian science evident. Later Clarke was Newton's parish priest at St. James's, Piccadilly, where they were associated in various charitable concerns. Clarke made an "official" translation of Newton's *Opticks* into Latin (1706) – though it is also reported that Abraham de Moivre had a hand in the work – for which (according to legend) Newton gave him the vast sum of £500. Thus it is certain that Newton and Clarke were on terms of close acquaintance and probably intimate friendship long before 1715. Because they were neighbors, no correspondence between them is known to survive.[22]

Moreover, so far as the Clarke-Leibniz letters are concerned, there exist no drafts by Newton relating to them. It is certainly the case that Clarke was very well informed upon Newton's thought, and that Newton was made aware of the letters that

Leibniz was writing to him via the Princess of Wales; possibly also Clarke saw papers that Newton was sketching out for other purposes, as will be seen later. Thus in a very important sense Newton did assist Clarke indirectly in his defense of Newtonianism against Leibniz, as Leibniz himself believed that he was doing, and other contemporaries also supposed, Whiston for example writing that "Dr Clarke's Philosophy was . . . generally no other than Sir Isaac Newton's Philosophy." But to say this is not to say that Newton actually guided Clarke's pen or stood behind his shoulder as he wrote; of this we have no evidence at all, and the declaration of the two most recent students of the matter that "there is no doubt that Newton took part in the fight between Leibniz and Clarke," literally read, goes beyond what was doubtless their intention to express. Certainly Newton was in Clarke's corner, but he did not on more than an isolated occasion at most put lead in Clarke's gloves.[23]

We know, in fact, that Newton consistently regarded Leibniz's attack upon his philosophy, or supposed philosophy, as an irrelevancy. Of Leibniz's letter to Conti, adumbrating the issues solemnly debated between Leibniz and Clarke, Newton wrote that it was "nothing but a piece of railery from the beginning to the end," and in writing to Conti himself (after declaring that Leibniz had pretended not to have seen the *Commercium Epistolicum* in order to avoid answering it), Newton protested that Leibniz was still refusing to respond to the historical evidence against him and instead was "endeavouring to engage me in dispute about Philosophy & solving of Problems, both which are nothing to the Question." That Newton should see the situation in this light three years after the publication of the *Commercium* is hardly surprising, because no one had yet provided an analytical rebuttal of its documents, Leibniz and his friends having striven only to discredit Newton as a mathematician or extend their quarrel in fresh ways.[24]

Although the evidence indicates that Clarke was a far more independent champion of Newton than was Keill, Newton did have to address himself to Leibniz's criticism of his philosophy as expressed in Leibniz's letter to Conti of November 1715. This was as a consequence of court pressure; as Newton wrote in a later draft:

> When Mr l'Abbé Conti had received a letter from Mr Leibnitz with a large Postscript against me full of accusations

forreign to the Question, & the Postscript was shewed to
the King, & I was Pressed for an answer to be also shewed to
his Majesty, . . . the same was afterwards sent to Mr Leibnitz . . .

Like Newton, one may see Leibniz's court influence behind the
scenes here, gaining a small though Pyrrhic victory for Leibniz for
which Newton never forgave Conti, whom he considered to have
acted as Leibniz's cat's-paw. The letter Newton then had to prepare (and which was later printed by Newton himself without the
knowledge of its addressee) is almost wholly concerned with the
mathematical history of the 1670s on lines already becoming familiar; only near the beginning – following the passage about the
irrelevancy of all this just mentioned – Newton wrote

> As to Philosophy [Leibniz] colludes in the signification of
> words, calling those things miracles which create no wonder
> & those things occult qualities whose causes are occult
> though the qualities themselves be manifest, & those things
> the souls of men which do not animate their bodies[.] His
> Harmonia praestabilita [preestablished harmony] is miracu-
> lous & contradicts the daily experience of all mankind, every
> man finding in himself a power of seeing with his eyes &
> moving his body by his will. He prefers Hypotheses to Ar-
> guments of Induction drawn from experiments, accuses me
> of opinions which are not mine, & instead of proposing
> Questions to be examined by Experiments before they are
> admitted into Philosophy he proposes Hypotheses to be ad-
> mitted & believed before they are examined. But all this is
> nothing to the *Commercium Epistolicum.*

As indeed it was not, but Newton seems here to have forgotten
that it was he himself who had first in scientific works written of
God and nature and made a number of statements that certainly
could not "be examined by Experiments."

No more is to be found in the published version of the letter to
Conti, but a rejected draft contains a much lengthier discussion
(of which the quotation above is a condensation) in which Newton also alludes to such matters as Leibniz's notions of space and
time, and his calling "the world Gods Watch," so that if the world
should not be able to last forever as it is without God's providential intervention to keep it stable (a belief Newton had rejected in
his *Opticks* and elsewhere), then, in Leibniz's view, this meant that
God like the watchmaker must be imperfect: "It would be Gods

fault if his Watch should ever decay & want an amendment." This, retorted Newton, was like blaming God because matter cannot think. It is by no means impossible, if rather unlikely, that Clarke read such drafts before writing to Leibniz, or more probable that Newton conveyed the same ideas to him in conversation.[25]

Only one more of the metaphysical questions, which Leibniz used as a stick with which to belabor Newton, need be mentioned now, because it had freshly come into Leibniz's power as he told Johann Bernoulli (with evident delight) in March 1715:

> When I was told that Newton says something extraordinary about God in the Latin edition of his *Opticks*, which until then I had not seen, I examined it and laughed at the idea that space is the *sensorium* of God, as if God from whom every thing comes, should have need of a *sensorium*. . . . And so this man has little success with metaphysics.

To the Princess of Wales, in fact, Leibniz wrote that "Sir Isaac Newton says, that Space is an Organ, which God makes use of to perceive things by," so justifying his fear that in England some men "make God himself a corporeal Being." This Clarke subsequently denied, asserting Newton's own opinion that God does not need a *sensorium*, or organ of sensation (as we might say, a nervous system), that Newton had only analogically compared God's perception of things with man's, and that (in any case) *sensorium* properly meant not the *organ* but the *place* of sensation.[26]

Two brief passages in Queries 28 and 31 in Newton's *Opticks* gave rise to Leibniz's derision; in the first, to which Leibniz particularly referred, Newton had asked:

> does it not appear from Phaenomena that there is a Being incorporeal, living, intelligent, omnipresent, who in infinite space, as it were in his Sensory, sees the things themselves intimately, and throughly perceives them, and comprehends them wholly by their immediate presence to himself: . . .

And in the second (using the Latin form now of the same word) Newton asks whether the fitness of organic nature can be other than the effect of

> the Wisdom and Skill of a powerful ever-living Agent, who being in all Places, is more able by his Will to move the Bodies within his boundless uniform Sensorium . . . than we are by our Will to move the Parts of our own Bodies.

In Leibniz's justification it may be said that in these two passages Newton pushes the analogy between the body-soul relationship in man and the infinite-space-God relationship rather hard and crudely, relying a little too literally, perhaps, on the *Genesis* statement (to which he refers) that Man was made in God's image. We may feel (with Leibniz) that any comparison at all between the complex pathway by which man's "soul" apprehends the material world (sense organs, nervous system, brain, memory, etc.) and God's *immediate* apprehension of things by his omnipresent universality is so dangerous as to be actually misleading. On the other hand Newton's point that a providential God (to be distinguished from Leibniz's divine artificer, who by the very perfection of his work renders himself redundant) must be both *aware* of events in the world and *influential* upon them seems obvious. Of course, Leibniz was entitled to argue that God the creator is not a providential God, or that in so far as God is providential he works miracles, but he could not justly maintain that Newton's alternative position necessarily led him into the fallacy that Leibniz proclaimed as Newton's, which seems in fact to have resulted simply from naïvete or inadvertence on Newton's part.[27]

While this metaphysical discussion with Clarke occupied the last year of Leibniz's life, evidently relished by him far more than a rebuttal of the historicism of the *Commercium Epistolicum*, further steps had been taken in England to strengthen or at any rate to publicize Newton's case. Two publications resulted in 1715: The first was Newton's own "Account of the Book entitled *Commercium Epistolicum* . . . published by order of the Royal Society," which appeared anonymously as a review article in the *Philosophical Transactions* for February 1715; the second, presumably, was Joseph Raphson's posthumous *History of Fluxions*, which I shall consider first.[28]

Raphson is an enigmatic figure whose role, despite his little book with its impressive title, was much less than that of other men like William Jones and especially John Keill. He had been elected to the Royal Society as far back as 1689, had published a book on algebra in the next year, and went to Cambridge as a Fellow-Commoner – clearly at an age beyond the normal – in 1692. From a statement of his own (see below) he met Newton in Cambridge in 1691. Newton's editor Roger Cotes certainly knew him and knew of his intention to write about fluxions early in 1711. He then expressed the hope that Raphson would "do justice to Sir Isaac," thus indicating that the inner circle of Newtonians already saw Raphson's book in the context of the calculus controversy, even though Leibniz's letter to the Royal Society complaining against Keill was still to be written. As Raphson lived in London, the absence of correspondence with Newton is easily explained; but Newton must surely have been acquainted with him, as he was with Jones and others who never appear in any distinct way in Newton's own surviving papers. Raphson died before his book appeared.

The *History of Fluxions* is both an apologia for Newton and an exposition of the methods of the calculus; "history," in the basic sense of a narrative of events through time, it is not. Raphson stated explicitly in his Preface that his object was "to assert the Principal Inventions of this Method, to their First and Genuine Authors; and especially those of Sir Isaac Newton, who has vastly the Advantage of others, as well in respect of Priority of Time, as the Great & Noble Nature of his Discoveries." For knowledge of these Raphson chiefly drew on the two mathematical treatises published by Newton with his *Opticks* in 1704 and his correspondence as published by Wallis; he also used the books of Craige and Cheyne but does not seem to mention William Jones's edition of some of Newton's mathematical work in 1711, so that perhaps this appeared too late to be of service to him. Raphson did explain that the *Commercium Epistolicum* had been published while his own book was in the press; he added a conclusion dealing with this and rehearsing from it the "Judgement of the Royal Society."

Considered as history in the modern sense of the word, Raphson made a number of statements that are false or at least very susceptible of criticism. For example, he asserted that Newton had communicated his incomparable method of fluxions "(by several Letters) . . . to some of the best Mathematicians of Europe"; no doubt he would have instanced Collins and Leibniz, but in the ordinary plain sense of words such a communication – so damaging to Leibniz if true – never occurred, and Raphson failed to make it clear that Leibniz was given no useful information about the "method of tangents" and so forth about which Newton wrote to him. Raphson indeed wrote, contrary to fact, that Newton "gave Notice" of fluxions to Leibniz. He further declared that the *Principia* "shews almost innumerable Applications of [the Algorithm of his Calculus] both to Geometry and Nature" a view justified by only a limited and special kind of truth and erroneous if taken as implying an *open employment* of the fluxional algorithm in the *Principia*. Raphson does add, however, the interesting historical detail that Newton entrusted to himself and Halley in 1691 the original manuscript of Newton's 1671 tract on fluxions "in order to bring it up [to London?] to be printed"; it was already "much worn by having been lent out" to Collins and others. Later, Raphson wrote, Newton decided that the tract was too much in need of revision to be printed.[29]

The writings of the Continental mathematicians were by no means neglected by Raphson, who usually analyzed them in order to make a point in favor of Newton. Thus he based an account of Leibniz's calculus upon his 1684 *Acta Eruditorum* paper "partly that [the reader] may see [in] how less apt and more laborious a Method of Notation, and far-fetch'd symbolizing insignificant Novelties, (perhaps on purpose to distinguish it from the plain and easy one it was communicated to him in) he has published it to the World." And again, when he describes the "exponential" calculus of Johann Bernoulli, Raphson takes pains to point out that the basic idea and its logarithmic equivalent had been familiar to Newton. It is hardly surprising that his conclusion unreservedly declares Newton "the first Author of this Method" and puts Leibniz far behind him "even upon supposition that he [Leibniz] had it not from his Correspondence with Sir Isaac, which many suspect . . . "[30]

Raphson's little book might well have made more stir had it not been overshadowed by the *Commercium Epistolicum*. It is enveloped in mystery. Was it to have been Newton's first covert retort to Leibniz, begun before Keill's return to England from America and rendered redundant first by the Leibniz-Keill dispute and then by Newton's decision to prepare the *Commercium*? Was Newton responsible for its coming out in Latin as well as English, obviously for the benefit of Continental readers, because Newton certainly saw and improved its Latin version? Why, as it was not suppressed altogether, was the *History of Fluxions* so long delayed? Newton later declared – with what degree of sincerity it is hard now to tell – that he had known nothing of the book before it was printed "& I stopt its coming abroad for three or four years," in fact until the publisher made a fuss about his unrecouped costs in preparing it. Against Newton's seeming noninvolvement, there is the certain fact that after Leibniz's death – probably in 1717 or 1718 – Newton had the *History of Fluxions* reissued, with the same title page and date as before but with errata noted on the opening pages and the last chapter (on the *Commercium Epistolicum*) added;

to this he also appended the correspondence between Leibniz and the Abbé Conti with his own "Observations" upon their letters. All this, of course, without any indication of Newton's hand or indeed that there were two (or more) versions of Raphson's *History*. Whatever the extent of Newton's interest in this book, he molded it firmly and secretly to his own ends.[31]

As for the "Account of the Book entituled *Commercium Epistolicum*," which appeared in the *Philosophical Transactions* for 1715 (to be reissued with the book itself in 1722, now in Latin), it has been widely believed since 1761 that Newton himself prepared it and the fact is now beyond doubt. Originating in the abortive riposte prepared by Newton in 1714 for the *Journal Literaire* of The Hague and developed through many subsequent drafts during the latter part of that year, it stands as Newton's only long, coherent narrative of his case against Leibniz and was given official status in publications authorized by the Royal Society.

The "Account" is, of course, advocacy – Newton had given up the hope that the documents he had printed would explain themselves. Its argument divides into five chief sections, along by now familiar lines. The first proves that Newton's method was already in a finished state by 1669, as demonstrated by *On Analysis* and his correspondence of immediately subsequent years; all that came after had been the development of ideas that he had fully grasped long before Leibniz had been heard of. This section of the argument is illustrated by examples of Newton's mathematics taken from *On Analysis* and elsewhere, but nowhere in the tract does Newton enter into any detailed comparison or identification of his method with that of Leibniz.

The second theme in the argument is that Leibniz did not devise his calculus before 1677, so that when he wrote about it to Oldenburg in June 1677 (in response to Newton's *Second Letter*), it was of fresh minting. Point after point is skillfully piled up by Newton to Leibniz's discredit, showing him (apparently) always limping along behind, seeming not to understand what he had already claimed to have discovered, continually begging for fresh enlightenment and so forth. What was Leibniz's great discovery in numerical series, Newton asks scornfully? "See the mystery!" It is that if you subtract all the terms after the first from the series

(1) $$1 + \tfrac{1}{2} + \tfrac{1}{3} + \tfrac{1}{4} + \tfrac{1}{5} \ldots \text{ etc.}$$

so forming another

(2) $(1 - \frac{1}{2}) + (\frac{1}{2} - \frac{1}{3}) + (\frac{1}{3} - \frac{1}{4}) + (\frac{1}{4} - \frac{1}{5}) \ldots = 1$

the terms can be rearranged to make yet another series

(3) $\frac{1}{2} + \frac{1}{6} + \frac{1}{12} + \frac{1}{20} \ldots = 1$

and so on.

As for the "geometrical" series boasted of by Leibniz in his letters, they had all (in one form or another) been sent him from Gregory or Newton himself, a fact he ever afterward conveniently forgot to mention. In one and the same letter (27 August 1676), Newton alleges, Leibniz "pretended to have found two series for the number whose logarithm was given, and yet desired Mr. Newton to explain to him the method of finding these very two series." This same letter provides Newton with another piece of chronological evidence; as Leibniz himself confessed later, he had not "polished" his demonstration of the arithmetical quadrature of the circle in 1675; he sent it "polished" to Oldenburg only in August 1676. Then came his removal from Paris, when he did no mathematical work. Then, again according to Leibniz himself, he did not think it worthwhile to publish the demonstration because his new Analysis would prove it more briefly ("*Analysis nostra paucis exhibet*"). Accordingly, Newton reasoned, Leibniz could not have found his new analysis *before* leaving Paris – because, seemingly, he was still pleased with the demonstration in August 1676 – therefore he must have found it only after his resettlement in Germany, that is in 1677.[32]

One must admire Newton's ingenuity. This is just the kind of argument that a literary scholar or historian might use (in the absence of more direct evidence) today in order to date a poem or other document. The satisfaction of Leibniz in August 1676 with a piece of mathematics that, he later says, the calculus taught him to view as unsatisfactory seems neatly to make possession of that analysis in August 1676 impossible. With hindsight one sees how subtly perverse and misleading this plausible piece of logic is, but Newton could not but view it as convincing. To the same effect, he reasoned about some remarks made by Leibniz concerning the method of tangents: How *could* he, so late as March 1677, have shown knowledge of nothing more than the processes of Hudde and Sluse and written of the desirability of constructing "an analytical table of tangents, of the greatest use both for many other problems and for my resolution of equations by means of series" (again, the seeming grab!) if, at that time, he understood how generally the differential calculus handled tangency? Indeed, the discussion of tangents in terms of the calculus appears in Leibniz's letters to England only in June 1677, and Leibniz's assertion of "for a long time" dealing with them by this method Newton regarded with justifiable skepticism, because it seemed to be flatly contradicted by the old-fashioned tone of the preceding letter. And what, after all, in Newton's eyes, was this new method of tangents to which Leibniz came so tardily? It was, he says, Isaac Barrow's "Method of Tangents exactly, excepting that he has changed the letters a and e of Dr Barrow into dx and dy . . . Well therefore did the Marquis de l'Hospital observe that where Dr Barrow left off Mr Leibniz began; for their methods of tangents are exactly the same," with the sole improvement that Leibniz showed how to obtain Sluse's tangent rule from his own.[33] The facts that Leibniz had followed Barrow and that he had not seen some potentialities of his own calculus, as swiftly as Newton had done ten years before, could only be understood, Newton was confident, if Leibniz had made no genuine discovery.

The third theme in the "Account" is the distinction between fluxions and differentials, combined with the contention (against Johann Bernoulli) that Newton had changed neither his concepts nor his procedure since the 1660s; because this contention was not completely accurate, it led Newton into an apparent skepticism about the value of any particular notation for mathematical processes:

> Mr Newton doth not place his method in forms of symbols, nor confine himself to any particular sort of symbols of fluents or fluxions.

What Newton preferred to emphasize was the intrinsic superiority of his fluxional method, as against Leibniz's calculus, in avoiding the use of infinitesimals as far as possible, employing moments in a rigorous way rather than differences in an approximate way:

> We have no ideas of infinitely little quantities, and therefore Mr Newton introduced fluxions into his method, that it might proceed by finite quantities as much as possible . . . ;

the basis of the fluxion being (as Newton rightly said) the "first ratio of nascent quantities, which have a being in geometry" rather than the "first nascent quantities" or Leibnizian differences, "which have no being either in geometry or nature." Moreover, Newton argued, Leibniz's calculus necessitated a "summing of indivisibles to compose an area or solid . . . never yet admitted into geometry" and hence was only suitable for analysis, whereas his own method admitted of demonstration.

Not only was the calculus logically inferior to the method of fluxions, it was incomplete and maimed without the method of series to which Leibniz had no possible claim. Newton's original mathematical discovery was accordingly, he claimed, a dual one: a method for the resolution of "finite equations into infinite ones, and applying these equations, both finite and infinite, to the solution of problems by the method of fluxions and moments." Newton's method was thus "incomparably more universal than that of Leibniz" – which had been limited to finite equations until he imitated the technique of series from Newton – and besides historically genuine because it had possessed this necessary duality from its inception in the 1660s.

Once more one can only admire Newton's forensic ingenuity, highly appropriate to a patent suit or any issue depending on circumstantial evidence. The fairly transparent casuistry about infinitesimals – for, though the fluxion is indeed a ratio, the constant variable or time-measure o has to be vanishingly small – did not trouble him, and he never paused to reflect that, to account for Leibniz's not following his own path exactly there might be other reasons than a desire to disguise a plagiarism, or still less that these very differences of concept and development in the calculus might, when regarded from a different point of view, be taken as indicating the genuine originality of Leibniz's discovery.

As a fourth argument, Newton insisted that from the first the method of fluxions had extended to second- and higher-order fluxions, thus implicity reacting to the Bernoullis' criticisms, and further that the method had never been modified:

> This was Mr Newton's way of working in those days, when he wrote this compendium of his analysis (*On Analysis*). And the same way of working he used in his Book of Quadratures, and still uses to this day.

Amplifying his last sentence, Newton specifically declared that he had never "changed o [the increment or moment of x] into $\dot{x}$," or employed $\dot{x}$ as a substitute for dx, because "pricked letters never signify moments"; furthermore, the method of fluxions had proved its practical utility because it had served Newton in the preparation of the *Principia*, as he declared in a famous and delusive passage:

> By the help of the new analysis Mr Newton found out most
> of the propositions in his *Principia Philosophiae*; but because
> the Ancients for making things certain admitted nothing into
> geometry before it was demonstrated synthetically, he dem-
> onstrated the propositions synthetically, that the system of
> the heavens might be founded upon good geometry. And this
> makes it now difficult for unskilful men to see the analysis by
> which those propositions were found out.

The claim is categorical, but it cannot literally be true, though it
has deceived many. The complete absence of a "fluxional *Principia*" from Newton's work sheets, the straightforward evolution
of the book from Newton's 1684 "Propositions on Motion" –
purely geometrical – and the inherent implausibility of the claim
to a "translation" of this sort have long puzzled scholars; we can
now – as indicated already – believe only that Newton resorted to
an analytical attack on a problem in one or two exceptional instances.

Fifth and last, the "Account" concludes with a defense of Newton's philosophy of nature: "We are not to fill this philosophy with
opinions which cannot be proved by phenomena." Honest confes-
sion of ignorance of causes best fits experimental science, Newton
argued, and therefore it was no "crime to content himself with
certainties and let uncertainties alone." The seeming certainty of
the mechanical philosophy (as understood by Leibniz) was a mere
illusion, one leading directly to the very confusion of true
religion with which Leibniz had falsely charged Newton:

> Must all the arguments for a God taken from the phenomena
> of nature be exploded by new hard names? And must Ex-
> perimental Philosophy be exploded as miraculous and absurd
> because it asserts nothing more than can be proved by ex-
> periments, and we cannot yet prove by experiments that all
> the phenomena in nature can be solved by mere mechanical
> causes?

Thus Newton ends on a solemn note in order to refute the "rail-
lery" of the Leibnizians who sought thereby to demonstrate that
Newton wanted "judgment, and was not able to invent the infinitesimal method."

What was the effect of the "Account" on the public mind? It
seems likely that the original English version in the *Philosophical
Transactions* was read only in England, where it was soon accepted
and followed as the authoritative statement of the case for New-

ton, even before Newton's authorship of it was widely admitted. But the English, barring a very few (John Flamsteed, the Astronomer Royal, for example, and John Woodward, the physician, antiquary, and "geologist," another of Newton's settled opponents), were already parti pris on Newton's side. The "Account" was a sermon for the converted. In Europe it was known only by rumor until a French version was published in the *Journal Literaire* of The Hague, again with the willing aid of John Keill, Newton's "avowed champion in this quarrel." That version appeared only in November 1715; if in England Conti, "more in love with Newton than ever," felt that it was now up to Leibniz, in face of so strong a challenge, to present his own evidence of independent discovery, in general the Continent was not greatly impressed. The "Account" did not prove the blockbuster Newton had expected it to be. Leibniz dismissed it as a stale réchauffé and replied to it in the rival journal, the *Nouvelles Literaires*, of the following month in a few dismissive lines: It was not the case, he wrote, that Leibniz was making any claim upon a discovery of Newton's, but that in the eyes of all the world Leibniz was already recognized as the discoverer, and had been so recognized from 1684 without dispute until the mathematicians of England surprised the world by thinking otherwise. And he again quoted the judgment of Johann Bernoulli.[34]

The "Account" settled nothing. It defeated nobody. The dispute went on, and Newton once more took the offensive.

11

WAR BEYOND DEATH:
1715–1722

O EXAMINE the last years of the calculus dispute does not increase one's admiration for some of the greatest of mankind. Leibniz never conceded an inch toward the recognition of Newton's mathematical precocity and remorselessly continued to the end his attrition of Newton's philosophical absurdities, as he saw them. Newton pushed his pursuit of Leibniz beyond the grave – for his death did not, as Conti once exclaimed, end the quarrel – until at least 1722. And subsidiary warfare broke out on no small scale which, however, I do not mean to explore in detail here. What was written in these last years, at least so far as the original point at issue is concerned, was all passion and tedious repetition. Very little that was new in fact or argument was made public after 1715 – for the essence even of the Clarke-Leibniz exchanges had all been stated before – and the weapons of polemic forged by either party seemed increasingly to be hurled, not at the chief opponents, but at the men of straw who, by now, had firmly assumed their places. It is no surprise to find the dispute concluding amid the futility of offensive wagers, or supposed wagers, and childish abuse. Had Newton, or had he not, publicly called Johann Bernoulli Leibniz's "skirmisher" (*enfant perdu*)? Who can care?

During the interval of less than three years between Newton's first reading of the *Charta Volans* and the death of Leibniz, his preoccupation with the demolition of Leibniz's claim to the calculus became obsessional, requiring hundreds of hours of paper work. The "Account," having failed in its effect, was to be followed in the next year by Newton's "Observations" upon Leibniz's exculpatory letter to Conti of March 1716, both of which Newton put into print in the second issue of Raphson's *History of Fluxions* (1717–18). This stage of the calculus dispute, with Conti as its luckless pivot (whether agent-provocateur, lay figure, or honest broker, think of him as one will, he won only the contempt of both the great contestants), was brought about by Leibniz's Court connections, which were, however, insufficiently strong to effect a summons to England as a loyal servant of George I. Conti at Newton's request (so he claimed) brought together a group of Hanoverian and other notables who were to verify with their own eyes the documents brought in evidence by Newton. This done satisfactorily, one of them, Baron von Kilmansegge, proclaimed that this was not enough: What was needed was a direct challenge from Newton to Leibniz, which must be answered. Conti went to Newton who within a few days of February 1716 produced a letter which George I much approved, saying it was difficult to answer facts. As usual, Newton had spent many hours in drafting this letter, even though he thought Leibniz had given him nothing but "railery"; yet it contains nothing new. It is a short version of the "Account," which contains some good specimens of Newton's English prose, as when he complains of Leibniz's "endeavouring to engage me in dispute about philosophy & about solving of problems, both which are nothing to the question." Not only had Leibniz thus failed to deal with the original documentary issue on which Newton always harped, but he had gone back on his own prior acknowledgments of Newton's attainments, which he ought "in candour to acknowledge still," and now contradicted statements of them made by Wallis and others, which formerly "he did not contradict nor found fault with. And I expect that he still forbear to contradict." But this iteration was to no purpose, except to produce another retort from Leibniz. It is, on the whole, dignified, denying the Newtonian charges and criticizing the *Commercium Epistolicum* as quite missing its aim because the letters in it were entirely concerned with series, the glosses upon the documents merely voicing "baseless suspicions which are sometimes absurd, and sometimes feigned contrary to the conscience of some of those" involved, and so not at all depreciating his own discovery of the calculus. And here at last Leibniz told something of that discovery, how he had come to London in 1673 almost innocent of higher geometry, and remained then unacquainted with Collins, how on his return to Paris he profited from reading Mercator and conversing with Huygens and so found "his" arithmetical quadrature of the circle, and how then, still largely innocent of the methods of infinite series (here Newton was right!):

> I entered at last into my calculus of differences, where the things I had already noticed when very young on the differ-

ences between series of numbers helped to open my eyes; because I came to it not by the flow of lines, but by the differences of numbers . . .

As for Barrow, he had nothing to do with it, and if anyone profited from him, it should surely be Newton who was his pupil.[1]

Concluding with the assurance that he had never charged Newton with plagiarism, blaming Newton's adherents for poisoning his mind, and trying to wriggle out of the accusation that the *Acta* review of 1705 had accused Newton of playing a Fabri to his own Cavalieri, Leibniz's defense, although prevaricating on occasion, is at least (as we know now) essentially accurate in describing his own mathematical evolution. But far from mollifying Newton, Leibniz's smooth assurance only irritated him the more and he was outraged at the suggestion that only a troublemaker would see harm in the words of the *Acta* review of *On Quadrature*. The process of drafting began again . . . The product appeared in the form of "Observations upon the preceding Epistle" in the reissue of Raphson's book, which Newton "caused to be printed" as soon as he heard that Leibniz was dead, "lest they [the letters and 'Observations'] should at any time come abroad imperfectly in France." His reason for not writing privately to Leibniz or Conti was that he had heard that Leibniz had sent to his friends in Paris copies of his supposedly private letters to Conti – the fine lines of moral conduct become very involved; thus, if the "Observations" were written in the summer of 1716, they first became public (probably) in the following year.

Once again Newton covers the familiar ground: Leibniz had been the aggressor, not only in the 1705 review but in demanding an apology from Keill in 1711; the Royal Society had not condemned him unheard, but had left him liberty to state his case historically, which he had always refused to do. Leibniz's writings against Newton had been nothing but defamation, whereas the *Commercium Epistolicum* had been nothing but facts, openly printed, unlike the clandestine, unavowed *Charta Volans*. The method of series and the method of fluxions were all one and should not be artificially distinguished:

> And if Mr Leibniz has been tearing this general method in pieces, and taking from me first one part, and then another part, whereby the rest is maimed, he has given a just occasion to the Committee [of the Royal Society] to consider the whole. It is also to be observed, that he is perpetually giving

testimony for himself, and it's allowed in all Courts of Justice
to speak to the credit of the witness.

As ever, Newton identifies Clio with the Goddess of Justice. Leibniz's charges that the committee had omitted from the papers in
the *Commercium Epistolicum* passages that told against Newton's
claim are dismissed; Leibniz had already withdrawn his first example of such an omission as a mistake, and now Newton shows
that his second is no better

> So you see that Mr Leibniz hath accused the Committee of
> the Royal Society without knowing the truth of his accusa-
> tion, and therefore is guilty of a misdemeanour.

The legal phraseology recurs. And so it goes on, Newton now
allowing no shred of independent discovery to cover the naked-
ness of Leibniz's supposed theft, until he embarks on yet another
narrative of his own early mathematical writings, which (in
Whiteside's words) "was to serve as the well-nigh unique *locus
classicus* for authentic information about the early stages of [New-
ton's] development of fluxional calculus," for all but the privileged
few who had knowledge of Newton's manuscripts, until the nineteenth century.[2]

The "Observations" were still not Newton's last words on the
calculus dispute, but before noting its final appearance and turning
(with relief now) to the last rosy glow of venerable respectability
that warmed the last years of Newton's life after this cold, titanic
struggle, a little should be said about the last sharp snappings of
the dogs of war, Johann Bernoulli and John Keill.

While these thunderclouds had been reverberating about Conti's
head, the English mathematicians had not neglected the "easy"
version of the mathematical problem that Leibniz had passed on
to them (from Bernoulli) in his letter to Conti of November 1715.
Their cries of triumph were cut short when Leibniz revised the
problem to increase its generality (or, as the English claimed, sent
a fresh, harder problem). This, effectively, beat them. Amid these
events Johann Bernoulli took notice of the attack made on him in
the *Philosophical Transactions* for the summer of 1714 by Keill, who
had written a paper on the inverse problem of central forces (the
problem of deriving the curve[s] from the force[s], rather than
vice versa). Bernoulli, as far back as the *Mémoires* of the French
Academy of Sciences for 1710, had claimed that Newton in the
Principia had not provided a satisfactory demonstration that, with
the inverse-square law assumed, only conic sections and no other

curves could furnish the resulting orbits. Keill had rejoined in
Newton's defense that Bernoulli's own demonstration to this effect was, in fact, only a slightly reworked version of Newton's
own Proposition 41 in Book I of the *Principia*: His criticism of
Newton, therefore, was out of place and his "improvement'" bo-
gus. To such a correction – given unsparingly – the Swiss mathematician could not submit: Too proud (let us suppose) to speak
out openly against the derided Keill, he wrote a letter to Christian
Wolf, which the latter transposed into indirect speech and printed,
thus anonymously, in the *Acta Eruditorum* for July 1716. At least
anonymity was their intention; with the quality of farce, which
often attached itself to the stern and biting pen of Bernoulli, a slip
in the printed text revealed all.

The "Letter on behalf of an eminent Mathematician" against
Keill contained nothing fresh; it restated Bernoulli's claim to be an
independent inventor of the integral calculus "if we mean to distinguish this calculus from the differential calculus which, even
according to Bernoulli himself, is beyond all controversy owed
entirely to the great Leibniz," abused Keill, and in general was not
at all about the original question in the great debate but about
Newton's and Keill's incompetence in the mathematical science of
mechanics. As for the inverse problem of central forces, let anyone examine how different Bernoulli's approach to a solution was
from Newton's

> and let them afterwards declare whether anyone but [Keill]
> himself can persuade himself that my formula [in the 1710
> *Mémoires*] was derived from Newton's.

An unnoticed "m" (*meam* for *eam*) revealed the author; denials
proved futile.

Keill retorted, Bernoulli primed one of his students to make a
reply under his own name, and much else went to and fro, which
may be passed over. Other mathematicians – Brook Taylor on the
English side, Rémond de Monmort on the French – were dragged
into the futile wrangling. What is more interesting is that Johann
Bernoulli was also at last revealed as the famous mathematician
quoted in the *Charta Volans*; this fact he refused publicly to confess
in Newton's lifetime, but Newton had no doubt inwardly of its
truth, and never forgave Bernoulli for his conduct.[3]

The French version of the pro-Leibniz letter by the "eminent
mathematician" had first appeared in the *Journal Literaire* of December 1713, in the translated reprint there of the *Charta Volans*.

It reappeared in the *Nouvelles Literaires,* just two years later, as part of Leibniz's protest against the *Commercium Epistolicum* addressed to the journal's editor, Du Sauzet. But *this* time, clearly from the pen of Leibniz himself and against Bernoulli's frequent insistence, this same extract was blandly headed "Lettre de M. Jean Bernoulli de Bâle, du 7 de Juin 1713." Probably Leibniz let the name out deliberately, though it would be idle to speculate about his reasons; certainly, with a little note of pleasure, he was able to tell Bernoulli, in April of the following year (1716), that Newton now knew him to be the "eminent mathematician," and had referred to him (in his reply to Conti, written at the abbé's request in February) as a "mathematician (or pretended mathematician)." The insult rankled with Bernoulli for many years, though how he could logically assert – as he always untruthfully did – that he was *not* that eminent mathematician, while at the same time objecting that Newton should not have called the eminent mathematician "pretended," it is hard to understand. In fact, Newton in February 1716 probably did not know that Bernoulli was the man (and certainly did not mention his name to Conti), but Leibniz ensured that he should not fail to know by again putting Bernoulli's name as the "eminent mathematician" in a letter he wrote to his friend the Baroness von Kilmansegge in April, only a day or two after telling Bernoulli of the predicament he was now in. That letter was intended for Newton's eyes; not surprisingly, in the "Observations" Newton for the first time makes Bernoulli the "eminent mathematician."

In parenthesis, Leibniz's letter to the baroness contains a passage that Newton could never have written, in a sense rather different from that in which he could not have written the whole letter. Leibniz tells the story of the Leiden shoemaker who was fond of attending the public disputations of the university students. An acquaintance asked him if he understood Latin: "No, I don't trouble to understand it." Why then did he come so often to the place? "Because I like to judge the strokes." And how could he be a judge of what was said? "I have another way to tell who is in the right of it." And what was that? said the acquaintance. "It is that when I see from someone's face that he is getting annoyed and becoming angry, I conclude that he lacks arguments." It is very true that Newton gave far more time and attention to the calculus dispute than Leibniz did, that it made him more angry, that he could not attain in his writings about it the assured, slightly supercilious, not wholly humorless tone that Leibniz achieved. One can see Newton as the red-faced angry disputant, Leibniz as the defendant calm and confident. Only an extraordinarily clever and subtle liar would have told this story – which would have, and was meant to have, more than one private reader – when confronted with a *justly* irate opponent. Leibniz could tell it because he did believe in the justice of his own cause. Unfortunately, two rights can all too easily make a wrong.[4]

Leibniz's death left Johann Bernoulli isolated and now, though not very old, the senior mathematician of the Continental school. He was willing to carry on the fight with Keill (and did so until the latter's death in August 1721), but he was eager to make peace with Newton. He even nourished the ambition – never to be realized – of receiving the gift of Newton's portrait. Accordingly, in the spring after Leibniz's death, he approached one of the French mathematicians, Rémond de Monmort, whom he knew to be in touch with the Newtonians, in order to get Monmort to assure Newton that his part in "testing the pulse of the English mathematicians" had been purely passive, and that

> I desire nothing so much as to live in good fellowship with
> him, and to find an opportunity of showing him how much
> I value his rare merits, indeed I never speak of him save with
> much praise . . .

if only, too, Newton could persuade Keill to live and let live! And in print a couple of years later, not having yet made a great deal of progress in private, Bernoulli lamented Keill's anger at Bernoulli's "correction" of Newton's mistakes in the *Principia* (in 1713):

> substituting the truth for what was false and supplying what
> was absent. This I did, too, with such modesty and candour
> that Newton himself (who at the end of his preface begs this
> from his reader) was not only not offended by my notes but,
> as I afterwards learned from a common friend who converses
> familiarly with Newton [Abraham de Moivre], seemed not a
> little obliged to me for them . . .

For, after all, the new edition of the *Principia* would have been spoiled if these blemishes noted by Bernoulli had remained in it; he is, of course, conveniently conflating Nikolaus Bernoulli's discreetly signifying a mistake privately to Newton with the imperfections once trumpeted in the Paris *Mémoires*. Keill, on the other hand, still carrying on a guerrilla war with Bernoulli (who had called him, among other things, "a certain person of Scottish extraction, who has become no less distinguished among his own people for his immorality than odious to all foreigners"), was not at all inclined to let Newton forget Bernoulli's role as Leibniz's lieutenant; only after he was removed from the scene could a pacification be effected.[5]

Its principal agent was Pierre Varignon, and it was in part a fruit of the improved Anglo-French scientific relations that followed upon the conclusion of peace in 1714 and the Hanoverian succession. Varignon, as an Academician, had distributed the three copies of the second English edition of *Opticks* (1717) sent by Newton to the Academy; one he sent to Johann Bernoulli with the following explanation to Newton:

> I add fresh gratitude to the thanks just offered because you have correctly judged me wholly innocent of acting against you in the Leibnizian controversy: quite the contrary, I have taken so little part in it that I have rather always kept silent about that controversy in my letters to Mr Leibniz and Mr Bernoulli, only lamenting to myself and in private that such great men are troubled by it, whom, if I had any influence, I would have restored to their former cordiality; and this was the sole object that I had in mind when I sent to Johann Bernoulli from yourself the new edition of your book on colours.

That was in November 1718, and Newton welcomed Varignon's action. Seven months later Bernoulli seized the opportunity to write his thanks to Newton directly for "his" gift and regret his own unwitting involvement in the dispute, which, against all his own inclination, had brought about the loss of Newton's goodwill. Even so, Bernoulli could not forbear to put a large share of the blame on the Newtonians, expressing his anger at the flatterers who abused all foreigners both innocent and guilty, out of a desire to build monuments for themselves upon the ruined reputations of others. Leibniz, Bernoulli wrote with remarkable disingenuity, had surely been misled by someone into naming himself as the author of the anonymous publications for which Newton had since held him responsible; and again he proclaimed that he was not. Although at this very moment Keill was once more busy with a vigorous "answer to Mr Bernoulli and the Leipzig rogues" who had introduced the new and barbarous practice of extending their abuse of Newton and himself into the index of the *Acta Eruditorum*, Newton was by now wearying of Keill's ceaseless championship and he accepted Bernoulli's olive branch without, surely, changing his opinion of the man. In September 1719 he wrote to Varignon that as Bernoulli had categorically denied anonymous support of Leibniz, "I readily welcome and court his friendship," while in an enclosed letter to Bernoulli himself Newton took the softest possible line; never had he stirred up quarrels with Bernoulli, whose esteem and friendship he greatly valued "on account of his enormous merit in mathematical matters."

> Now that I am old [he went on] I take very little pleasure in mathematical studies, nor have I ever taken the trouble of spreading opinions through the world, but rather I take care not to allow them to involve me in disputes.

Newton was also able (in due course) to assure Bernoulli that he had not been expelled from the Royal Society, and behind the scenes he delayed publication of Pierre Des Maizeaux's book on the calculus quarrel; yet he was not absolutely satisfied about Bernoulli (according to de Moivre), because the Swiss continually harped on his old grievances, made new threats about publishing things damaging to the British mathematicians, and demanded that Newton restrain them from attacking him. So the correspondences went on and on through 1720 and 1721 until Des Maizeaux's *Collection* at last appeared, and now again Bernoulli complained that Newton had called him a "pretended mathematician," "Leibniz's knight-errant," and so forth, though in fact Newton had only used such sarcasm in writing of the anonymous "eminent mathematician." And Newton on his side forbade Bernoulli to print to his own advantage any expressions of cordiality or conviction of his innocence, though he allowed them to be circulated privately. In the end, to conclude the business, Varignon drafted a letter stating his understanding of Newton's lack of animosity to Bernoulli, which was to be sent to him, with freedom to print it, if Newton approved, as, reluctantly and only through Abraham de Moivre (for Newton had now fallen ill) he did. Varignon duly passed the authorized statement to Bernoulli in June 1722.

That should have been the end of the matter. Keill was out of the way and had no successor. But it was not, quite. To Varignon's distress, Bernoulli (difficult as ever, and on the whole the more difficult as concessions were made to him) replied that he was not satisfied; he had received no apology from Newton for the harsh things said about him, and the attributions of things to him on Leibniz's word. Besides, Newton had not promised to put a leash on Brook Taylor and other English mathematicians who might try to offend him. This was too much even for the kindly, equable Varignon: He told Bernoulli his demands were impossible and washed his hands of the matter. He too died in the following December, possibly before receiving Newton's latest, not wholly unprickly, literary gifts, that is to say the revised *Commercium Epistolicum*, now enlarged by a Latin translation of the 1715 "Account" (*Recensio*) and his *Arithmetica Universalis*, or algebra, written forty years before. Consequently, Johann Bernoulli himself made one last approach to Newton, having received (as Varignon's last gesture) copies of the French version of *Opticks*, published at Paris that autumn. Again, though one may suppose in a rough way he was trying to be pleasant, Bernoulli touched the comic, for after assuring Newton of his admiration for Newton's work "concerning light and the system of colours . . . a discovery more enduring than any bronze [monument], and one which will be even more highly prized by posterity than it is now," Bernoulli went on to explain how Nicolas Hartsoeker, "a foolish man," had claimed priority in this discovery: "So that I am astonished that no one comes forward from your fellow-countrymen to defend your reputation." And he could not let the matter of Des Maizeaux's *Collection* alone either. To this last letter Newton never replied, and it marks the end of the calculus dispute in Newton's *Correspondence*, if not in his thoughts. Indeed, Newton seems to have written few letters during the last four years of his life, apart from the necessary minimum number on official business.

He was now very old, not in good health, and increasingly reliant upon trusted friends like de Moivre and his half nephew by marriage, John Conduitt. Yet he was by no means lacking in the mental power to control, almost up to his last months, the third edition of his *Principia*, which was seen through the press by Henry Pemberton, even if he had as long ago as 1718 ironically written to Des Maizeaux

> ever since I wrote that book I have been forgetting the methods by which I wrote it.[6]

He now excluded from that edition the mathematical additions he had toyed with through the years since 1713 – a reimpression of his treatises already in print, perhaps, with or without a historical introduction, or some entirely new statement about the whole business of fluxions and calculus – and left the book in essence as it had been before. Because this project was abandoned as Newton moved on into his eighties, Des Maizeaux's *Collection* remains the last publication in the calculus dispute with which Newton himself was associated – excepting, of course, the reissued *Commercium Epistolicum* two years later (1722), doctored – much to the indignation of nineteenth-century students – by those same capable but anonymous hands that had prepared it in the first place.

Des Maizeaux's *Collection (Recueil de diverses pièces sur la philosophie, la religion naturelle, l'histoire, les mathématiques &c,* Amsterdam 1720) was for long one of the most useful and convenient source books for the philosophers' war, reprinted in 1740 and again in 1759, and only recently supplanted by the publication of Newton's letters and papers. Further, some of Newton's most interesting autobiographical statements were addressed (in draft) to Des Maizeaux, who, like de Moivre, was a Huguenot refugee from the revocation of the Edict of Nantes (1685), first in Switzerland, then in England from 1699 onward. He was a professional writer, who (like Newton himself) won the patronage of Charles Montagu, Earl of Halifax, and it was through him or through de Moivre that he may first have met Newton. It is not impossible that he encountered Leibniz also during his European travels, and certain (because Leibniz himself tells us this) that the two men were exchanging letters "both in Mr Bayle's lifetime and after his death" in 1706, for Des Maizeaux was desirous of publishing a correspondence that had occurred between Pierre Bayle and Leibniz. Through his acquaintance with Conti in London, ten years later, Des Maizeaux formed the similar idea of printing the letters exchanged between Leibniz and Dr Samuel Clarke about Newton's philosophy of nature and, having once more approached Leibniz, had received from him copies of the first four letters and their replies, by August 1716. His plan was, however, anticipated by Clarke's own edition (1717), leaving Des Maizeaux only the option of arranging a Continental printing, which in fact forms the first volume of his *Collection*. To bolster it, either he or Conti thought of adding Leibniz's other "English" correspondence of the same period, including, besides the letters to Conti (now made over to him), letters from Leibniz to the Baroness von Kilmansegge and others, French correspondents and members of the German Court of George I, together with whatever could be got together from Newton's side. By July or August 1718 the texts had been assembled, a long introduction written, and the whole set in type in Amsterdam by the publisher of the *Nouvelles Literaires,* Du

Sauzet. Des Maizeaux then submitted the proof sheets to Newton, who was to assure Varignon three years or so later that "I
knew nothing of the design of printing them [Leibniz's letters] till
I saw them in print." But this recollection may not have been
wholly accurate, because Des Maizeaux had probably already obtained from Newton a French translation of his "Observations"
upon Leibniz's second letter to Conti, which he was in due course
to publish (after getting this French version set in type in 1717 or
thereabouts, Newton had decided not to proceed with it at that
time). Moreover, the earliest extant letter from Newton to Des
Maizeaux, in which Newton reported his views of the "printed
papers" left with him, does not quite read like the first in a series.
To various correspondents, not least Johann Bernoulli, who were
ruffled by the tone of Des Maizeaux's own prose in the *Collection*,
Newton was afterward to protest that he had had no hand in it,
and that Des Maizeaux had been Leibniz's friend (though there
seems to be no independent evidence of this), who had spoiled
rather than improved his own case against Leibniz. He told Bernoulli that he "did not in the least wish" the publication of Des
Maizeaux's book, then still standing in type – Newton paid the
bookseller – while in another draft (perhaps prepared for de
Moivre to communicate to Bernoulli) he wrote

> Mr Leibniz kept a general correspondence & has friends in
> England & France & other countries as well as Germany.
> Some of those in England have been collecting his remains in
> honour of his memory & the two letters above mentioned [in
> which Bernoulli was revealed by Leibniz as the "eminent
> mathematician" of the *Charta Volans*] are in this collection; &
> I have no hand in what they do.[7]

As so often with Newton's utterances of this sort, his assertion
was half true at best. Newton immediately organized Des Maizeaux's proposed book to his own advantage, reordering the
documents and correcting the editorial comments upon them. In
one draft of a letter to Des Maizeaux containing a list of errata in
the printed book – by which the editor corrected the copy he presented to the Royal Society – Newton actually wrote: "since you
have the originals I think it is right to let them come abroad"
though in other drafts he seems to blame Des Maizeaux for giving publicity to Leibniz's side of the story in the many letters from
his pen:

> in some of those letters he tells his story at large; and though
> I will not write an answer to those letters since he is dead,
> yet, . . . I think I may be allowed to tell the story myself and
> leave it to be compared with his narrations.

Of this clear threat to add to the *Commercium Epistolicum*, the "Account" and the "Observations" yet another version of his case built around Leibniz's correspondence, there are other signs such as a draft headed "To the Reader" in which Newton speaks of receiving the proof sheets of Des Maizeaux's book "a few days before" and finding the documents "put together in the wrong order . . . I have caused them to be reprinted in due order of time together with a paper published about the same time in the *Acta Lipsiensia* by a nameless author who called a solution of Mr John Bernoulli's *solutionem meam*." Fortunately Newton changed his mind, perhaps because he found Des Maizeaux compliant, but one may readily link Newton's abortive intention with the *Commercium Epistolicum* reissue a few years later.

As always, pen in hand, putting words on paper that perhaps would find no reader for three hundred years, Newton reverted to his constant view of his quarrel with Leibniz:

> The proper question is, Who was the first inventor? Let it be
> proved that Mr Leibniz had the method before he had any
> notice of it from England . . .

This was the question Leibniz had always refused to answer:

> If Mr Leibniz could have made a good objection against the
> *Commercium Epistolicum*, he might have done it in a short let-
> ter without writing another book as big. But this book being
> matter of fact and unanswerable he treated it with opprob-
> rious language and avoided answering it by several excuses,
> and then endeavoured to lay it aside by appealing to the
> judgement of his friend Mr Bernoulli and by writing to his
> friends at Court and by running the dispute into a squabble
> about a Vacuum and Atoms, and universal gravity and occult
> qualities, and miracles, and the sensorium of God, and the
> perfection of the world, and the nature of time and space, and
> the solving of problems, and the question whether he did not
> find the differential method *proprio marte* [off his own bat]: all
> of which are nothing to the purpose.

That was Newton's view of Liebniz's conduct of the calculus quarrel in 1718; and there we may leave it because there is no reason to believe that Newton ever changed his mind during the last years of his life. Why should he? It was clear to him that the *Commercium Epistolicum* was unshaken and unshakable, and that against its documentary evidence Leibniz had produced nothing but assertions and irrelevancies. And though we now know Newton to have been in error as to the fundamental justice of the case, were there not grounds for his opinion? For it was indeed only in the nineteenth century that the documents supporting Leibniz's independent discovery became public property. Having said this, one should not also forget Newton's tergiversations and distortions, as when he claimed (also to Des Maizeaux) that he had written the "Book of Quadratures" as early as 1676, and found the "demonstration of Keplers Astronomical Proposition [the second law of planetary motion] by the inverse method of fluxions in the year 1677"; we may forgive these lapses of memory – if such they were – half a century after the events but so also we must forgive Leibniz's refusal to fight the calculus battle with the weapons chosen by Newton.[8]

At this point we may abandon the narrative – for there was never any grand crisis in the philosophers' war, still less a final resolution, but rather a decline into inanition – in order to compare the public appearance of the two chief contestants toward its close. Leibniz died in the full heat of battle, his last letter to Clarke unwritten. His friends, though numerous all over Europe, were scattered and disunited; some, like Conti, perhaps disaffected. The preeminence of his calculus had been gloriously vindicated by many leading mathematicians despite the attacks of Keill and Taylor, and was still to be upheld successfully in the next few years, so that Leibniz never had cause to doubt for a moment that European mathematics would bear his mark forever, but in the few years after Leibniz's death that same hand of Johann Bernoulli that did so much for the calculus also helped to render the whole case against Newton absurd. By the 1720s no intelligent Continental could doubt that the great Leibniz really had devised his calculus independently, but equally he could not but allow that in some sense Newton had been first in that field. It was necessary – except for the English who had no doubt about Leibniz's utter guilt – to award something of victory to either contestant. Thus, the result of the calculus quarrel was that Leibniz certainly lost something, and Newton gained a great deal. If this seems an opinion contrary to what may be a prevalent supposition that Newton in some sense "lost" the calculus debate, both because (in the longer run) the method of differentials triumphed over that of fluxions, so that treatises on the fluxional calculus became mere antiquarian curiosities even in England after about 1820, and because the honest independence of Leibniz's discovery was never doubted save in England (and there, too, from early in Victoria's reign opinion began to alter), two points must be made in its defense. The first is that though always asserting the superior merits of fluxions, Newton recognized that there was really only one new infinitesimal calculus: Which notation should be used was not an important question, nor even the choice between moments and fluxions. Therefore the preference of later generations for differentials – made inevitable by the enormous development of Continental mathematics during the eighteenth century – has no bearing on the calculus quarrel as it went on between 1699 and 1720. It is certainly not to be read as indicating a vote by subsequent mathematicians – English as well as Continental European – against Newton's case. Second, it has to be realized that Newton's progress in the infinitesimal calculus would have been completely forgotten and obliterated had not he and his friends revived and asserted it. Until Wallis, in defense of Newton's priority, extracted elucidations from him, nothing of Newton's work in the calculus was available in print save the fluxions Lemma in the *Principia*; until after Fatio de Duillier had made his protest in 1699, no early mathematical paper by Newton was published. The growth of the controversy and the emergence of Newton's studies to the light of day went along together; but for the controversy, stirred up by Newton's few partisans, no one would ever have conceived of an alternative to Leibniz's right as the unique founder of the calculus. It would have remained for the nineteenth and twentieth centuries to illuminate that obscure lemma from the mass of Newton's papers (presuming their preservation) just as Leibniz's own first steps have been reconstructed. Even such men as Cheyne, Craige, David Gregory, and Keill might never have learned anything of the mathematical discoveries of Isaac Newton, and certainly there would have been no systematic development and exposition of a fluxional calculus such as occurred in Britain during the eighteenth century.

Would it have mattered if Newton had remained in the language of pure mathematics a mute Milton or a village Hampden? Is the fluxional calculus more than a closed loop, a *huit clos*, leading no-where? No major book was ever written in it; it withered until the "Cambridge Analysts" abandoned it with relief. The fluxional calculus seems now to have been a product of intense youthful brilliance, never fully finished, never matured into a ripe middle age, not unlike the poetry of Keats or Rimbaud. When the trea-tises of Newton's early years belatedly saw print, they were al-ready equaled or outstripped. Whiteside has drawn attention to the "complex skein of discovery of the algorithmic mode of analy-sis by limit-infinitesimals which Newton knew from the middle 1660s as the method of fluxions and which Leibniz a decade later christened by its now standard name of differential calculus," both formulations drawing heavily on preceding tradition, both utiliz-ing the "insights of so many who had gone before," so that the simultaneity and similarity between the two discoveries seems to us more impressive than their difference in time and form: "The priority in time of creation of his fluxional method which Newton indubitably has must seem of mininal significance." Here one en-counters the paradox of history: That all personal human effort seems, on the one hand, in terms of the moment, vital, crucial, revolutionary and yet, on the other hand, in secular terms, ines-sential, inevitable, a mere part of the ceaseless flux of events. If one path of development had remained closed, another would have opened: So Newton's abortive discovery of fluxions seems vain, and any discussion of it pointless.

Somewhat analogously, Clifford Truesdell – unlike Whiteside, no sympathetic critic of Newton – has written of Book II of New-ton's *Principia*, containing a large fraction of the whole work and some of its most difficult mathematical argument:

> Almost all of the results are original, and but few correct. Newton's alternating regard of a fluid as a "rare medium, consisting of equal particles freely disposed at equal distances from each other", or as "compressed", while sometimes praised as evidence of great physical insight, did not lead to trustworthy conclusions.

It is certainly the case that in rational fluid mechanics much sounder foundations were laid during the eighteenth century by Continental mathematicians using the Leibnizian calculus – yet through a dialogue with the *Principia*, which, again employing the "insights of so many who had gone before," first gave a coherent

mathematical structure to the subject. It is surely a historical simplification to see one man's achievement as furnishing a unique
channel of continuity between past and future, because if that
channel had remained closed, others would have opened; but – as
with Newton's work – the channel actually followed does condi-
tion future explorations.[9]

Thus, it seems to me, the Newtonian method of fluxions did
have a real importance, the importance of all singular historical
events. If Newton's route was not unique, neither (because it existed) can it be negligible. If Newton had not been a great pure
mathematician, if he had not been a master of infinitesimal methods, he could not have written the *Principia*. To that considerable
extent, history would have been the less rich; and the particular
history of mathematics would have been poorer without the twin
conceptions of fluxion and differential. Nor can the historian iso-
late the philosophers' war from the broader history of Newtonian
science. In a perhaps unexpected way, the emergence of Newton
as a pure mathematician seems to have enhanced his reputation in
applied mathematics. When the *Principia* was first published in
1687, Newton enjoyed only a very modest Continental reputation
as an optical experimenter with extraordinary theoretical notions,
whereas before he died, Continental Europe had largely accepted
Newton's theory of light and color, was rapidly reconciling itself
to the mathematical theory of universal gravitation, and had reg-
istered as a historical monument his discoveries in the infinitesimal
calculus. The philosophers' war was one aspect of the rising tide
of Newtonianism, which forever dimmed the scientific reputation
of Leibniz without touching his brilliance in philosophy and pure
mathematics.

Because the eighteenth century was to be, above all, the French
century in European and world history, we may (rather briefly)
consider the sunset glow of French recognition that warmed
Newton's last years. Not that this was uniquely a French phenomenon: The Dutch – as already noted – under the leadership of
'sGravesande, jettisoned Cartesianism rapidly from about 1720
onward, and the North Italians were not far behind. One of these
at a later date, Francesco Algarotti, was to become author of a
widely read popularization of Newtonian science, translated into
many languages, *Il Newtonianismo per le Dame* (Naples, 1737).
Only the German world of learning and philosophy, partly
through the lingering influence of Leibnizian ideas, never submit-

ted entirely to the experimental charms and mathematical rigors of Newtonian science; in time German criticism would find an eloquent exponent in Goethe.

Although it was the *Principia* – despite the unacceptability of its theory of gravitation – that brought Newton (along with Leibniz) into the reorganized French Academy of Sciences in 1699, as a foreign associate, the solid growth of Newton's reputation in France (especially among the intellectuals associated with the philosopher of the Oratory, Nicole Malebranche) was nourished by interest in his early experimental study of light and colors. This work had been passed over in France during the later seventeenth century, first because Huygens had found Newton's theoretical interpretation of his experiments unintelligible, and second because Edmé Mariotte, the chief authority on experimental physics in the Academy of Sciences at that time, had found it impossible to duplicate the results reported by Newton in the 1670s. In any case, Newton seems to have failed to react warmly to the honor done him by the French, and through the following years of war nothing resulted from his election, nor did he present either the English (1704) or Latin (1706) edition of *Opticks* to the academy. Other men, despite the war between France and the Allies, maintained scientific relations between the two countries, notably Etienne-Francois Geoffroy and Hans Sloane, secretary of the Royal Society. Geoffroy had visited England during the brief interval of peace between the wars of Louis XIV and understood the language; during the autumn and winter of 1706-7 he read at meetings of the Academy of Sciences a French version of *Opticks*, having received a copy of the book from Sloane in 1705. Shortly afterward the Malebranchistes accepted it as convincing, and the first successful repetitions of Newton's experiments were made, overthrowing Mariotte's earlier skepticism.[10]

The calculus dispute made Newton eager to win foreign friends. In 1713 he sent the Abbé Jean-Paul Bignon, the great factotum of French academic life at this time, some copies of the *Commercium Epistolicum* for distribution, which were succeeded by others of the second edition of the *Principia*. These gifts were the occasion of Newton's first extant letters from Fontenelle, the secretary of the Academy of Sciences, who assured Newton (a little formally) of his

> admiration for all your works, which I have in common with everyone in the world who possesses some hint of geometry,

and I greatly regret that I am not skilful enough to admire you as I ought.

Newton's much warmer correspondence with Varignon began about the same time – Varignon clearly had a real leaning toward Newtonian physics, which Fontenelle did not share at all – and on his side Newton recognized his own membership of the academy. A number of Frenchmen also read the *Principia* at this time, and one Academician at least, the Chevalier de Louville, was converted to Newtonianism. Curiosity about intellectual life in England and the occurrence in 1715 of an eclipse visible at London but not at Paris brought a party of Academicians on a visit: Rémond de Monmort, Claude-Joseph Geoffroy (younger brother of Etienne-François), and Louville. This party witnessed Newton's optical experiments, were elected Fellows of the Royal Society, and found Newton a delightful host. They returned to Paris the first Anglophiles of the Enlightenment. In due course Monmort repaid Newton's hospitality by sending him a present of fifty bottles of champagne, to which Newton responded with gifts, perhaps the second English *Opticks*, and ornaments for his wife from Catherine Barton, who had won Monmort's admiration, and as for Newton himself "I confine myself at present to assure you Sir that I have for you not only all the respect that your great name inspires, but that nothing can equal the tender and perfect attachment with which I have the honour to be . . . (etc)." Having been involved in the Conti affair from the beginning, Monmort had been, and was to remain, neutral in the calculus dispute, playing his part in the attempt to reconcile Newton and Johann Bernoulli after the death of Leibniz, and protesting against the continuing hostility against Bernoulli displayed by Keill and Brook Taylor. He never was reconciled to Newtonian physics.

The greatest service to Newton's reputation in France at this time was performed by Pierre Varignon, who to his regret was never able to visit Newton but received from him a present of Newton's portrait. In 1720 a Parisian bookseller wished to reprint the French version of Newton's *Opticks* (made by Pierre Coste, another Huguenot refugee in London) recently published at Amsterdam; the task of giving an official "Approbation" to the book was entrusted to Varignon, who thereupon volunteered to revise the translation and see it through the press in Paris on Newton's behalf. The Paris *Traité d'Optique* (1722), though slow in emergence, was to be far the most elegant edition of Newton's book, embellished by a vignette of the celebrated "crucial experiment" with two prisms based on a sketch from Newton's pen. The text, and the experiments of more than half a century before on which it was based, aroused the enthusiasm of some of the greatest in France, including the chancellor, Henri François Daguesseau; Newton's faithfulness as a reporter of the behavior of light was vindicated at every test. Daguesseau in particular used his authority to spur on the dilatory publisher

> not for your sake [as he told Newton] but my own, or rather to speak more truly not for my own sake but for that of all France, who freely and with applause offers her tongue and her speech to you; nor for the sake of France alone, but on behalf of all philosophers and mathematicians of every nation everywhere, who already look up to you as obvious master of them all.

Did the French honey of these last years do something to sweeten the recollection of the German gall, which had been so bitter in the past? One can but hope so; and Newton was certainly full of attention and courtesy to his French correspondents. He too, in old age, had softened (to all but Johann Bernoulli) and now relished letters of compliment that he would once have thrown aside unanswered, and as president of the Royal Society found it possible – though such duties forced him to lay aside his still active quill – to be properly deferential to foreign academicians and ambassadors.[11]

Almost in the scriptural spirit, the meek had been exalted and the rich sent hungry away. Twenty years before, the sometime Cambridge professor, whom disillusionment had driven to take up a civil-service post in London, had enjoyed an uncertain international reputation: His optical experiments discredited by many as impossible, his *Mathematical Principles of Natural Philosophy* accused of making a return to occult qualities. Newton's had been the *last* name of those chosen to be Foreign Associates of the French Academy of Sciences; and Fatio's declaration of his mathematical preeminence had been contemptuously silenced. Already near the end of the sixth decade of his life, Newton (busy at the Mint in pursuit of clippers and coiners, anxious lest he be thought to waste time on mathematics that he ought to devote to the King's business) was not yet a great man. All the greatness came after (though not on account of) the publication of *Opticks* in 1704 – presidency of the Royal Society, knighthood, unquestioned leadership of British science – and analogously it was the Paris *Traité d'Optique* (1722) that brought him welcome symphonies of international praise and prestige.

If the day of Newton the experimenter was already dawning bright in Europe, the day of Newton the mathematical philosopher was not far behind, though Newton did not live to see recognition of the *Principia* as the greatest achievement of the human intellect in natural science. The first semipopular treatment of the Newtonian philosophy of nature to appear on the Continent (and, simultaneously, in English translation) was 'sGravesande's *Mathematical Elements of Physics confirmed by Experiments* in 1720. The naturalization of this philosophy in France begun by Maupertuis's *Discourse on the different shapes of the Stars* (1732) was extended by Voltaire's *Letters on the English* two years later, followed by his enormously successful *Elements of Newton's Philosophy* (1738). Voltaire had spent three years in England (1726–9), where he witnessed Newton's burial in Westminster Abbey and met some of Newton's close friends, whence the celebrated story of Newton and the apple, which Voltaire was the first to tell in print, and the scandalous (though gallant) suggestion that Newton's advancement owed less to his intellect than to his niece's bright eyes. Maupertuis had also made a brief visit and become a Fellow of the Royal Society in 1728 (Voltaire followed in 1743). In the 1730s the official French geodesic expeditions to Lapland and Peru definitively pronounced in favor of the Newtonian theory of the earth's shape. From that decade French mathematical science fully assimilated Newtonianism and wedded it to the Continental tradition of mathematical science, even to the dynamics of Leibniz and Bernoulli. Here at last, in the work of D'Alembert and others, thesis and antithesis were reconciled in a new synthesis.[12]

Simultaneously, though the method of fluxions was never to find a welcome outside England, Newton's long-neglected mathematical writings found publishers abroad. Rémond de Monmort had privately circulated a small edition of *On Quadrature* as early as 1708. After a long interval, Newton's widely read algebra, the *Universal Arithmetic* (first printed by Whiston in 1707), was published in no fewer than three cities (Leiden, Milan, and Paris) in 1732. But the most interesting of all these Continental printings is the French version of the *Method of Fluxions and Infinite Series* made by the Comte de Buffon in 1738 and printed in 1740. Buffon, to be above all distinguished for his voluminous universal natural history of later years, was at this time active in spreading English science in France: He was also the translator of Stephen Hales's *Vegetable Staticks*. Now the *Method of Fluxions* had only just seen the light of day in the English version produced by John Colson in 1736 and therefore could not possibly have appeared much earlier in French. Buffon recognized, naturally, that Newton's treatise (whose origin he dates back to 1664) contained nothing that was novel in 1740: Because its procedures were now familiar, and in any case were developed with wonderful clarity, he thought it needed none of the heavy commentary Colson had wrapped around it. With complete frankness he gives priority in the discovery to Newton:

> It will be very agreeable to see the differential and integral calculus with all their applications in a single small volume; in the manner with which topics are treated the hand of the great Master will be recognized, and the genius of the discoverer; and the conviction will persist that Newton alone is the author of these marvellous modes of calculus, as he is also of many other achievements which are quite as wonderful.

As for Leibniz, "who wished to share the glory of the discovery," many, says Buffon drily, still give him at least the right of second inventor. How far Buffon's partisanship could carry him in historical narrative may be seen from the following quotations taken from his Newtonian account of the origins of the calculus, based probably on the *Commercium Epistolicum* and "Account" of 1722:

> . . . thus as early as 1669 Newton had discovered infinite series, the differential calculus, and the integral calculus; all this was sent by Barrow to Collins who took copies and communicated it to Brouncker and Oldenburg, who sent it to Sluse; moreover Collins had sent it in his letters to James Gregory, Bertet, Borelli, Vernon, Strode and many other geometers; these letters are printed in the *Commercium Epistolicum* and from these letters one sees that Newton had discovered all these things . . . as early as 1664 or 1665; in these letters, too, it may be read that Newton meant to print the work of which we now give the translation as early as 1671.

The misapprehensions or exaggerations of Buffon's history may be left unexplored, because they are of interest now only to demonstrate how far the *Commercium Epistolicum* could succeed with a French reader, and how far the genuine and the spurious might blend together and reinforce each other in the rapid, and be it said thoroughly justified, increase of Newton's reputation to heights of eminence during the middle years of the eighteenth century.[13]

Thus the philosophers' war did the cause of Newtonianism no harm. Rather, it made Newton a more fascinating figure, almost (one might say) in accord with the immoral maxim that bad publicity is better than no publicity. Newton's name had been widely canvassed in the French-language journals issued from Amsterdam and, not for the last time, denunciation of false ideas introduced them to readers who would otherwise never have encountered them. The calculus dispute brought to light whole new aspects of Newton's achievement, which might have remained concealed, while incidentally giving his natural philosophy also an exposure to the public eye, which Newton unprovoked would never have sought. Just as we can never understand Newton the man if we forget that he was first and foremost a mathematician, so we shall not understand Newtonianism if we overlook its genesis in the philosophers' war.

Simultaneous discovery is a phenomenon that has greatly interested sociologists of science: Some hundreds of cases have been noted and such instances as Faraday and Henry, Darwin and Wallace, are familiar. Effectively, the discoveries of the method of fluxions and of the differential calculus were simultaneous, because Newton, first discoverer in order of time, did not publish. The fact that so long a period of latency occurred (say, twenty years from 1664 to 1684) added bitterness to the dispute, whereas the fact that the first to publish (Leibniz) could never deny seniority in the area to the rival claimant (Newton) gave an additional twist, made the more acute by the irrefutable evidence of communication between the two claimants before the first publication. How different the situation of Charles Darwin in Kent and Alfred Russel Wallace safely in Borneo in 1858! But tension (at least) is a common result of simultaneous discovery in the opinion of the sociologist Robert K. Merton:

> Two or more scientists quietly announce a discovery. Since it is often the case that these are truly independent discoveries, with each scientist having separately exhibited originality of mind, the process is sometimes stabilized at that point . . .

This was briefly true of Newton and Leibniz, as we now know:

But since the situation is often ambiguous with the role of
each not easy to demonstrate and since each *knows* that he had
himself arrived at the discovery, and since the institutional-
ized stakes of reputation are high and the joy of discovery
immense, this is often not a stable solution . . . Then begins
the familiar deterioration of standards . . . Reinforced by
group-loyalties and often by chauvinism, the controversy
gains force, mutual recriminations of plagiary abound, and
there develops an atmosphere of thorough-going hostility
and mutual distrust.[14]

Merton's phrases here might have been specifically descriptive of
the dispute between Newton and Leibniz.

Yet this did not originate in simultaneous *announcements*, and
therefore it is in some ways more closely comparable to those
wrangles in which an author (like Leibniz, to his surprise) has
found himself accused of publishing as recent and original a discovery that had been made long before, but lain hid. So, in Newton's own lifetime, the English seized on the obscure physician
George Joyliffe as the true discoverer of the lymphatic system of
vessels, and in later times Lavoisier's theory of the role of oxygen
in combustion was said to have been anticipated by John Mayow,
or Charles Darwin was accused of silently appropriating the ideas
of others, even his own grandfather. In some of these cases, of
course, what was said to be prior was already in the public domain, and the gravest charge to be laid at the door of Lavoisier or
Darwin was that of a careless inattention to the meritorious ideas
of their predecessors (though such a charge in either case now
seems totally unjustified, it did lead Darwin in later editions of
The Origin of Species to survey the history of the theory of bio-
logical evolution). It is when the accusation ceases to be one of
inadequate scholarship and begins to carry the implication of
moral guilt, of a criminal suppression of truth, that the affair becomes bitter: Even as late as 1699 the main grievance of Newton's
friends against Leibniz was that he had been inattentive to Newton's claims and the dispute worsened only after charges of "imitation" were bandied about. In the end Newton seems to have set
almost no limit to his image of Leibniz's turpitude, although Leibniz for his part seems to have restricted his charges of malicious
conduct and slander to Newton's supporters.[15]

Simultaneous discovery is a historical "accident," or rather the
product of convergence in research, and therefore disputes occa-

sioned by it may be regarded (as by Merton) in much the same light as simultaneous discovery itself, as phenomena of the sociology of science. The philosophers' war between Newton and Leibniz may be seen in this light, but it must be also seen as the outcome of the combination of a peculiar psychological characteristic in a discoverer, with the sociological phenomenon of convergence. If Newton had brought himself to publish his early mathematical discoveries, the calculus dispute could hardly have occurred (although Newton might well have thereby involved himself in other polemics). Reluctance to publish before some possibly never-to-be-attained perfection has been achieved, in accordance with the Horatian precept and natural fear of criticism encountered unprepared, has been common enough: Copernicus worked upon *De revolutionibus orbium coelestium* for thirty years, Darwin on *The Origin* for twenty (and it would have been a longer period still but for Wallace's letter), though in both these instances the object of ultimate publication was more purposefully held in mind throughout the long years than it ever was by Newton. Clearly, the longer a "first discoverer" holds his hand, the greater the probability that the process of convergence will operate and (as with Leibniz) the "second discovery" through the maturation of the subject may well be made in a form more convincing and assimilable than was the first. Contrary to Newton's own assertion of the plenary rights of the "first discoverer" – and one cannot but wonder what his reaction would have been had Marcus Marci's claim to priority in the discovery of the constant refrangibility of light been asserted against him[16] – one might almost jump to an opposite ethical position that the "first discoverer" who deliberately suppresses his work, who takes no "reasonable" steps to make it known (however these might be defined!) loses his right to immediate fame, and must be content with whatever antiquarian priority subsequent generations may allow him. Of course, Newton was cognizant of this weakness in his position, which he attempted to defend by assertions that his discoveries had been broadcast in correspondence (equivalent to the circulation of preprints in our own day) and had actually been known to Leibniz. And it is precisely here, in the claim to "prepublication" – not in the justified proclamations of his own genuine leadership in the calculus discovery – that the greatest distortions of the Newtonian record reside.

Newton, then, magnanimous (or idle?) at first in his response to Leibniz, then increasingly unscrupulous in his attempts to strengthen his last feeble hold upon the calculus discovery, which he had almost resigned by inertia, was inevitably ungenerous to the rights of the "second discoverer." What of Leibniz during the years when he was willing to admit to Newton some carefully unformulated "analogous method," which admission in times of stress was also to be disavowed? One may feel, perhaps, that Leibniz was less greathearted than Alfred Russel Wallace who, having certainly known nothing of Darwin's speculations and notebooks, nevertheless accepted the Linnean Society's actions in 1858 in good part. Conscious always of his own great powers, driven by the forces of his own inventiveness whenever some stimulus brought them into play, Leibniz was little inclined toward any critical self-analysis of his own creativity. A great intellectual steamroller smoothing the way of the clear and distinct concept, he had little enough respect for the under-laborers who had blazed the trees and cleared the scrub. Some general acknowledgment of their distinguished work should be enough – it was what *he* could do in a very superior way that interested him. If Newton was totally ignorant of the convergence of research and the way in which it must affect the position of a dilatory "first discoverer," Leibniz no less ignored convergence as enforcing the need for scholarship as a part of the process of discovery.

I end this story with a question: Did Newton and Leibniz discover the same thing? Obviously, in a straightforward mathematical sense they did: Calculus and fluxions are not identical, but they are certainly equivalent. Equally, we may say that the conservative Tycho Brahe and the revolutionary Kepler, if they could watch the same sunset side by side, see the same thing: a glowing red ball slowly approaching the rim of the horizon. Or do they? My friend Norwood Russell Hanson raised that question long ago: And, of course, "seeing" is always more than watching the gap between sun and horizon-rim lessen, even though it is almost impossible to see the horizon rising up to meet the motionless sun.[17] Believing, one may well say, is seeing. The mathematical case of Newton and Leibniz is not at all the same, a mathematician might argue, because a fluxional proposition can always be re-written as a proposition in differential calculus by applying strict and invariant rules of translation: There is no question of seeing anything one way or the other. Yet one wonders whether some more subtle element may not remain, concealed, for example, in that word "equivalent." I hazard the guess that unless we obliterate the distinction between "identity" and "equivalence," then if two sets of propositions are logically equivalent, but not identical, there must be some distinction between them of a more than trivial symbolic character. Perhaps this is in some sense metaphysical rather than operational, of psychological interest rather than cash value. So – in nineteenth-century chemistry – propositions in terms of equivalent weights are not the same as propositions in terms of elemental atoms, though operationally each does the same work as the other. Calculus and the method of fluxions do the same work, but they were products of very different minds, adjuncts of unlike systems of thought. Let me release from this Pandora's box no more than the simplistic affirmation that Leibniz's was a calculus of discontinuity, of monads, while Newton was concerned with the continuity of flow, with time; or, one might say, differentials belong to the relative, fluxions to the absolute. Does not this involve seeing different things? In the world of ideas, one term of a pair of equivalents has tended always to prevail, for reasons not always easy to describe; Leibnizian differentials prevailed. But may one not speculate that the development of mathematics would have been different if the fluxional method had swept the board of European mathematics in the 1670s and 1680s?

How deep the difference was between the minds of the two philosophers, and how cogently their conceptions of infinitesimals were related to this difference, may perhaps be best seen in that piece of Newton's metaphysical writing so distressing to Leibniz, the Scholium on Space and Time in the *Principia*, and especially the following lines in it, written long before Newton had begun to insist upon the distinction between fluxion (absolute) and differential (relative) but already foreshadowing it:

> Accordingly, the relative quantities are not those actual quantities whose names they bear but those physical quantifications of them (true or false) which people use instead of the actual quantities, measured. But if the meaning of words is to be defined from their uses, these [physical] quantifications will be properly understood under the terms time, space, place and motion and an expression will be unusual and purely mathematical if the actual quantities, measured, are intended by them. And hence they abuse language who understand these words as applying to the actual quantities,

measured. Nor do they less defile mathematics and philosophy who confuse the actual quantities with their relationships
and the customary quantifications of them.[18]
How ironical and how prescient that Newton should, by implication, make the fluxion an unattainable Platonic ideal and
the differential the common currency among men that it was in
fact to be!

APPENDIX:
NEWTON'S "ACCOUNT
OF THE BOOK ENTITULED
COMMERCIUM EPISTOLICUM"

II. *An Account of the Book entituled* Commercium Epiſtolicum Collinii & aliorum, De Analyſi promota ; *publiſhed by order of the* Royal-Society, *in relation to the Diſpute between Mr.* Leibnitz *and Dr.* Keill, *about the Right of Invention of the Method of* Fluxions, *by ſome call'd the* Differential Method.

SEveral Accounts having been publiſhed abroad of this *Commercium*, all of them very imperfect : It has been thought fit to publiſh the Account which follows.

This *Commercium* is compoſed of ſeveral ancient Letters and Papers, put together in order of Time, and either copied or tranſlated into Latin from ſuch Originals as are deſcribed in the Title of every Letter and Paper; a numerous Committee of the Royal-Society being appointed to examin the Sincerity of the Originals, and compare therewith the Copies taken from them. It relates to a general Method of reſolving finite Equations into infinite ones, and applying theſe Equations, both finite and infinite, to the Solution of Problems by the Method of Fluxions and Moments. We will firſt give an Account of that Part of the Method which conſiſts in reſolving finite Equations into infinite ones, and ſquaring curvilinear Figures thereby. By Infinite Equations are meant ſuch as involve a Series of Terms converging or approaching the Truth nearer and nearer *in infinitum*, ſo as at length to differ from the Truth leſs than by any given Quantity, and if continued *in infinitum*, to leave no Difference.

G g   Dr.

(174)

Dr. *Wallis* in his *Opus Arithmeticum* publiſhed *A. C.* 1657: *Cap.* 33. *Prop.* 68. reduced the Fraction $\frac{A}{1-R}$ by perpetual Diviſion into the Series $A + AR + AR^2 + AR^3 + AR^4 + \&c.$

Viſcount *Brounker* ſquared the Hyperbola by this Series $\frac{1}{1 \times 2} + \frac{1}{3 \times 4} + \frac{1}{5 \times 6} + \frac{1}{7 \times 8} + \&c.$ that is by this, $1 - \frac{1}{2} + \frac{1}{3} - \frac{1}{4} + \frac{1}{5} - \frac{1}{6} + \frac{1}{7} - \frac{1}{8} + \&c.$ conjoyning every two Terms into one. And the Quadrature was publiſhed in the *Philoſophical Tranſactions* for *April* 1668.

Mr. *Mercator* ſoon after publiſhed a Demonſtration of this Quadrature by the Diviſion of Dr. *Wallis*; and ſoon after that Mr. *James Gregory* publiſhed a Geometrical Demonſtration thereof. And theſe Books were a few Months after ſent by Mr. *John Collins* to Dr. *Barrow* at *Cambridge*, and by Dr. *Barrow* communicated to Mr. *Newton* (now Sir *Iſaac Newton*) in *June* 1669. Whereupon Dr. *Barrow* mutually ſent to Mr. *Collins* a Tract of Mr. *Newton*'s entituled *Analyſis per aquationes numero terminorum infinitas.* And this is the firſt Piece publiſhed in the *Commercium*, and contains a general Method of doing in all Figures, what my Lord *Brounker* and Mr. *Mercator* did in the Hyperbola alone. Mr. *Mercator* lived above ten Years longer without proceeding further than to the ſingle Quadrature of the Hyperbola. The Progreſs made by Mr. *Newton* ſhews that he wanted not Mr. *Mercator*'s Aſſiſtance. However, for avoiding Diſputes, he ſuppoſes that my Lord *Brounker* invented, and Mr. *Mercator* demonſtrated, the Series for the Hyperbola ſome Years before they publiſhed it, and, by conſequence, before he found his general Method.

The aforeſaid Treatiſe of *Analyſis* Mr. *Newton*, in his Letter to Mr. *Oldenburgh*, dated *Octob.* 24. 1676, mentions in the following Manner. *Eo ipſo tempore quo* Mercatoris *Logarithmotechnia prodiit, communicatum eſt per amicum* D. Barrow *(tunc* Matheſeos Profeſſorem *Cantab.) cum* D. Collinio *Compendium quoddam harum Serierum, in quo ſignificaveram Areas & Longitudines Curvarum omnium, & Solidorum ſuperficies & contenta ex datis.*

(175)

datis rectis ; & vice verſa ex his datis rectas determinari poſſe : & methodum indicatam illuſtraveram diverſis ſeriebus.
Mr. *Collins* in the Years 1669, 1670, 1671 and 1672 gave notice of this Compendium to Mr *James Gregory* in *Scotland*, Mr. *Bertet* and Mr. *Vernon* then at *Paris*, Mr. *Alphonſus Borelli* in *Italy*, and Mr. *Strode*, Mr. *Townſend*, Mr. *Oldenburgh*, Mr. *Dary* and others in *England*, as appears by his Letters. And Mr. *Oldenburg* in a Letter, dated *Sept.* 14. 1669. and entred in the Letter-Book of the Royal-Society, gave notice of it to Mr. *Francis Sluſius* at *Liege*, and cited ſeveral Senten-ces out of it. And particularly Mr. *Collins* in a Letter to Mr. *James Gregory* dated *Novemb.* 25. 1669. ſpake thus of the Method contained in it. Barrovius *Provinciam ſuam publicè prælegendi remiſit cuidam nomine* Newtono Cantabrigienſi, *cujus tanquam viri acutiſſimo ingenio præditi in Præfatione Prælectionum Opticarum, meminit : quippe antequam ederetur* Mercatoris *Lo-garithmotechnia, eandem methodum adinvenerat, eamque ad om-nes Curvas generaliter & ad Circulum diverſimode applicârat.*
And in a Letter to Mr. *David Gregory* dated *Auguſt* 11. 1676. he mentions it in this manner. *Paucos poſt menſes quam editi ſunt hi Libri* [viz. *Mercatoris* Logarithmotechnia & Exercita-tiones Geometricæ *Gregorii*] *miſſi ſunt ad* Barrovium Cantabri-giæ. *Ille autem reſponſum dedit hanc inſinitarum Serierum Doctri-nam à* Newtono *biennium ante excogitatam fuiſſe quam ederetur* Mercatoris *Logarithmotechnia & generaliter omnibus figuris ap-plicatam, ſimulque tranſmiſit* D. Newtoni *opus manuſcriptum.*
The laſt of the ſaid two Books came out towards the End of the Year 1668, and Dr. *Barrow* ſent the ſaid Compendi-um to Mr. *Collins* in *July* following, as appears by three of Dr. *Barrow*'s Letters. And in a Letter to Mr. *Strode*, dated *July* 26. 1672, Mr. *Collins* wrote thus of it. *Exemplar ejus* [Logarithmotechniæ] *miſi* Barrovio Cantabrigiam, *qui quaſdam* Newtoni *chartas extemplo remiſit : E quibus & aliis quæ prius ab authore cum* Barrovio *communicata fuerant, patet illam methodum à dicto* Newtono *aliquot annis antea excogitatam & modo univer-*

Gg 2 *ſali*

fali applicatam fuiſſe : Ita ut ejus ope, in quavis Figura Curvilinea propoſita, quæ una vel pluribus proprietatibus definitur, Quadratura vel Area dictæ figuræ, accurata ſi poſſibile ſit, ſin minus infinitè verò propinqua, Evolutio vel longitudo Lineæ Curvæ, Centrum gravitatis figuræ, Solida ejus rotatione genita & eorum ſuperficies ; ſine ulla radicum extractione obtineri queant. Poſtquam intellexerat D. Gregorius *hanc methodum à* D. Mercatore *in Logarithmotechnia uſurpatam & Hyperbolæ quadrandæ adhibitam, quamque adauxerat ipſe* Gregorius, *jam univerſalem redditam eſſe, omnibuſque figuris applicatam; acri ſtudio eandem acquiſivit multumque in ea enodanda deſudavit. Uterque* D. Newtonus & Gregorius *in animo habet hanc methodum exornare :* D. Gregorius *autem* D. Newtonum *primum ejus inventorem anticipare haud integrum ducit.* And in another Letter written to Mr. *Oldenburgh* to be communicated to Mr. *Leibnitz,* and dated *June* 14 1676. Mr. *Collins* adds : *Hujus autem methodi ea eſt præſtantia ut cum tam late pateat ad nullam hæreat difficultatem.* Gregorium *autem alioſque in ea fuiſſe opinione arbitror, ut quicquid uſpiam antea de hac re innotuit, quaſi dubia diluculi lux fuit ſi cum meridiana claritate conferatur.*

This Tract was firſt printed by Mr. *William Jones,* being found by him among the Papers and in the Hand-writing of Mr. *John Collins,* and collated with the Original which he afterwards borrowed of Mr. *Newton.* It contains the above-mention'd general Method of *Analyſis,* teaching how to reſolve finite Equations into infinite ones, and how by the method of Moments to apply Equations both finite and infinite to the Solution of all Problems. It begins where Dr. *Wallis* left off, and founds the method of Quadratures upon three Rules.

Dr. *Wallis* publiſhed his *Arithmetica infinitorum* in the Year 1655, and by the 59*th* Propoſition of that Book, if the *Abſciſſa* of any curvilinear Figure be called *x,* and *m* and *n* be Numbers, and the Ordinates erected at right Angles be $x^{\frac{m}{n}}$, the Area of the Figure ſhall be $\frac{n}{m+n} x^{\frac{m+n}{n}}$. And this is aſſumed by by Mr. *Newton* as the firft Rule upon which he founds his Quadrature of Curves. Dr. *Wallis* demonftrated this Propofition by Steps in many particular Propofitions, and then collected all the Propofitions into One by a Table of the Cafes. Mr. *Newton* reduced all the Cafes into One, by a Dignity with an indefinite Index, and at the End of his Compendium demonftrated it at once by his method of Moments, he being the firft who introduced indefinite Indices of Dignities into the Operations of *Analyfis*.

By the 108*th* Propofition of the faid *Arithmetica Infinitorum*, and by feveral other Propofitions which follow it; if the Ordinate be compofed of two or more Ordinates taken with their Signes $+$ and $—$, the Area fhall be compos'd of two or more Areas taken with their Signes $+$ and $—$ refpectively. And this is affumed by Mr. *Newton* as the fecond Rule upon which he founds his Method of Quadratures.

And the third Rule is to reduce Fractions and Radicals, and the affected Roots of Equations into converging Series, when the Quadrature does not otherwife fucceed; and by the firft and fecond Rules to fquare the Figures, whofe Ordinates are the fingle Terms of the Series. Mr *Newton*, in his Letter to Mr. *Oldenburgh* dated *June* 13. 1676. and communicated to Mr. *Leibnitz*, taught how to reduce any Dignity of any Binominal into a converging Series, and how by that Series to fquare the Curve, whofe Ordinate is that Dignity. And being defired by Mr. *Leibnitz* to explain the Original of this Theoreme, he replied in his Letter dated *Octob.* 24. 1676, that a little before the Plague (which raged in *London* in the Year 1665) upon reading the *Arithmetica Infinitorum* of Dr. *Wallis*, and confidering how to interpole the Series x, $x - \frac{1}{3}x^3$, $x - \frac{2}{3}x^3 + \frac{1}{5}x^5$, $x - \frac{3}{3}x^3 + \frac{3}{5}x^5 - \frac{1}{7}x^7$, &c.

he found the Area of a Circle to be $x - \dfrac{\frac{1}{2}x^3}{3} - \dfrac{\frac{1}{8}x^5}{5} - \dfrac{\frac{1}{16}x^7}{7}$

$- \dfrac{\frac{5}{128}x^9}{9} - $ &c. And by purfuing the Method of Interpolation he found the Theoreme abovemention'd, and by means of this Theoreme he found the Reduction of Fractions and Surds into converging Series, by Division and Extraction of Roots; and then proceeded to the Extraction of affected Roots. And these Reductions are his third Rule.

When Mr. *Newton* had in this Compendium explained these three Rules, and illustrated them with various Examples, he laid down the Idea of deducing the Area from the Ordinate, by considering the Area as a Quantity, growing or increasing by continual Flux, in proportion to the Length of the Ordinate, supposing the Abscissa to increase uniformly in proportion to Time. And from the Moments of Time he gave the Name of Moments to the momentaneous Increases, or infinitely small Parts of the Abscissa and Area, generated in Moments of Time. The Moment of a Line he called a Point, in the Sense of *Cavallerius*, tho' it be not a geometrical Point, but a Line infinitely short, and the Moment of an Area or Superficies he called a Line, in the Sense of *Cavallerius*, tho' it be not a geometrical Line, but a Superficies infinitely narrow. And when he consider'd the Ordinate as the Moment of the Area, he understood by it the Rectangles under the geometrical Ordinate and a Moment of the Abscissa, tho' that Moment be not always expressed. *Sit ABD,* faith he, *Curva*

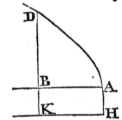

quævis, & AHKB rectangulum cujus latus AH vel *KB eft unitas. Et cogita rectam DBK uniformiter ab AH motam areas ABD & AK defcribere; & quod* [recta] *BK* (1)*fit momentum quo* [area] *AK (x), &* [recta] *BD (y) momentum quo* [area curvilinea] *ABD gradatim augetur; & quod ex momento BD perpetim dato poffis, per præcedentes* [tres] *Regulas, aream ABD ipfo defcriptam inveftigare, five cum area AK (x) momento* 1 *defcripta conferre.* This is his Idea of the Work in squaring of Curves, and how he applies this to other Problems, he expreffes in the next Words. *Jam qua ratione,* faith he, *fuperficies ABD ex momento fuo perpetim*

petim dato per præcedentes [tres] *Regulas elicitur, eâdem quæli-*
bet alia quantitas ex momento suo fic dato elicietur. Exemplo res
fiet clarior. And after some Examples he adds his Method
of Regreffion from the Area, Arc, or folid Content, to the
Abfciffa ; and fhews how the fame Method extends to Mechanical Curves, for determining their Ordinates, Tangents,
Areas, Lengths, *&c.* ·And that by affuming any Equation
expreffing the Relation between the Area and Abfcilfa of a
Curve, you may find the Ordinate by this Method. And
this is the Foundation of the Method of Fluxions and Moments, which Mr. *Newton* in his Letter dated *Octob.* 24, 1676
comprehended in this Sentence. *Data æquatione quotcunque*
fluentes quantitates involvente, invenire Fluxiones ; & vice verfa.

In this Compendium Mr. *Newton* reprefents the uniform
Fluxion of Time, or of any Exponent of Time by an Unit ;
the Moment of Time or of its Exponent by the Letter *o* ; the
Fluxions of other Quantities by any other Symbols; the Moments of thofe Quantities by the Rectangles under thofe
Symbols and the Letter *o* ; and the Area of a Curve by the
Ordinate inc'ofed in a Square, the Area being put for a Fluent and the Ordinate for its Fluxion. When he is demonftrating anyPropofition he ufes the Letter *o* for a finite Moment
of Time, or of its Exponent, or of any Quantity flowing
uniformly, and performs the whole Calculation by the Geometry of the Ancients in finite Figures or Schemes without
any Approximation *:* and fo foon as the Calculation is at an
End, and the Equation is reduced, he fuppofes that the
Moment *o* decreafes *in infinitum* and vanifhes. But when he
is not demonftrating but only inveftigating a Propofition, for
making Difpatch he fuppofes the Moment *o* to be infinitely
little, and forbears to write it down, and ufes all manner of
Approximations which he conceives will produce no Error in
the Conclufion. An Example of the firft kind you have in
the End of this Compendium, in demonftrating the firft of
the three Rules laid down in the Beginning of the Book.

G g 4   Exam-

Examples of the fecond kind you have in the fame Compendium, in finding the Length of Curve Lines *p.* 15. and in finding the Ordinates, Areas and Lengths of Mechanical Curves *p.* 18. 19. And he tells you, that by the fame Method, Tangents may be drawn to mechanical Curves *p.* 19. And in his Letter of *Decemb.* 10. 1672. he adds, that Problems about the Curvature of Curves Geometrical or Mechanical are refolv'd by the fame Method. Whence its manifeft, that he had then extended the Method to the fecond and third Moments. For when the Areas of Curves are confidered as Fluents (as is ufual in this *Analyfis*) the Ordinates exprefs the firft Fluxions, the Tangents are given by the fecond Fluxions, and the Curvatures by the third, And even in this *Analyfis* *p.* 16. where Mr. *Newton* faith, *Momentum eft fuperficies cum de folidis,& Linea cum de fuperficiebus,& Punctum cum de lineis agitur,* it is all one as if he had faid, that when Solids are confidered as Fluents, their Moments are Superficies, and the Moments of thofe Moments (or fecond Moments) are Lines, and the Moments of thofe Moments (or third Moments) are Points, in the Senfe of *Cavallerius.* And in his *Principia Philofophiæ,* where he frequently confiders Lines as Fluents defcribed by Points, whofe Velocities increafe or decreafe, the Velocities are the firft Fluxions, and their Increafe the fecond. And the Probleme, *Data æquatione fluentes quantitates involvente fluxiones invenire & vice verfa,* extends to all the Fluxions, as is manifeft by the Examples of the Solution thereof, publifhed by Dr. *Wallis Tom.* 2. *p* 391, 392, 396. And in *Lib.* II. *Princip. Prop.* xiv. he calls the fecond Difference the Difference of Moments.

Now that you may know what kind of Calculation Mr. *Newton* ufed in, or before the Year 1669. when he wrote this Compendium of his *Analyfis*, I will here fet down his Demonftration of the firft Rule abovementioned. *Sit Curvæ*

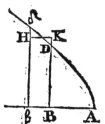

Sit Curvæ alicujus $AD\delta$ *Basis* AB = x, *perpendiculariter applicata* BD = y, *& area* $ABD = z$, *ut prius. Item fit* $B\beta = o$, $BK = v$, *& Rectangulum* $B\beta HK$ (o,v) *æquale spatio* $B\beta\delta D$. *Est ergo* $A\beta = x + o$, *&* $A\delta\beta = z + ov$, *His præmissis, ex relatione inter* x *&* z *ad arbitrium assumpta, quæro* y *ut sequitur.*

Pro lubitu sumatur [æquatio] $\frac{2}{3}x^{\frac{3}{2}} = z$, *sive* $\frac{4}{9}x^{3} = zz$. *Tum* $x + o$ (Aβ) *pro* x, *&* $z + ov$ (A$\delta\beta$) *pro* z *substitutis, prodibit* $\frac{4}{9}$ *in* $x^{3} + 3x^{2}o + 3xo^{2} + o^{3} =$ (*ex natura Curvæ*) $z^{2} + $ $+ 2zov + o^{2}v^{2}$. *Et sublatis* $\frac{4}{9}x^{3}$ *&* zz *æqualibus, reliquisque per* o *divisis, restat* $\frac{4}{9}$ *in* $3x^{2} + 3xo + o^{2} = 2zv + ov^{2}$. *Si jam supponamus* B$\beta$ *in infinitum diminui & evanescere, sive* o *esse nihil, erunt* v *&* y *æquales, & termini per* o *multiplicati evanescent; ideoque restabit* $\frac{4}{9} \times 3xx = 2zv$, *sive* $\frac{2}{3}xx (= zy) = \frac{2}{3}x^{\frac{3}{2}}y$, *sive* $x^{\frac{1}{2}} (= \frac{x^{2}}{x^{\frac{3}{2}}}) = y$. *Quare è contra, si* $x^{\frac{1}{2}} = y$, *erit* $\frac{2}{3}x^{\frac{3}{2}} = z$.

Vel generaliter, Si $\frac{n}{m+n} \times ax^{\frac{m+n}{n}} = z$; *sive ponendo* $\frac{na}{m+n}$ $= c$, *&* $m + n = p$, *Si* $cx^{\frac{p}{n}} = z$, *sive* $c^{n}x^{p} = z^{n}$: *Tum* $x + o$ *pro* x, *&* $z + ov$ *sive (quod perinde est)* $z + oy$ *pro* z *substitutis, prædit* c^{n} *in* $x^{p} + pox^{p-1}$ &c. $= z^{n} + noyz^{n-1}$ &c. *reliquis nempe* [Serierum] *terminis, qui tandem evanescerent, omissis. Jam sublatis* $c^{n}x^{p}$ *&* z^{n} *æqualibus, reliquisque per* o *divisis, restat* $c^{n}px^{p-1} = nyz^{n-1} (= \frac{nyz^{n}}{z} = \frac{nyc^{n}x^{p}}{cx^{\frac{p}{n}}})$ *sive dividendo per* $c^{n}x^{p}$, *exit* $p.x^{-1} = \frac{ny}{c x^{\frac{p}{n}}}$ *sive* $pcx^{\frac{p-n}{n}} = ny$; *vel restituendo* $\frac{na}{m+n}$ *pro* c *&* $m + n$ *pro* p, *hoc est* m *pro* $p - n$, *&* na

2   H h   *pro*

(182)

pro p c, *fiet* a $x^{\frac{m}{n}}=y$. *Quare è contra,fi* a$x^{\frac{m}{n}}=y$ *erit* $\dfrac{n}{m+n}$ a $x^{\frac{m+n}{n}}$

$=z$ Q. E. D.

By the fame way of working the fecond Rule may be alfo demonftrated. And if any Equation whatever be affumed expreffing the Relation between the Abfciffa and Area of a Curve, the Ordinate may be found in the fame manner, as is mentioned in the next Words of the *Analyfis*. And if this Ordinate drawn into an Unit be put for the Area of a new Curve, the Ordinate of this new Curve may be found by the fame Method: And fo on perpetually. And thefe Ordinates reprefent the firft, fecond, third, fourth and following Fluxions of the firft Area.

This was Mr. *Newton's* Way of working in thofe Days, when he wrote this Compendium of his *Analyfis*. And the fame Way of working he ufed in his Book of Quadratures, and ftill ufes to this Day.

Among the Examples with which he illuftrates the Method of Series and Moments fet down in this Compendium, are thefe. Let the Radius of a Circle be 1, and the Arc z, and the Sine x, the Equations for finding the Arc whofe Sine is given, and the Sine whofe Arc is given, will be

$$z = x + \tfrac{1}{6}x^3 + \tfrac{3}{40}x^5 + \tfrac{5}{112}x^7 + \tfrac{35}{1152}x^9 + \&c.$$
$$x = z - \tfrac{1}{6}z^3 + \tfrac{1}{12}z^5 - \tfrac{1}{5040}z^7 + \tfrac{1}{362880}z^9 - \&c.$$

Mr. *Collins* gave Mr. *Gregory* notice of this Method in Autumn 1669, and Mr. *Gregory*, by the Help of one of Mr. *Newton's* Series, after a Year's Study, found the Method in *December* 1670; and two Months after, in a Letter dated *Feb.* 15 1671. fent feveral Theorems, found thereby, to Mr. *Collins*, with leave to communicate them freely. And Mr. *Collins* was very free in communicating what he had received both from Mr. *Newton* and from Mr. *Gregory*, as appears by his Letters printed in the *Commercium*. Amongft the Series which Mr *Gregory* fent in the faid Letter, were thefe

(183)

thefe two. Let the Radius of a Circle be r, the Arc a, and the Tangent t, the Equations for finding the Arc whofe Tangent is given, and the Tangent whofe Arc is given, will be thefe.

$$a = t - \frac{t^3}{3r^2} + \frac{t^5}{5r^4} - \frac{t^7}{7r^5} + \frac{t^9}{9r^8} - \&c.$$

$$t = a + \frac{a^3}{3r^2} + \frac{2a^5}{15r^4} + \frac{17a^7}{315r^5} + \frac{62a^9}{2835r^8} + \&c.$$

In this Year (1671) Mr. *Leibnitz* publifhed two Tracts at *London*, the One dedicated to the Royal-Society, the Other dedicated to the Academy of Sciences at *Paris*; and in the Dedication of the Firft he mentioned his Correfpondence with Mr. *Oldenburgh*.

In *February* 167$\frac{1}{7}$ meeting Dr. *Pell* at Mr. *Boyle's*, he pretended to the differential Method of *Mouton*. And notwithftanding that he was fhewn by Dr. *Pell* that it was *Mouton's* Method, he perfifted in maintaining it to be his own Invention, by reafon that he had found it himfelf without knowing what *Mouton* had done before, and had much improved it.

When one of Mr. *Newton's* Series was fent to Mr. *Gregory*, he tried to deduce it from his own Series combined together, as he mentions in his Letter dated *December* 19. 1670. And by fome fuch Method Mr. *Leibnitz*, before he left *London*, feems to have found the Sum of a Series of Fractions decreafing *in Infinitum*, whofe Numerator is a given Number and Denominators are triangular or pyramidal or triangulo-triangular Numbers, &c. See the Myftery! From the Series $\frac{1}{1} + \frac{1}{2} + \frac{1}{3} + \frac{1}{4} + \frac{1}{5} + \&c.$ fubduct all the Terms but the firft (*viz* $\frac{1}{2} + \frac{1}{3} + \frac{1}{4} + \frac{1}{5} \&c$) and there will remain $1 = 1 - \frac{1}{2} + \frac{1}{2} - \frac{1}{3} + \frac{1}{3} - \frac{1}{4} + \frac{1}{4} - \frac{1}{5} + \frac{1}{5} \&c) = \frac{1}{1 \times 2} + \frac{1}{2 \times 3} + \frac{1}{3 \times 4} + \frac{1}{4 \times 5} + \&c.$ And from this Series take all the Terms but the firft, and there will remain $\frac{1}{2} = \frac{2}{1 \times 2 \times 3} + \frac{2}{2 \times 3 \times 4} + \frac{2}{3 \times 4 \times 5} + \frac{2}{4 \times 5 \times 6} + \&c.$ And from the firft Series take all the Terms but the two firft, and there will remain $\frac{3}{2} = \frac{2}{1 \times 3} + \frac{2}{2 \times 4} + \frac{2}{3 \times 5} + \frac{2}{4 \times 6} + \&c.$

Hh 2

In

(184)

In the End of *February* or beginning of *March* 167$\frac{2}{3}$. Mr. *Leibnitz* went from *London* to *Paris*, and continuing his Correspondence with Mr. *Oldenburg* and Mr. *Collins*, wrote in *July* 1674. that he had a wonderful Theoreme, which gave the Area of a Circle or any Sector thereof exactly in a series of rational Numbers; and in *October* following, that he had found the Circumference of a Circle in a Series of very simple Numbers, and that by the same Method (so he calls the said Theoreme) any Arc whose Sine was given might be found in a like Series, though the Proportion to the whole Circumference be not known. His Theoreme therefore was for finding any Sector or Arc whose Sine was given. If the Proportion of the Arc to the whole Circumference was not known, the Theoreme or Method gave him only the Arc; if it was known it gave him also the whole Circumference: and therefore it was the first of Mr. *Newton's* two Theoremes above-mention'd. But the Demonstration of this Theoreme Mr *Leibnitz* wanted. For in his Letter of *May* 12. 1676. he desired Mr. *Oldenburgh* to procure the Demonstration from Mr *Collins*, meaning the Method by which Mr. *Newton* had invented it.

In a Letter compos'd by Mr. *Collins* and dated *April* 15. 1675. Mr. *Oldenburgh* sent to Mr. *Leibnitz* Eight of Mr *Newton's* and Mr. *Gregory's* Series, amongst which were Mr. *Newton's* two Series above-mention'd for finding the Arc whose Sine is given, and the Sine whose Arc is given; and Mr. *Gregory's* two Series above mentioned for finding the Arc whose Tangent is given, and the Tangent whose Arc is given. And Mr. *Leibnitz* in his Answer, dated *May* 20. 1675. acknowledged the Receipt of this Letter in these Words. *Literas tuas multa fruge Algebraica refertas accepi, pro quibus tibi & doctissimo* Collinio *gratias ago. Cum nunc præter ordinarias curas Mechanicis imprimis negotiis distrahar, non potui examinare Series quas misistis ac cum meis comparare. Ubi fecero perscribam tibi sententiam meam: nam aliquot jam anni sunt quod inveni meas via quadam sic satis singulari.* But

(185)

But yet Mr. *Leibnitz* never took any further notice of his having received thefe Series, nor how his own differed from them, nor ever produced any other Series then thofe which he received from Mr *Oldenburgh*, or numeral Series deduced from them in particular Cafes. And what he did with Mr. *Gregory*'s Series for finding the Arc whofe Tangent is given, he has told us in the *Acta Eruditorum menfis Aprilis* 1691. pag. 178. *Jam anno 1675, faith he, compofitum habebam opufculum Quadratura Arithmetica ab amicis ab illo tempore lectum,* &c. By a Theoreme for tranfmuting of Figures, like thofe of Dr. *Barrow* and Mr. *Gregory*, he had now found a Demonftration of this Series, and this was the Subject of his *Opufculum*. But he ftill wanted a Demonftration of the reft: and meeting with a Pretence to ask for what he wanted, he wrote to Mr. *Oldenburg* the following Letter, dated at *Paris May 12. 1676.*

Cum Georgius Mohr *Danus nobis attulerit communicatam fili à Doctiffimo* Collinio *veftro expreffionem rationis inter arcum & finum per infinitas Series fequentes; pofito finu* x, *arcu* z, *radio* 1,

$$z = x + \tfrac{1}{6}x^3 + \tfrac{3}{40}x^5 + \tfrac{15}{112}x^7 + \tfrac{35}{1152}x^9 + \&c.$$

$$x = z - \tfrac{1}{6}z^3 + \tfrac{1}{120}z^5 - \tfrac{1}{5040}z^7 + \tfrac{1}{362880}z^9 - \&c.$$

Hac, INQUAM, cum nobis attulerit ille, qua mihi valde ingeniofa videntur, & pofterior imprimis Series elegantiam quandam fingularem habeat: ideo rem gratam mihi feceris, Vir clariffime, fi demonftrationem tranfmiferis. Habebis viciffim mea ab his longe diverfa circa hanc rem meditata, de quibus jam aliquot abhinc annis ad te perfcripfiffe credo, demonftratione tamen non addita, quam nunc polio. Oro ut Clariffimo Collinio *multam a me falutem dicas: is facile tibi materiam fuppeditabit fatisfaciendi defiderio meo.* Here, by the Word *INQUAM*, one would think that he had never feen thefe two Series before, and that his *diverfa circa hanc rem meditata* were fomething elfe than one of the Series which he had received from

H h 3   Mr.

(186)

Mr. *Oldenburgh* the Year before, and a Demonstration thereof which he was now polishing, to make the Present an acceptable Recompence for Mr. *Newton*'s Method.

Upon the Receipt of this Letter Mr. *Oldenburg* and Mr. *Collins* wrote pressingly to Mr. *Newton*, desiring that he himself would describe his own Method, to be communicated to Mr. *Leibnitz*. Whereupon Mr. *Newton* wrote his Letter, dated *June* 13. 1676, describing therein the Method of Series, as he had done before in the Compendium above-mentioned; but with this Difference : Here he described at large the Reduction of the Dignity of a Binomial into a Series, and only touched upon the Reduction by Division and Extraction of affected Roots : There he described at large the Reduction of Fractions and Radicals into Series by Division and Extraction of Roots, and only set down the two first Terms of the Series into which the Dignity of a Binomial might be reduced. And among the Examples in this Letter, there were Series for finding the Number whose Logarithm is given, and for finding the Versed Sine whose Arc is given : This Letter was sent to *Paris, June* 26. 1676. together with a MS. drawn up by Mr. *Collins*, containing Extracts of Mr. *James Gregory*'s Letters.

For Mr. *Gregory* died near the End of the Year 1675 ; and Mr. *Collins*, at the Request of Mr *Leibnitz* and some other of the Academy of Sciences, drew up Extracts of his Letters, and the Collection is still extant in the Hand Writing of Mr. *Collins* with this Title; *Extracts of Mr. Gregory's Letters, to be lent to Mr.* Leibnitz *to peruse, who is desired to return the same to you.* And that they were sent is affirmed by Mr. *Collins* in his Letter to Mr. *David Gregory* the Brother of the Deceas'd, dated *August* 11. 1676. and appears further by the Answers of Mr. *Leibniz* and Mr. *Tschurnhause* concerning them.

The Answer of Mr. *Leibnitz* directed to Mr *Oldenburgh* and dated *August* 27. 1676, begins thus ; *Literæ tuæ die* Julii 26. *datæ plura ac memorabiliora circa rem Analyticam continent quam*

2

multa

276

multa volumina spissa de his rebus edita. Quare tibi pariter ac clarissimis viris Newtono *ac* Collinio *gratias ago, qui nos participes tot meditationum egregiarum esse voluistis.* And towards the End of the Letter, after he had done with the Contents of Mr. *Newton's* Letter, he proceeds thus. *Ad alia tuarum Literarum venio quæ doctissimus* Collinius *communicare gravatus non est. Vellem adjecisset appropinquationis* Gregorianæ *linearis demonstrationem. Fuit enim his certe studiis promovendis aptissimus.* And the Answer of Mr. *Tschurnhause*, dated *Sept.* 1. 1676, after he had done with Mr. *Newton's* Letter about Series, concludes thus. *Similia porro quæ in hac re præstitit eximius ille Geometra* Gregorius *memoranda certe sunt. Et quidem optime famæ ipsius consulturi, qui ipsius relicta Manuscripta luci publicæ ut exponantur operam navabunt.* In the first Part of this Letter, where Mr. *Tschurnhause* speaks of Mr. *Newton's* Series, he saith that he looked over them cursorily, to see if he could find the Series of Mr. *Leibnitz* for squaring the Circle or Hyperbola. If he had searched for it in the Extracts of *Gregory's* Letters he might have found it in the Letter of *Febr.* 15. 1671. above-mentioned. For the MS. of those Extracts with that Letter therein is still extant in the Hand-Writing of Mr. *Collins.*

And tho' Mr *Leibnitz* had now received this Series twice from Mr. *Oldenburgh*, yet in his Letter of *August* 27. 1676. he sent it back to him by way of Recompence for Mr. *Newton's* Method, pretending that he had communicated it to his Friends at *Paris* three Years before or above; that is, two Years before he received it in Mr. *Oldenburgh's* Letter of *April* 15. 1675; at which Time he did not know it to be his own, as appears by his Answer of *May* 20. 1675 above-mentioned. He might receive this Series at *London*, and communicate it to his Friends at *Paris* above three Years before he sent it back to Mr. *Oldenburg*: but it doth not appear that he had the Demonstration thereof so early. When he found the Demonstration, then he compos'd it in his *Opusculum*, and communicated that also to his Friends; and he himself has told

Hh 4

told us that this was in the Year 1675. However, it lies upon him to prove that he had this Series before he received it
from Mr. *Oldenburgh*. For in his Answer to Mr. *Oldenburgh* he
did not know any of the Series then sent him to be his own;
and concealed from the Gentlemen at *Paris* his having received it from Mr. *Oldenburgh* with several other Series, and his
having seen a Copy of the Letter in which Mr. *Gregory* had
sent it to Mr. *Collins* in the Beginning of the Year 1671.

In the same Letter of *August* 27. 1676, after Mr. *Leibnitz*
had described his Quadrature of the Circle and Equilateral
Hyperbola, he added : *Vicissim ex seriebus regressuum pro Hyperbola hanc inveni. Si sit numerus aliquis unitate minor* $1 - m$,
ejusque logarithmus Hyperbolicus l. *Erit* $m = \frac{1}{1} - \frac{1^2}{1 \times 2} + \frac{1^3}{1 \times 2 \times 3}$
$- \frac{1^4}{1 \times 2 \times 3 \times 4} +$ &c. *Si numerus sit major unitate, ut* $1 + n$, *tunc
pro eo inveniendo mihi etiam prodiit Regula quae in* Newtoni *Epi-
stola expressa est* : *scilicet erit* $n = \frac{1}{1} + \frac{1^2}{1 \times 2} + \frac{1^3}{1 \times 2 \times 3} + \frac{1^4}{1 \times 2 \times 3 \times 4}$
$+$ &c. ——— *Quod regressum ex arcubus attinet, incideram ego
directe in Regulam quae ex dato arcu sinum complementi exhibet.
Nempe sinus complementi* $= 1 - \frac{a^2}{1 \times 2} + \frac{a^4}{1 \times 2 \times 3 \times 4} -$ &c. *Sed po-
stea quoque deprehendi ex ea illam nobis communicatam pro inveni-
endo sinu recto, qui est* $\frac{a}{1} - \frac{a^3}{1 \times \times 3} + \frac{a^5}{1 \times 2 \times 3 \times 4 \times 5} -$ &c. *posse de-
monstrari.* Thus Mr. *Leibnitz* put in his Claim for the Co-inven-
tion of these four Series, tho' the Method of finding them
was sent him at his own Request, and he did not yet understand it. For in this same Letter of *August* 27 1676. he desired
Mr. *Newton* to explain it further. His Words are. *Sed desi-
deraverim ut Clarissimus* Newtonus *nonnulla quoque amplius ex-
plicet* ; *ut originem Theorematis quod initio ponit* : *Item modum
quo quantitates* p, q, r, *in suis Operationibus invenit* : *Ac denique
quomodo in Methodo regressuum se gerat, ut cum ex Logarithmo quae-
rit Numerum. Neque enim explicat quomodo id ex methodo sua de-
rivetur.* He pretended to have found two Series for the Number
whose Logarithm was given, and yet in the same Letter desired

278

fired Mr. *Newton* to explain to him the Method of finding those very two Series.

When Mr. *Newton* had received this Letter, he wrote back that all the said four Series had been communicated by him to Mr. *Leibnitz*; the two first being one and the same Series in which the Letter *l* was put for the Logarithm with its Sign $+$ or $-$; and the third being the Excess of the Radius above the versed Sine, for which a Series had been sent to him. Whereupon Mr. *Leibnitz* defifted from his Claim. Mr. *Newton* alfo in the fame Letter dated *Octob.* 24. 1676. further explained his Methods of Regreffion, as Mr. *Leibnitz* had defired. And Mr. *Leibnitz* in his Letter of *June* 21. 1677. defired a further Explication : but foon after, upon reading Mr. *Newton's* Letter a fecond time, wrote back *July* 12. 1677. that he now underftood what he wanted ; and found by his old Papers that he had formerly ufed one of Mr. *Newton's* Methods of Regreffion, but in the Example which he had then by chance made ufe of, there being produced nothing elegant, he had, out of his ufual Impatience, neglected to ufe it any further. He had therefore feveral direct Series, and by confequence a Method of finding them, before he invented and forgot the inverfe Method. And if he had fearched his old Papers diligently, he might have found this Method alfo there ; but having forgot his own Methods he wrote for Mr. *Newton's*.

When Mr. *Newton* in his Letter dated *June* 13. 1676. had explained his Method of Series, he added : *Ex his videre eft quantum fines Analyfeos per hujufmodi infinitas aquationes ampliantur : quippe quæ earum beneficio ad omnia pene dixerim problemata (fi numeralia* Diophanti *& fimilia excipias) fefe extendit. Non tamen omnino univerfalis evadit, nifi per ulteriores quafdam Methodos eliciendi Series infinitas. Sunt enim quædam Problemata in quibus non licet ad Series infinitas per Divifionem vel Extraĉionem radicum fimplicium affeĉarumve pervenire. Sed quomodo in iftis cafibus procedendum fit jam non vacat dicere ; ut neque alia quædam tradere, quæ circa Reduĉionem infinitarum Serierum*

2   I i   *in*

in finitas, ubi rei natura tulerit, excogitavi. Nam parcius scribo, quod hæ speculationes diu mihi fastidio esse cœperunt ; adeo ut ab iisdem jam per quinque fere annos abstinuerim. To this Mr. *Leibnitz* in his Letter of *August* 27. 1676. answered: *Quod dicere videmini plerasque difficultates (exceptis Problematibus* Diophantæis*) ad series Infinitas reduci ; id mihi non videtur. Sunt enim multa usque adeo mira & implexa ut neque ab æquationibus pendeant neque ex Quadraturis. Qualia sunt (ex multis aliis) Problemata methodi Tangentium inversæ.* And Mr. *Newton* in his Letter of *Octob.* 24. 1676, replied : *Ubi dixi omnia pene Problemata solubilia existere ; volui de iis præsertim intelligi circa quæ Mathematici se hactenus occuparunt, vel saltem in quibus Ratiocinia Mathematica locum aliquem obtinere possunt. Nam alia sane adeo perplexis conditionibus implicata excogitare liceat, ut non satis comprehendere valeamus : & multo minus tantarum computationum onus sustinere quod ista requirerent. Attamen ne nimium dixisse videar, inversa de Tangentibus Problemata sunt in potestate, aliaque illis difficiliora. Ad quæ solvenda usus sum duplici methodo, una concinniori, altera generaliori. Utramque visum est impræsentia literis transpositis consignare, ne propter alios idem obtinentes, institutum in aliquibus mutare cogerer.* 5 a c c d æ 10 e ff h, &c. id est, *Una methodus consistit in extractione fluentis quantitatis ex æquatione simul involvente fluxionem ejus : altera tantum in assumptione seriei pro quantitate qualibet incognita, ex qua cætera commode derivari possunt; & in collatione terminorum homologorum æquationis resultantis ad eruendos terminos assumptæ seriei.* By Mr. *Newton*'s two Letters, its certain that he had then (or rather above five Years before) found out the Reduction of Problems to fluxional Equations and converging Series : and by the Answer of Mr. *Leibnitz* to the first of those Letters, its as certain that he had not then found out the Reduction of Problems either to differential Equations or to converging Series.

And the same is manifest also by what Mr. *Leibnitz* wrote in the *Acta Eruditorum, Anno* 1691, concerning this Matter.

Jam

(191)

Jam anno 1675, faith he, *compofitum habebam opufculum Quadraturæ Arithmeticæ ab amicis ab illo tempore lectum, fed quod, materia fub manibus crefcente, limare ad Editionem non vacavit, poftquam aliæ occupationes fupervenere; præfertim cum nunc prolixius exponere vulgari more quæ Analyfis noftra paucis exhibet, non fatis operæ pretium videatur.* This Quadrature compofed *vulgari more* he began to communicate at *Paris* in the Year 1675. The nextYear he was polifhing theDemonftration thereof, to fend it to Mr. *Oldenburgh* in Recompence for Mr. *Newton's* Method, as he wrote to him *May* 12. 1676; and accordingly in his Letter of *Auguft.* 27. 1676. he fent it compofed and polifhed *vulgari more.* The Winter following he returned into *Germany*, by *England* and *Holland*, to enter upon publick Bufinefs, and had no longer any Leifure to fit it for the Prefs, nor thought it afterwards worth his while to explain thofe Things prolixly in the vulgar manner which his new *Analyfis* exhibited in fhort. He found out this new *Analyfis* therefore after his Return into *Germany*, and by confequence not before the Year 1677.

The fame is further manifeft by the following Confideration. Dr. *Barrow* publifhed his *Method of Tangents* in the Year 1670. Mr. *Newton* in his Letter dated *December* 10. 1672. communicated his *Method of Tangents* to Mr. *Collins*, and added : *Hoc eft unum particulare vel Corollarium potius Methodi generalis, quæ extendit fe citra moleftum ullum calculum, non modo ad ducendum Tangentes ad quafvis Curvas five Geometricas five Mechanicas, vel quomodocunque rectas Lineas aliafve Curvas refpicientes; verum etiam ad refolvendum alia abftrufiora Problematum genera de Curvitatibus, Areis, Longitudinibus, Centris Gravitatis Curvarum,* &c. *Neque (quemadmodum* Huddenii *methodus de* Maximis & Minimis) *ad folas reftringitur æquationes illas, quæ quantitatibus furdis funt immunes. Hanc methodum intertexui alteri ifti quâ Æquationum Exegefin inftituo, reducendo eas ad feries infinitas.* Mr. *Slufius* fent his Method of Tangents to Mr. *Oldenburgh Jan.* 17. 167$\frac{2}{3}$, and the fame was

2   I i 2   foon

soon after published in the *Transactions*. It proved to be the same with that of Mr. *Newton*. It was founded upon three *Lemmas*, the first of which was this, *Differentia duarum digni-tatum ejusdem gradus applicata ad differentiam laterum dat partes singulas gradus inferioris ex binomio laterum, ut* $\dfrac{y^3 - x^3}{y - x} = yy$

$+ yx + xx$, that is, in the Notation of Mr. *Leibnitz* $\dfrac{dy^3}{dy} =$

$= 3yy$. A Copy of Mr. *Newton*'s Letter of *Decemb.* 10. 1672 was sent to Mr. *Leibnitz* by Mr. *Oldenburg* amongst the Papers of Mr. *James Gregory*, at the same time with Mr. *Newton*'s Letter of *June* 13. 1676. And Mr. *Newton* having described in these two Letters that he had a very general *Analysis*, consisting partly of the Method of converging Series, partly of another Method, by which he applied those Series to the Solution of almost all Problems (except perhaps some nume-ral ones like those of *Diophantus*) and found the Tangents, Areas, Lengths, solid Contents, Centers of Gravity, and Curvities of Curves, and curvilinear Figures Geometrical or Mechanical, without sticking at Surds; and that the Method of Tangents of *Slusius* was but a Branch or Corollary of this other Method : Mr. *Leibnitz* in his returning Home through *Holland*, was meditating upon the Improvement of the Me-thod of *Slusius*. For in a Letter to Mr. *Oldenburgh*, dated from *Amsterdam Nov.* ⅟ 1676, he wrote thus. *Methodus Tangentium à Slusio publicata nondum rei fastigium tenet. Potest aliquid amplius præstari in eo genere quod maximi foret usus ad omnis generis Problemata : etiam ad meam (sine extractionibus) Æquationum ad series reductionem. Nimirum posset brevis qua-dam calculari circa Tangentes Tabula, eousque continuanda donec progressio Tabulæ apparet ; ut eam scilicet quisque quousque libuerit sine calculo continuare possit.* This was the Improvement of the Method of *Slusius* into a general Method, which Mr. *Leibnitz* was then thinking upon, and by his Words,
Potest

(193)

Potest aliquid amplius præstari in eo genere quod maximi foret usus ad omnis generis Problemata, it seems to be the only Improvement which he had then in his Mind for extending the Method to all sorts of Problems. The Improvement by the differential Calculus was not yet in his Mind, but must be referred to the next Year.

Mr. *Newton* in his next Letter, dated *Octob.* 24. 1676, mentioned the *Analysis* communicated by Dr. *Barrow* to Mr. *Collins* in the Year 1669, and also another Tract written in 1671. about converging Series, and about the other Method by which Tangents were drawn after the Method of *Slusius*, and *Maxima* and *Minima* were determined, and the Quadrature of Curves was made more easy, and this without sticking at Radicals, and by which Series were invented which brake off and gave the Quadrature of Curves in finite Equations when it might be. And the Foundation of these Operations he comprehended in this Sentence exprest enigmatically as above. *Data æquatione fluentes quotcunque quantitates involvente fluxiones invenire,& vice versa.* Which puts it past all Dispute that he had invented the Method of Fluxions before that time. And if other things in that Letter be considered, it will appear that he had then brought it to great Perfection, and made it exceeding general ; the Propositions in his Book of Quadratures, and the Methods of converging Series and of drawing a Curve Line through any Number of given Points, being then known to him. For when the Method of Fluxions proceeds not in finite Equations, he reduces the Equations into converging Series by the binomial Theoreme, and by the Extraction of Fluents out of Equations involving or not involving their Fluxions. And when finite Equations are wanting, he deduces converging Series from the Conditions of the Probleme, by assuming the Terms of the Series gradually, and determining them by those Conditions. And when Fluents are to be derived from Fluxions, and the Law of the Fluxions is wanting, he finds

I i 3

that Law *quam proxime,* by drawing a Parabolick Line through any Number of given Points. And by thefe Improvements Mr. *Newton* had in thofe Days made his Method of Fluxions much more univerfal than the Differential Method of Mr. *Leibnitz* is at prefent.

This Letter of Mr. *Newton's,* dated *Octob.* 24. 1676, came to theHands of Mr. *Leibnitz* in the End of the Winter or Beginning of the Spring following; and Mr. *Leibnitz* foon after, *viz.* in a Letter dated *June* 21. 1677, wrote back : *Clariffimi* Slufii *methodum Tangentium nondum effe abfolutam Celeberrimo* Newtono *affentior. Et jam à multo tempore rem Tangentium generalius tractavi, fcilicet per differentias Ordinatarum.* ——— *Hinc nominando,* in pofterum,dy *differentiam duarum proximarum* y *&c.* Here Mr. *Leibnitz* began firft to propofe his Differential Method, and there is not the leaft Evidence that he knew it before the Receipt of Mr. *Newton's* laft Letter. He faith indeed, *Jam à multo tempore rem Tangentium generalius tractavi, fcilicet per differentias Ordinatarum* : and fo he affirmed in other Letters, that he had invented feveral converging Series direct and inverfe before he had the Method of inventing them ; and had forgot an inverfe Method of Series before he knew what ufe to make of it. But no Man is a Witnefs in his own Caufe. A Judge would be very unjuft, and act contrary to the Laws of all Nations, who fhould admit any Man to be a Witnefs in his own Caufe. And therefore it lies upon Mr. *Leibnitz* to prove that he found out this Method long before the Receipt of Mr. *Newton's* Letters. And if he cannot prove this, the Queftion, Who was the firft Inventor of the Method, is decided.

The Marquifs *De l' Hofpital* (a Perfon of very great Candour) in the Preface to his Book *De Analyfi quantitatum infinitè* 'parvarum,* publifhed *A. C.* 1696. tells us, that a little after 'the Publication of the Method of Tangents of *Des Cartes,* ' Mr. *Fermat* found alfo a Method, which *Des Cartes* himfelf 'at length allowed to be, for the moft part, more fimple than his

'his own. But it was not yet fo fimple as Mr. *Barrow* after-
'wards made it, by confidering more nearly the nature of Po-
'lygons, which offers naturally to the Mind a little Triangle,
'compos'd of a Particle of the Curve lying between two Or-
'dinates infinitely near one another, and of the Difference of
'thefe two Ordinates, and of that of the two correfpondent
'*Abfciffa*'s. And this Triangle is like that which ought to be
'made by the Tangent, the Ordinate, and the Sub-tangent:
'fo that by one fimple Analogy, this laft Method faves all
'the Calculation which was requifite either in the Method of
'*Des Cartes*, or in this fame Method before.   Mr. *Barrow*
'ftopt not here, he invented alfo a fort of Calculation proper
'for this Method.   But it was neceffary in this as well as in
'that of *Des Cartes*, to take away Fractions and Radicals for
'making it ufeful. Upon the Defect of this Calculus, that of
'the celebrated Mr. *Leibnitz* was introduced, and this learned
'Geometer began where Mr. *Barrow* and others left off.   This
'his *Calculus* led into Regions hitherto unknown, and there
'made Difcoveries which aftonifhed the moft able Mathema-
'ticians of *Europe*,' &c. Thus far the Marquifs.   He had not
feen Mr. *Newton's Analyfis*, nor his Letters of *Decem.* 10. 1672.
June 13. 1676, and *Octob.* 24. 1676 : and fo not knowing that
Mr. *Newton* had done all this and fignified it to Mr. *Leibnitz,*
he reckoned that Mr. *Leibnitz* began where Mr. *Barrow* left
off, and by teaching how to apply Mr. *Barrow's* Method
without fticking at Fractions and Surds, had enlarged the
Method wonderfully.   And Mr. *James Bernoulli*, in the *Acta
Eruditorum* of *January* 1691 *pag.* 14. writes thus : *Qui calculum* Barrovianum *(quem in Lectionibus fuis Geometricis adum-
bravit Auctor, cujufque Specimina funt tota illa Propofitionum
inibi contentarum farrago,) intellexerit,* [calculum] *alterum à*
Domino Leibnitio *inventum, ignorare vix poterit; utpote qui
in priori illo fundatus eft, & nifi forte in Differentialium notatione
& operationis aliquo compendio ab eo non differt.*

Now

Now Dr. *Barrow*, in his Method of Tangents, draws two Ordinates indefinitely near to one another, and puts the Letter *a* for the Difference of the Ordinates, and the Letter *e* for the Difference of the *Abscissa*'s, and for drawing the Tangent gives thefe Three Rules. 1. *Inter computandum,* faith he, *omnes abjicio terminos in quibus ipfarum* a *vel* e *poteftas habeatur, vel in quibus ipfæ ducuntur in fe. Etenim ifti termini nihil valebunt.* 2. *Poft æquationem conftitutam omnes abjicio terminos literis conftantes quantitates notas feu determinatas fignificantibus, aut in quibus non habentur* a *vel* e. *Etenim illi termini femper ad unam æquationis partem adducti nihilum adæquabunt.* 3. *Pro* a *Ordinatam, & pro* e *Subtangentem fubftituo. Hinc demum Subtangentis quantitas dignofcetur.* Thus far Dr. *Barrow.*

And Mr. *Leibnitz* in his Letter of *June* 21. 1677 above-mentioned, wherein he firft began to propofe his Differential Method, has followed this Method of Tangents exactly, excepting that he has changed the Letters *a* and *e* of Dr. *Barrow* into *dx* and *dy.* For in the Example which he there gives, he draws two parallel Lines and fets all the Terms below the under Line, in which *dx* and *dy* are (feverally or jointly) of more than one Dimenfion, and all the Terms above the upper Line, in which *dx* and *dy* are wanting, and for the Reafons given by Dr. *Barrow,* makes all thefe Terms vanifh. And by the Terms in which *dx* and *dy* are but of one Dimenfion, and which he fets between the two Lines, he determines the Proportion of the Subtangent to the Ordinate. Well therefore did the Marquifs *de l'Hofpital* obferve that where Dr. *Barrow* left off Mr. *Leibnitz* began : for their Methods of Tangents are exactly the fame.

But Mr. *Leibnitz* adds this Improvement of the Method, that the Conclufion of this Calculus is coincident with the Rule of *Slufius,* and fhews how that Rule prefently occurs to any one who underftands this Method. For Mr. *Newton* had reprefented in his Letters, that this Rule was a Corollary of his general Method.

And

(197)

And whereas Mr. *Newton* had faid that his Method in drawing of Tangents, and determining *Maxima* and *Minima*, &c.
proceeded without fticking at Surds : Mr. *Leibnitz* in the
next Place, fhews how this Method of Tangents may be improved fo as not to ftick at Surds or Fractions, and then adds:
Arbitror quæ celare voluit Newtonus *de Tangentibus ducendis ab
his non abludere. Quod addit, ex hoc eodem fundamento Quadraturas quoque reddi faciliores me in hac fententia confirmat ;
nimirum femper figuræ illæ funt quadrabiles quæ funt ad æquationem
differentialem.* By which Words, compared with the preceding
Calculation, its manifeft that Mr. *Leibnitz* at this time underftood that Mr. *Newton* had a Method which would do all thefe
things, and had been examining whether Dr. *Barrow's* Differential Method of Tangents might not be extended to the
fame Performances.

In *November* 1684 Mr. *Leibnitz* publifhed the Elements of
this Differential Method in the *Acta Eruditorum,* and illuftra-
ted it with Examples of drawing Tangents and determining
Maxima and *Minima,* and then added. *Et hæc quidem initia
funt Geometriæ cujufdam multo fublimioris, ad difficillima & pulcherrima quæque etiam miftæ* Mathefeos *Problemata pertingentis,
quæ fine calculo differentiali* AUT SIMILI *non temere quif-
quam pari facilitate tractabit.* The Words AUT SIMILI
plainly relate to Mr. *Newton's* Method. And the whole Sentence contains nothing more than what Mr. *Newton* had affirmed of his general Method in his Letters of 1672 and 1676.

And in the *Acta Eruditorum* of *June* 1686, *pag.* 297.
Mr. *Leibnitz* added: *Malo autem* dx *& fimilia adhibere quam
literas pro illis, quia iftud* dx *eft modificatio quædam ipfius* x,
&c. He knew that in this Method he might have ufed
Letters with Dr. *Barrow,* but chofe rather to ufe the new
Symbols *dx* and *dy,* though there is nothing which can be
done by thefe Symbols, but may be done by fingle Letters
with more brevity.

2   . K k   The

(198)

The next Year Mr. *Newton's Principia Philosophiæ* came abroad, a Book full of such Problemes as Mr. *Leibnitz* had called *difficillima & pulcherrima etiam mistæ Mathescos problemata, quæ sine calculo differentiali aut* SIMILI *non temere quisquam pari facilita e tractabit.* And the Marquess *de L' Hospital* has represented this Book *presque tout de ce calcul*; composed almost wholly of this Calculus. And Mr. *Leibnitz* himself in a Letter to Mr. *Newton*, dated from *Hannover, March* ⁷⁄ 1693, and still extant in his own Hand-writing, and upon a late Occasion communicated to the Royal Society, acknowledged the same thing in these Words : *Mirifice ampliaveras Geometriam tuis Seriebus, sed edito* Principiorum *opere ostendisti patere tibi etiam quæ Analysi receptæ non subsunt. Conatus sum ego queque, notis commodis adhibitis quæ differentias & summas exhibeant, Geometriam illam quam Transcendentem appello Analysi quodammodo subjicere, nec res male processit* ; And again in his Answer to Mr. *Fatio,* printed in the *ActaEruditorum* of *May* 1700. *pag.*203. *lin.*21. he acknowledged the same thing. In the second Lemma of the second Book of these *Principles,* the Elements of this Calculus are demonstrated synthetically, and at theEnd ofthe Lemma there is a Scholium in these Words. *In Literis quæ mihi cum Geometra peritissimo* G. G. Leibnitio *annis al hinc decem intercedebant, cum significarem me compotem esse methodi determinandi* Maximas & Minimas, *ducendi Tangentes & similia peragendi,quæ in terminis surdis æque ac in rationalibus procederet; & literis transpositis hanc sententiam involventibus* [Data æquatione quotcunque fluentes quantitates involvente,fluxiones invenire, & vice versa] *eandem celarem : rescripsit Vir clarissimus se quoque in ejusmodi methodum incidisse, & methodum suam communicavit à mea vix abludentem præterquam in verborum & notarum formulis. Utriusque fundamentum continetur in hoc Lemmate.* In those Letters, and in another dated *Decem.* 10. 1672, a Copy of which, at that time, was sent to Mr. *Leibnitz* by Mr. *Oldenburgh,* as is mentioned above, Mr. *Newton* had so far explained his Method, that it was not difficult for

Mr.

(199)

Mr. *Leibnitz*, by the Help of Dr. *Barrow*'s Method of Tangents, to collect it from those Letters. And its certain, by the Arguments above-mentioned, that he did not know it before the writing of those Letters.

Dr. *Wallis* had received Copies of Mr. *Newton*'s two Letters of *June* 13. and *Octob.* 24. 1676 from Mr. *Oldenburgh*, and published several things out of them in his *Algebra*, printed in *English* 1683, and in Latin 1693 ; and soon after had Intimation from *Holland* to print the Letters entire, because Mr. *Newton*'s Notions of Fluxions passed there with Applause by the Name of the Differential Method of Mr. *Leibnitz*. And thereupon he took notice of this Matter in the Preface to the first Volume of his Works published *A. C.* 1695. And in a Letter to Mr. *Leibnitz* dated *Decemb.* 1. 1696, he gave the Account of it *Cum Præfationis (præfigendæ) postremum folium erat sub prælo, ejusque typos jam posuerant Typothetæ ; me monuit amicus quidam (harum rerum gnarus) qui peregre fuerat, tum talem methodum in Belgio prædicari, tum illam cum* Newtoni *methodo Fluxionum quasi coincidere. Quod fecit ut (translatis typis jam positis) id monitum interseruerim.* And in a Letter dated *April* 10. 1695, and lately communicated to the Royal-Society, he wrote thus about it. *I wish you would print the two large Letters of* June *and* August [he means *June* and *October*] 1676. *I had intimation from* Holland, *as desired there by your Friends, that somewhat of that kind were done ; because your Notions (of Fluxions) pass there with great Applause by the Name of* Leibnitz's *Calculus Differentialis. I had* * *this intimation when all but part of the Preface to this Volume was printed off ; so that I could only insert (while the Press stay'd) that short Intimation thereof which you there find. You are not so kind to your Reputation (and that of the Nation) as you might be, when you let things of worth lye by you so long, till others carry away*

---

* Extat hæc Epistola in tertio volumine operum *Wallisii.*

2   K k 2   *the*

the Reputation that is due to you. I have endeavoured to do you Juſtice in that Point, and am now ſorry that I did not print thoſe two Letters verbatim.

The ſhort Intimation of this Matter, which Dr. *Wallis* inſerted into the ſaid Preface, was in theſe Words. *In ſecundo Volumine (inter alia) habetur* Newtoni *Methodus de Fluxionibus (ut ille loquitur) conſimilis naturæ cum* Leibnitii *(ut hic loquitur) Calculo Differentiali (quod qui utramque methodum contulerit ſatis animadvertat, ut ut ſub loquendi formulis diverſis) quam ego deſcripſi (Algebræ* cap. 91. &c. *præſertim* cap 95) *ex binis* Newtoni *Literis, aut earum alteris,* Junii 13. *&* Octob. 24. 1676 *ad* Oldenburgum *datis, cum* Leibnitio *tum communicandis (iiſdem fere verbis, ſaltem leviter mutatis, quæ in illis literis habentur,) ubi* METHODUM HANC *LEIBNITIO* EXPONIT, *tum ante* DECEM ANNOS *nedum plures* [id eſt, anno 1666 vel 1665] *ab ipſo excogitatam. Quod moneo, nequis cauſetur de hoc Calculo Differentiali nihil à nobis dictum eſſe.*

Hereupon the Editors of the *Acta Lipſienſia,* the next Year in *June,* in the Style of Mr. *Leibnitz,* in giving an Account of theſe two firſt Volumes of Dr. *Wallis,* took notice of this Clauſe of the Doctor's Preface, and complained, not of his ſaying that Mr. *Newton* in his two Letters above-mentioned explained to Mr *Leibnitz* the Method of Fluxions found by him Ten Years before or above ; but that while the Doctor mentioned the Differential Calculus, and ſaid that he did it *nequis cauſetur de calculo differentiali nihil ab ipſo dictum fuiſſe,* he did not tell the Reader that Mr. *Leibnitz* had this Calculus at that time when thoſe Letters paſſed between him and Mr. *Newton,* by means of Mr. *Oldenburgh.* And in ſeveral Letters which followed hereupon, between Mr. *Leibnitz* and Dr. *Wallis,* concerning this Matter, Mr. *Leibnitz* denied not that Mr. *Newton* had the Method Ten Years before the writing of thoſe Letters, as Dr. *Wallis* had affirmed; pretended not that he himſelf had the Method ſo early ; brought no Proof that he had it before the Year 1677;

(201)

1677; no other Proof besides the Conceffion of Mr. *Newton* that he had it fo early; affirmed not that he had it earlier; commended Mr. *Newton* for his Candour in this Matter; allowed that the Methods agreed in the main, and faid that he therefore ufed to call them by the common Name of his *Infinitifimal Analyfis*; reprefented, that as the Methods of *Vieta* and *Cartes* were called by the common Name of *Analyfis Speciofa*, and yet differed in fome things; fo perhaps the Methods of Mr. *Newton* and himfelf might differ in fome things, and challenged to himfelf only thofe things wherein, as he conceived, they might differ, naming the Notation, the differential Equations and the Exponential Equations. But in his Letter of *June* 21. 1677 he reckon'd differential Equations common to Mr. *Newton* and himfelf.

This was the State of the Difpute between Dr. *Wallis* and Mr. *Leibnitz* at that time And Four years after, when Mr. *Fatio* fuggefted that Mr. *Leibnitz*, the fecond Inventor of this Calculus, might borrow fomething from Mr. *Newton*, the oldeft Inventor by many Years: Mr. *Leibnitz* in his Anfwer, publifhed in the *Acta Eruditorum* of *May* 1700, allowed that Mr. *Newton* had found the Method apart, and did not deny that Mr. *Newton* was the oldeft Inventor by many Years, nor afferted any thing more to himfelf, than that he alfo had found the Method apart, or without the Affiftance of Mr. *Newton*, and pretended that when he firft publifhed it, he knew not that Mr. *Newton* had found any thing more of it than the Method of Tangents. And in making this Defence he added: *Quam* [methodum] *ante Dominum* Newtonum & Me *nullus quod fciam Geometra babuit; uti ante hunc maximi nominis Geometram N E M O fpecimine publice dato fe habere probavit, ante Dominos* Bernoullios & Me *nullus communicavit.* Hitherto therefore Mr. *Leibnitz* did not pretend to be the firft Inventor. He did not begin to put in fuch a Claim till after the Death of Dr. *Wallis*, the laft of the old Men who were acquainted with what had paffed between the *Englifh* and Mr. *Leibnitz*

K k 3   Forty

Forty Years ago. The Doctor died in *October A. C.* 1703, and Mr. *Leibnitz* began not to put in this new Claim before *January* 1705.

Mr. *Newton* publifhed his *Treatife of Quadratures* in the Year 1704. This Treatife was written long before, many things being cited out of it in his Letters of *Octob.* 24. and *Novemb.* 8. 1676. It relates to the Method of Fluxions, and that it might not be taken for a new Piece, Mr. *Newton* repeated what Dr. *Wallis* had publifhed Nine Years before without being then contradicted, namely, that this Method was invented by Degrees in the Years 1665 and 1666. Hereupon the Editors of the *Acta Lipfienfia* in *January* 1705, in the Style of Mr. *Leibnitz,* in giving an Account of this Book, reprefented that Mr. *Leibnitz* was the firft Inventor of the Method, and that Mr. *Newton* had fubftituted Fluxions for Differences. And this Accufation gave a Beginning to this prefent Controverfy.

For Mr. *Keill,* in an Epiftle publifhed in the *Philofophical Tranfactions* for *Sept.* and *Octob.* 1708, retorted the Accufation, faying : *Fluxionum Arithmeticam fine omni dubio primus invenit D.* Newtonus, *ut cuilibet ejus Epiftolas à* Wallifio *editas legenti facile conftabit. Eadem tamen Arithmetica poftea mutatis nomine & notationis modo à Domino* Leibnitio *in* Actis Eruditorum *edita eft.*

Before Mr. *Newton* faw what had been publifhed in the *Acta Leipfica,* he exprefs'd himfelf offended at the printing of this Paragraph of Mr. *Keill*'s Letter, leaft it fhould create a Controverfy. And Mr. *Leibnitz,* underftanding it in a ftronger Senfe than Mr. *Keill* intended it, complain'd of it as a Calumny, in a Letter to Dr. *Sloane* dated *March* 4. 1711 *N.S.* and moved that the Royal-Society would caufe Mr. *Keill* to make a publick Recantation. Mr. *Keill* chofe rather to explain and defend what he had written ; and Mr. *Newton,* upon being fhewed the Accufation in the *Acta Lipfica,* gave him leave to do fo. And Mr. *Leibnitz* in a fecond Letter to Dr. *Sloane,* dated *Decem.* 29. 1711, inftead of making good his Accufation

fation, as he was bound to do that it might not be deem'd a Calumny, infifted only upon his own Candour, as if it would be Injuftice to queftion it ; and refus'd to tell how he came by the Method ; and faid that the *Acta Lipfica* had given every Man his due, and that he had concealed the Invention above Nine Years, (he fhould have faid Seven Years) that No body might pretend (he means that Mr. *Newton* might not pretend) to have been before him in it ; and called Mr. *Keill* a Novice unacquainted with things paft, and one that acted without Authority from Mr. *Newton*, and a clamorous Man who deferved to be filenced, and defired that Mr. *Newton* himfelf would give his Opinion in the Matter. He knew that Mr. *Keill* affirmed nothing more than what Dr. *Wallis* had publifhed thirteen Years before, without being then contradicted. He knew that Mr. *Newton* had given his Opinion in this matter in the Introduction to his Book of *Quadratures*, publifhed before this Controverfy began : but Dr. *Wallis* was dead ; the Mathematicians which remained in *England* were Novices ; Mr. *Leibnitz* may Queftion any Man's Candour without Injuftice, and Mr. *Newton* muft now retract what he had publifhed or not be quiet.

The Royal-Society therefore, having as much Authority over Mr. *Leibnitz* as over Mr. *Keill*, and being now twice preffed by Mr. *Leibnitz* to interpofe, and feeing no reafon to condemn or cenfure Mr. *Keill* without enquiring into the matter ; and that neither Mr. *Newton* nor Mr. *Leibnitz* (the only Perfons alive who knew and remembred any thing of what had paffed in thefe matters Forty Years ago) could be Witneffes for or againft Mr. *Keill* ; appointed a numerous Committee to fearch old Letters and Papers, and report their Opinion upon what they found ; and ordered the Letters and Papers, with the Report of their Committee to be publifhed. And by thefe Letters and Papers it appear'd to them, that Mr. *Newton* had the Method in or before the Year 1669, and it did not appear to them that Mr. *Leibnitz* had it before the Year 1677.

For

For making himfelf the firft Inventor of the Differential Method, he has reprefented that Mr. *Newton* at firft ufed the Letter *o* in the vulgar manner for the given Increment of *x*, which deftroys the Advantages of the Differential Method; but after the writing of his Principles, changed *o* into $\dot{x}$, fubftituting $\dot{x}$ for *dx*. It lies upon him to prove that Mr. *Newton* ever changed *o* into $\dot{x}$, or ufed $\dot{x}$ for *dx*, or left off the Ufe of the Letter *o*. Mr. *Newton* ufed the Letter *o* in his *Analyfis* written in or before the Years 1669, and in his Book of *Quadratures*, and in his *Principia Philofophiæ*, and ftill ufes it in the very fame Senfe as at firft. In his Book of Quadratures he ufed it in conjunction with the Symbol $\dot{x}$, and therefore did not ufe that Symbol in its Room. Thefe Symbols *o* and $\dot{x}$ are put for things of a different kind. The one is a Moment, the other a Fluxion or Velocity as has been explained above. When the Letter *x* is put for a Quantity which flows uniformly, the Symbol $\dot{x}$ is an Unit, and the Letter *o* a Moment, and $\dot{x}o$ and *dx* fignify the fame Moment. Prickt Letters never fignify Moments, unlefs when they are multiplied by the Moment *o* either expreft or underftood to make them infinitely little, and then the Rectangles are put for Moments.

Mr. *Newton* doth not place his Method in Forms of Symbols, nor confine himfelf to any particular Sort of Symbols for Fluents and Fluxions. Where he puts the Areas of Curves for Fluents, he frequently puts the Ordinates for Fluxions, and denotes the Fluxions by the Symbols of the Ordinates, as in his *Analyfis*. Where he puts Lines for Fluents, he puts any Symbols for the Velocities of the Points which defcribe the Lines, that is, for the firft Fluxions; and any other Symbols for the Increafe of thofe Velocities, that is, for the fecond Fluxions, as is frequently done in his *Principia Philofophiæ*. And where he puts the Letters *x*, *y*, *z* for Fluents, he denotes their Fluxions, either by other Letters as *p*, *q*, *r*; or by the fame Letters in other Forms as *X*, *Y*, *Z* or $\dot{x}$, $\dot{y}$, $\dot{z}$; or by
any

any Lines as *D E*, *FG*, *H I*, confidered as their Exponents.
And this is evident by his Book of *Quadratures*, where he
reprefents Fluxions by prickt Letters in the firft Propofition,
by Ordinates of Curves in the laft Propofition, and by other
Symbols, in explaining the Method and illuftrating it with Examples, in the Introduction. Mr. *Leibnitz* hath no Symbols
of Fluxions in his Method, and therefore Mr. *Newton's* Symbols of Fluxions are the oldeft in the kind. Mr. *Leibnitz* began to ufe the Symbols of Moments or Differences *dx, dy, dz*
in the Year 1677. Mr. *Newton* reprefented Moments by the
Rectangles under the Fluxions and the Moment *o*, when he
wrote his *Analyfis*, which was at leaft Forty Six Years ago.
Mr. *Leibnitz* has ufed the Symbols *ʃx, ʃy, ʃz* for the Sums
of Ordinates ever fince the Year 1686; Mr. *Newton* reprefented the fame thing in his *Analyfis*, by infcribing the Ordinate in a Square or Rectangle. All Mr. *Newton's* Symbols
are the oldeft in their feveral Kinds by many Years.

And whereas it has been reprefented that the ufe of the
Letter *o* is vulgar, and deftroys the Advantages of the Differential Method: on the contrary, the Method of Fluxions,
as ufed by Mr. *Newton*, has all the Advantages of ths Differential, and fome others. It is more elegant, becaufe in his
Calculus there is but one infinitely little Quantity reprefented
by a Symbol, the Symbol *o*. We have no Ideas of infinitely
little Quantities, and therefore Mr. *Newton* introduced Fluxions into his Method, that it might proceed by finite Quantities as much as poffible. It is more Natural and Geometrical,
becaufe founded upon the *primæ quantitatum nafcentium rationes*, which have a Being in Geometry, whilft *Indivifibles*, upon
which the Differential Method is founded, have no Being either in Geometry or in Nature. There are *rationes primæ quantitatum nafcentium*, but not *quantitates primæ nafcentes*. Nature
generates Quantities by continual Flux or Increafe; and the
ancient Geometers admitted fuch a Generation of Areas and
Solids, when they drew one Line into another by local Motion

L l

to

to generate an Area, and the Area into a Line by local Motion to generate a Solid. But the ſumming up of Indiviſibles to compoſe an Area or Solid was never yet admitted into Geometry. Mr. *Newton*'s Method is alſo of greater Uſe and Certainty, being adapted either to the ready finding out of a Propoſition by ſuch Approximations as will create no Error in the Concluſion, or to the demonſtrating it exactly: Mr. *Leibnitz*'s is only for finding it out. When the Work ſucceeds not in finite Equations Mr. *Newton* has recourſe to converging Series, and thereby his Method becomes incomparably more univerſal than that of Mr. *Leibnitz*, which is confin'd to finite Equations : for he has no Share in the Method of infinite Series. Some Years after the Method of Series was invented, Mr. *Leibnitz* invented a Propoſition for tranſmuting curvilinear Figures into other curvilinear Figures of equal Areas, in order to ſquare them by converging Series: but the Methods of ſquaring thoſe other Figures by ſuch Series were not his. By the help of the new *Analyſis* Mr. *Newton* found out moſt of the Propoſitions in his *Principia Philoſophiæ*: but becauſe the Ancients for making things certain admitted nothing into Geometry before it was demonſtrated ſynthetically, he demonſtrated the Propoſitions ſynthetically, that the Syſteme of the Heavens might be founded upon good Geometry. And this makes it now difficult for unskilful Men to ſee the Analyſis by which thoſe Propoſitions were found out.

It has been repreſented that Mr. *Newton*, in the Scholium at the End of his Book of Quadratures, has put the third, fourth, and fifth Terms of a converging Series reſpectively equal to the ſecond, third, and fourth Differences of the firſt Term, and therefore did not then underſtand the Method of ſecond, third, and fourth Differences. But in the firſt Propoſition of that Book he ſhewed how to find the firſt, ſecond, third and following Fluxions *in infinitum*; and therefore when he wrote that Book, which was before the Year 1676, he did underſtand the Method of all the Fluxions, and by con-

2   ſequence

fequence of all the Differences. And if he did not understand it when he added that Scholium to the End of the Book,
which was in the Year 1704, it muſt have been becauſe he
had then forgot it. And ſo the Queſtion is only whether
he had forgot the Method of ſecond and third Differences
before the Year 1704.

In the Tenth Propoſition of the ſecond Book of his *Princi-*
pia Philoſophiæ, in deſcribing ſome of the Uſes of the Terms
of a converging Series for ſolving of Problemes, he tells us
that if the firſt Term of the Series re-
preſents the Ordinate *B C* of any Curve
Line *A C G*, and *C B D I* be a Paralle-
logram infinitely narrow, whoſe Side
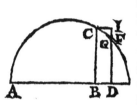
D I cuts the Curve in *G* and its Tan-
gent *C F* in *F*, the ſecond Term of the
Series will repreſent the Line *I F*, and
the third Term the Line *F G*. Now the
Line *FG* is but half the ſecond Difference of the Ordinate :
and therefore Mr. *Newton* when he wrote his *Principia*, put
the third Term of the Series equal to half of the ſecond Difference of the firſt Term, and by conſequence had not then forgotten the Method of ſecond Differences.

In writing that Book, he had frequent occaſion to conſider
the Increaſe or Decreaſe of the Velocities with which Quantities are generated, and argues right about it. That Increaſe
or Decreaſe is the ſecond Fluxion of the Quantity, and therefore he had not then forgotten the Method of ſecond Fluxions.

In the Year 1692, Mr. *Newton*, at the Requeſt of Dr.*Wallis*,
ſent to him a Copy of the firſt Propoſition of the Book of
Quadratures, with Examples thereof in firſt, ſecond and third
Fluxions : as you may ſee in the ſecond Volume of the Doctor's Works, *pag.* 391, 392, 393 and 396. And therefore he
had not then forgotten the Method of ſecond Fluxions.

L l 2   Nor

Nor is it likely, that in the Year 1704. when he added the aforesaid Scholium to the End of the Book of Quadratures, he had forgotten not only the first Proposition of that Book, but also the last Proposition upon which that Scholium was written. If the Word [ut], which in that Scholium may have been accidentally omitted between the Words [erit] and [ejus,] be restor'd, that Scholium will agree with the two Propositions and with the rest of his Writings, and the Objection will vanish.

Thus much concerning the Nature and History of these Methods, it will not be amiss to make some Observations thereupon.

In the *Commercium Epistolicum,* mention is made of three Tracts written by Mr. *Leibnitz.* after a Copy of Mr. *Newton's Principia Philosophiæ* had been sent to *Hannover* for him, and after he had seen an Account of that Book published in the *Acta Eruditorum* for *January* and *February* 1689. And in those Tracts the principal Propositions of that Book are composed in a new manner, and claimed by Mr. *Leibnitz* as if he had found them himself before the publishing of the said Book. But Mr. *Leibnitz* cannot be a Witness in his own Cause. It lies upon him either to prove that he found them before Mr. *Newton,* or to quit his claim.

In the last of those three Tracts, the 20th Proposition (which is the chief of Mr. *Newton's* Propositions) is made a Corollary of the 19th Proposition, and the 19th Proposition has an erroneous Demonstration adapted to it. It lies upon him either to satisfy the World that the Demonstration is not erroneous, or to acknowledge that he did not find that and the 20th Proposition thereby, but tried to adapt a Demonstration to Mr. *Newton's* Proposition to make it his own. For he represents in his 20th Proposition that he knew not how Mr. *Newton* came by it, and by consequence that he found it himself without the Assistance of Mr. *Newton.*

By the Errors in the 15th and 19th Proposition of the third Tract, Dr. *Keill* hath shewed that when Mr. *Leibnitz* wrote these

(209)

thefe three Tracts, he did not well underftand the Ways of working in fecond Differences. And this is further manifeft by the 10*th*, 11*th*, and 12*th* Propofitions of this third Tract. For thefe he lays down as the Foundation of his infinitefimal Analyfis in arguing about centrifugal Forces, and propofes the firft of them with relation to the Center of Curvity of the Orb, but ufes this Propofition in the two next, with Relation to the Center of Circulation. And by confounding thefe two Centers with one another in the fundamental Propofitions upon which he grounds this *Calculus*, he erred in the Superftructure, and for want of Skill in fecond and third Differences, was not able to extricate himfelf from the Errors. And this is further confirmed by the fixth Article of the fecond Tract. For that Article is erroneous, and the Error arifes from his not knowing how to argue well about fecond and third Differences. When therefore he wrote thofe Tracts he was but a Learner, and this he ought in candour to acknowledge.

It feems therefore that as he learnt the Differential Method by means of Mr. *Newton*'s aforefaid three Letters compared with Dr. *Barrow*'s Method of Tangents; fo Ten Years after, when Mr. *Newton*'s *Principia Philofophiæ* came abroad, he improved his Knowledge in thefe Matters, by trying to extend this Method to the Principal Propofitions in that Book, and by this means compofed the faid three Tracts. For the Propofitions contained in them (Errors and Trifles excepted) are Mr. *Newton*'s (or eafy Corollaries from them) being publifhed by him in other Forms of Words before. And yet Mr. *Leibnitz* publifhed them as invented by himfelf long before they were publifhed by Mr. *Newton*. For in the End of the firft Tract, he reprefents that he invented them all before Mr. *Newton*'s *Principia Philofophiæ* came abroad, and fome of them before he left *Paris*, that is before *October* 1676. And the fecond Tract he concludes with thefe Words: *Multa ex his deduci poffent praxi accommodata, fed nobis*

*bis nunc fundamenta Geometrica jeciſſe ſuffecerit, in quibus maxi-
ma conſiſtebat difficultas. Et ſortaſſis attente conſideranti vias
quaſdam novas ſatis antea impeditas aperuiſſe videbimur. Omnia
autem reſpondent noſtræ Analyſi Infinitorum, hoc eſt calculo Sum-
marum & Differentiarum (cujus elementa quædam in his Aƈtis de-
dimus) communibus quoad licuit verbis hic expreſſo.* He pretends
here that the *Fundamenta Geometrica in quibus maxima conſiſtebat difficultas* were firſt laid by himſelf in this very Traƈt, and
that he himſelf had in this very Traƈt opened *vias quaſdam
novas ſatis antea impeditas.* And yet Mr. *Newton's Principia
Philoſophiæ* came abroad almoſt two Years before, and gave
occaſion to the Writing of this Traƈt, and was written *communibus quoad licuit verbis,* and contains all theſe Principles
and all theſe new Ways. And Mr. *Leibnitz,* when he publiſhed that Traƈt, knew all this, and therefore ought then
to have acknowledged that Mr. *Newton* was the firſt who laid
the *Fundamenta Geometrica in quibus maxima conſiſtebat Difficultas,* and opened the *vias novas ſatis antea impeditas.* In his
Anſwer to Mr. *Fatio* he acknowledged all this, ſaying *Quam*
[methodum] *ante Dominum Newtonum & me nullus quod
ſciam Geometra habuit; uti ante hunc maximi nominis Geometram,
NEMO SPECIMINE PUBLICE DATO ſe ha-
bere PROBAVIT.* And what he then acknowledged he
ought in Candour and Honour to acknowledge ſtill upon all
Occaſions.

Mr. *Leibnitz* in his Letter of *May* 28. 1697, wrote thus to
Dr. *Wallis. Methodum Fluxionum profundiſſimi* Newtoni *cogna-
tam eſſe methodo meæ differentiali non tantum animadverti poſtquam opus ejus* [Principiorum ſcilicet] *& tuum prodiit; ſed etiam profeſſus ſum in Aƈtis Eruditorum, & alias qu que monui.
Id enim candori meo convenire judicavi, non minus quam ipſius
merito. Itaque communi nomine deſignare ſoleo Analyſeos infini-
teſimalis; quæ latius quam Tetragoniſtica patet. Interim quemadmodum &* Vietæa *&* Carteſiana *methodus Analyſeos ſpecioſæ
nomine venit; diſcrimina tamen nonnulla superſunt : ita fortaſſe*
                                                              &

(211)

& Newtoniana *&* Mea *differunt in nonnullis.* Here alfo Mr. *Leibnitz* allows that when Mr. *Newton's* Principles of Philofophy came abroad, he underftood thereby the Affinity that there was between the Methods, and therefore called them both by the common Name of the infinitefimal Method, and thought himfelf bound in candour to acknowledge this Affinity : and there is ftill the fame Obligation upon him in point of Candour. And befides this Acknowledgment, he here gives the Preference to Mr. *Newton's* Method in Antiquity. For he reprefents that as the vulgar Analyfis in *Species* was invented by *Vieta,* and augmented by *Cartes,* which made fome Differences between their Methods : fo Mr. *Newton's* Method and his own might differ in fome things. And then he goes on to enumerate the Differences by which he had improved Mr. *Newton's* Method as we mentioned above. And this Subordination of his Method to Mr. *Newton's,* which he then acknowledged to Dr. *Wallis,* he ought ftill to acknowledge.

In enumerating the Differences or Improvements which he had added to Mr. *Newton's* Method ; he names in the fecond Place Differential Equations : but the Letters which paffed between them in the Year 1676, do fhow that Mr. *Newton* had fuch Equations at that time, and Mr. *Leibnitz* had them not. He names in the third Place Exponential Equations : but thefe Equations are owing to his Correfpondence with the *Englifh.* Dr. *Wallis,* in the Interpolation of Series, confidered Fract and Negative Indices of Dignities. Mr. *Newton* introduced into his Analytical Computations, the Fract, Surd, Negative and Indefinitive Indices of Dignities ; and in his Letter of *October* 24. 1676, reprefented to Mr. *Leibnitz* that his Method extended to the Refolution of affected Equations involving Dignities whofe Indices were Fract or Surd. Mr. *Leibnitz* in his Anfwer dated *June* 21 1677, mutually defired Mr. *Newton* to tell him what he thought of the Refolution of Equations involving
Dig-

, (212)

Dignities whofe Indices were undetermined, fuch as were
thefe $x^y + y^x = xy$, $x^x + y^y = x + y$. And thefe Equations
he now calls Exponential, and reprefents to the World that
he was the firft Inventor thereof, and magnifies the Invention as a great Difcovery. But he has not yet made a publick
Acknowledgment of the Light which Mr. *Newton* gave him
into it, nor produced any one Inftance of the ufe that he has
been able to make of it where the Indices of Dignities are
Fluents. And fince he has not yet rejected it with his ufual
Impatience for want of fuch an Inftance, we have reafon
to expect that he will at length explain its Ufefulnefs to the
World.

Mr. *Newton* in his Letter of *October* 24. 1676 wrote that
he had two Methods of refolving the Inverfe Problems of
Tangents, and fuch like difficult ones ; one of which con-
fifted *in affuming a Series for any unknown Quantity from which
all the reft might conveniently be deduced, and in collating the
homologous Terms of the refulting Equation, for determining the
Terms of the affumed Series.* Mr. *Leibnitz* many Years after
publifhed this Method as his own, claiming to himfelf the
firft Invention thereof. It remains that he either renounce
his Claim publickly, or prove that he invented it before
Mr. *Newton* wrote his faid Letter.

It lies upon him alfo to make a publick Acknowledgment
of his Receipt of Mr. *Oldenburgh's* Letter of *April* 15. 1675,
wherein feveral converging Series for fquaring of Curves,
and particularly that of Mr. *James Gregory* for finding the
Arc by the given Tangent, and thereby fquaring the Circle,
were communicated to him. He acknowledged it privately
in his Letter to Mr. *Oldenburg* dated *May* 20. 1675 ftill extant
in his own Hand-writing, and by Mr *Oldenburg* left entred in
the Letter-Book of the *Royal-Society*. But he has not yet
acknowledged it publickly, as he ought to have done when
he publifhed that Series as his own.

It

302

(213)

It lies upon him also to make a publick Acknowledgment of his having received the Extracts of Mr. *James Gregory's* letters, which, at his own Request, were sent to him at *Paris* in *June* 1676 by Mr. *Oldenburgh* to peruse : amongst which was Mr. *James Gregory's* Letter of *Feb.* 15. 1671, concerning that Series, and Mr. *Newton's* Letter of *December* 10. 1672 concerning the Method of Fluxions.

And whereas in his Letter of *Decem.* 28. 1675 he wrote to Mr. *Oldenburgh,* that he had communicated that Series above two Years before to his Friends at *Paris,* and had written to him sometimes about it ; and in his Letter of *May* 12. 1676 said to Mr. *Oldenburgh* that he had written to him about that Series some Years before ; and in his Letter to Mr. *Oldenburgh* dated *Aug.* 27. 1676, that he had communicated that Series to his Friends above three Years before ; that is, upon his first coming from *London* to *Paris* : He is desired to tell us how it came to pass, that when he received Mr. *Oldenburgh's* Letter of *Apr.* 15. 1675 he did not know that Series to be his own.

In his Letters of *July* 15. and *Octob.* 26. 1674, he tells us of but one Series for the circumference of a Circle, and saith that the Method which gave him this Series, gave him also a Series for any Arc whose Sine was given, tho' the Proportion of the Arc to the whole Circumference be not known. This Method therefore, by the given Sine of 30 Degrees, gave him a Series for the whole Circumference. If he had also a Series for the whole Circumference deduced from the Tangent of 45 Degrees, he is desired to tell the World what Method he had in those Days, which could give him both those Series. For the Method by the Transmutation of Figures will not do it. He is desired also to tell us why in his said Letters he did not mention more Quadratures of the Circle than one.

And if in the Year 1674 he had the Demonstration of a Series for finding any Arc whose Sine is given, he is desired

M m

to

to tell the World what it was ; and why in his Letter of *May*
12. 1676 he defired Mr. *Oldenburgh* to procure from
Mr. *Collins* the Demonftration of Mr. *Newton's* Series for
doing the fame thing ; and wherein his own Series differed
from Mr. *Newton's*.   For upon all thefe Confiderations there
is a Sufpicion that Mr. *Newton's* Series for finding the Arc
whofe Sine is given, was communicated to him in *England* ;
and that in the Year 1673 he began to communicate it as his
own to fome of his Friends at *Paris*, and the next Year wrote
of it as his own in his Letters to Mr. *Oldenburgh*, in order to
get the Demonftration or Method of finding fuch Series.
But the Year following, when Mr. *Oldenburgh* fent him this
Series and the Series of Mr. *Gregory* and Six other Series,
he dropt his Pretence to this Series for want of a Demonftration, and took time to confider the Series fent him, and to
compare them with his own, as if his Series were others
different from thofe fent him   And when he had found a
Demonftration of *Gregory's* Series by a Tranfmutation of Figures, he began to communicate it as his own to his Friends
at *Paris*, as he reprefents in the *Acta Eruditerum* for *April*
1691. *pag.* 178, faying; *Jam Anno* 1675 *compofitum habebam
opufculum Quadratura Arithmetica ab Amicis ab illo tempore
lectum*, &c.   But the Letter by which he had received this
Series from Mr. *Oldenburgh* he concealed from his Friends,
and pretended to Mr. *Oldenburgh* that he had this Series a
Year or two before the Receipt of that Letter.   And the
next Year, upon receiving two of Mr. *Newton's* Series again,
by one *George Mohr*, he wrote to Mr. *Oldenburgh* in fuch a
manner as if he had never feen them before, and upon Pretence of their Novelty, defired Mr. *Oldenburgh* to procure
from Mr. *Collins* Mr. *Newton's* Method of finding them.   If
Mr. *Leibnitz* thinks fit to obviate this Sufpicion, he is in the
firft Place to prove that he had Mr. *Gregory's* Series before he
received it from Mr. *Oldenburgh*.

It

(215)

It lies upon him alfo to tell the World what was the Method by which the feveral Series of Regreffion for the Circle and Hyperbola, fent to him by Mr. *Newton June* 13. 1676, and claimed as his own by his Letter of *Auguft* 27. following, were found by him before he received them from Mr. *Newton*.

And whereas Mr. *Newton* fent him, at his own Requeft, a Method of Regreffion, which upon the firft reading he did not know to be his own, nor underftood it ; but fo foon as he underftood it he claimed as his own, by pretending that he had found it long before, and had forgot it, as he perceived by his old Papers : it lies upon him, in point of Candor and Juftice, either to prove that he was the firft Inventor of this Method, or to renounce his Claim to it for preventing future Difputes.

Mr. *Leibnitz* in his Letter to Mr. *Oldenburgh* dated *Feb.* 3. 167¾ claimed a Right to a certain Property of a Series of Numbers Natural, Triangular, Pyramidal, Triangulo-Triangular, *&c.* and to make it his own, reprefented that he wondred that Monfieur *Pafcal*, in his Book entituled *Triangulum Arithmeticum*, fhould omit it. That Book was publifhed in the Year 1665, and contains this Property of the Series ; and Mr. *Leibnitz* has not yet done him the Juftice to acknowledge that he did not omit it. It lies upon him therefore in Candor and Juftice, to renounce his Claim to this Property, and acknowledge Mr. *Pafchal* the firft Inventor.

He is alfo to renounce all Right to the Differential Method of *Mouton* as fecond Inventor : for fecond Inventors have no Right. The fole Right is in the firft Inventor until another finds out the fame thing apart. In which cafe to take away the Right of the firft Inventor, and divide it between him and that other, would be an Act of Injuftice.

In his Letter to Dr. *Sloane* dated *Decem.* 29. 1711. he has told us that his Friends know how he came by the Differential Method. It lies upon him, in point of Candor, openly. and plainly, and without further Hefitation, to fatisfy the World how he came by it.   M m 2   In

(216)

In the same Letter he has told us that he had this Method above Nine Years before he published it, and it follows from thence that he had it in the Year 1675 or before. And yet its certain that he had it not when he wrote his Letter to Mr. *Oldenburgh* dated *Aug.* 27. 1676, wherein he affirmed that Problems of the Inverse Method of Tangents and many others, could not be reduced to infinite Series, nor to Equations or Quadratures. It lies upon him therefore, in point of Candor, to tell us what he means by pretending to have found the Method before he had found it.

We have shewed that Mr. *Leibnitz* in the End of the Year 1676, in returning home from *France* through *England* and *Holland*, was meditating how to improve the Method of *Slusius* for Tangents, and extend it to all sorts of Problems, and for this end proposed the making of a general Table of Tangents; and therefore had not yet found out the true Improvement. But about half a Year after, when he was newly fallen upon the true Improvement, he wrote back ; *Clarifs.* Slusii *Methodum Tangentium nondum effe abfolutam Cele-berrimo* Newtono *affentior. Et jam A MULTO TEM-PORE rem Tangentium generalius tractavi, fcilicet per diffe-rentias Ordinatarum.* Which is as much as to say that he had this Improvement long before those Days. It lies upon him, in point of Candor, to make us understand that he pretended to this Antiquity of his Invention with some other Design than to rival and supplant Mr. *Newton*, and to make us believe that he had the Differential Method before Mr. *Newton* explained it to him by his Letters of *June* 13. and *Octob.* 24. 1676, and before Mr. *Oldenburgh* sent him a Copy of Mr. *Newton*'s Letter of *Decem* 10. 1672 concerning it.

The Editors of the *Acta Eruditorum* in *June* 1696, in giving an Account of the two first Volumes of the Mathematical Works of Dr. *Wallis*, wrote thus, in the Style of Mr. *Leibnitz: Caterum ipfe* Newtonus, *non minus Candore quam praclaris in rem Mathematicam meritis infignis, publice &*

I   *pr.*

(217)

privatim agnovit Leibnitium, *tum cum (interveniente celeberri-mo Viro* Henrico Oldenburgo *Bremenfi, Societatis Regiæ* Anglicanæ *tunc Secretario) inter ipfos (ejufdem jam tum Societatis Socios) Commercium intercederet, id eft jam fere ante annos viginti & amplius, Calculum fuum differentialem, Seriefque infinitas, & pro iis quoque Methodos generales habuiffe; quod* Wallifius *in Præfatione Operum, facta inter eos communicationis mentionem faciens, prætertit, quoniam de eo fortaffe non fatis ipfi conftabat. Cæterum Differentiarum confideratio* Leibnitiana, *cujus mentionem facit* Wallifius *(ne quis fcilicet, ut ipfe ait, cauferetur de Calculo Differentiali nihil ab ipfo dictum fuiffe) meditationes apernit, quæ aliunde non æque nafcebantur, &c.* By the Words here cited out of the Preface to the two firſt Volumes of Dr. *Wallis*'s Works, it appears that Mr. *Leibnitz* had feen that Part of the Preface, where Mr. *Newton* is faid to have explained to him (in the Year 1676) the Method of Fluxions found by him Ten Years before or above. Mr. *Newton* never allowed that Mr. *Leibnitz* had the Differential Method before the Year 1677. And Mr. *Leibnitz* himfelf in the *Acta Eruditorum* for *April* 1691. pag. 178. acknowledged that he found it after he returned home from *Paris* to enter upon Bufinefs, that is, after the Year 1676. And as for his pretended general Method of infinite Series, it is fo far from being general, that it is of little or no ufe. I do not know that any other Ufe hath been made of it, than to colour over the Pretence of Mr. *Leibnitz* to the Series of Mr. *Gregory* for fquaring the Circle.

Mr. *Leibnitz*, in his Anfwer to Mr. *Fatio* printed in the *Acta Eruditorum* for the Year 1700. pag. 203. wrote thus. *Ipfe* [Newtonus] *fcit unus omnium optime, fatifque indicavit publice cum fua* Mathematica Naturæ Principia *publicaret, Anno* 1687, *nova quædam inventa Geometrica, quæ ipfi communia mecum fuere,* NEUTRUM LUCI AB ALTERO ACCEPTÆ, *fed meditationibus quemque fuis debere, & a me decennio ante* [i. e. anno 1677] *expofita fuiffe.* In the

M m 3   Book

Book of *Principles* here referred unto, Mr. *Newton* did not acknowledge that Mr. *Leibnitz* found this Method without receiving Light into it from Mr. *Newton's* Letters above-mentioned ; and Dr. *Wallis* had lately told him the contrary without being then confuted or contradicted And if Mr. *Leibnitz* had found the Method without the Affiftance of Mr. *Newton,* yet fecond Inventors have no Right.

Mr. *Leibnitz* in his aforefaid Anfwer to Mr. *Fatio,* wrote further : *Certe cum elementa Calculi mea edidi anno* 1684, *ne conftabat quidem mihi aliud de inventis ejus* [fc. Newtoni] *in hoc genere, quam quod ipfe olim fignificaverat in literis, poffe fe Tangentes invenire non fublatis irrationalibus,* quod Hugenius *quoque fe poffe mihi fignificavit poftea, etfi cæterorum ejus Calculi adhuc expers. Sed majora multo confecutum* Newtonum, *vifo demum libro* Principiorum *ejus, fatis intellexi.* Here he again acknowledged that the Book of *Principles* gave him great Light into Mr. *Newton's* Method: and yet he now denies that this Book contains any thing of that Method in it. Here he pretended that before that Book came abroad he knew nothing more of Mr. *Newton's* Inventions of this kind, than that he had a certain Method of Tangents. and that by that Book he received the firft Light into Mr. *Newton's* Method of Fluxions : but in his Letter of *June* 21. 1677 he acknowledged that Mr. *Newton's* Method extended alfo to Quadratures of curvilinear Figures, and was like his own. His Words are ; *Arbitror quæ celare voluit* Newtonus *de Tangentibus ducendis ab his non abludere. Quod addit, ex hoc eodem fundamento Quadraturas quoque reddi faciliores me in fententia hac confirmat ; nimirum femper figuræ illæ funt quadrabiles quæ funt ad æquationem differentialem.*

Mr. *Newton* had in his three Letters above-mentioned (copies of which Mr. *Leibnitz* had received from Mr. *Oldenbergh*) reprefented his Method fo general, as by the Help of Equations, finite and infinite, to determin *Maxima* and *Minima,* Tangents, Areas, folid Contents, Centers of Gravity,

2   Lengths

(219)

Lengths and Curvities of curve Lines and curvilinear Figures,
and this without taking away Radicals, and to extend to
the like Problems in Curves uſually called Mechanical,
and to inverſe Problems of Tangents and others more difficult, and to almoſt all Problems, except perhaps ſome Numeral ones like thoſe of *Diophantus*. And Mr. *Leibnitz* in
his Letter of *Aug.* 27. 1676, repreſented that he could not
believe that Mr. *Newton*'s Method was ſo general.
Mr. *Newton* in the Firſt of his three Letters ſet down his
Method of Tangents deduced from this general Method,
and illuſtrated it with an Example, and ſaid that this Method of Tangents was but a Branch or Corollary of his General Method, and that he took the Method of Tangents of
Sluſius to be of the ſame kind : and thereupon Mr. *Leibnitz*,
in his Return from *Paris* through *England* and *Holland* into
Germany, was conſidering how to improve the Method of
Tangents of *Sluſius*, and extend it to all ſorts of Problems,
as we ſhewed above out of his Letters. And in his third
Letter Mr. *Newton* illuſtrated his Method with Theorems
for Quadratures and Examples thereof. And when he had
made ſo large an Explanation of his Method, that Mr. *Leib-*
nitz had got Light into it, and had in his Letter of *June* 21.
1677 explained how the Method which he was fallen into
anſwered to the Deſcription which Mr. *Newton* had given of
his Method, in drawing of Tangents giving the Method of
Sluſius, proceeding without taking away Fractions and Surds,
and facilitating Quadratures ; for him to tell the *Germans* that
in the Year 1684, when he firſt publiſhed his Differential
Method, he knew nothing more of Mr. *Newton*'s Invention,
than that he had a certain Method of Tangents, is very extraordinary and wants an Explanation.

At that time he explained nothing more concerning his
own Method, than how to draw Tangents and determin
Maxima and *Minima* without taking away Fractions or Surds.
He certainly knew that Mr. *Newton*'s Method would do all
this

(220)

this, and therefore ought in Candor to have acknowledged
it. After he had thus far explained his own Method, he
added that what he had there laid down were the Principles of a much sublimer Geometry, reaching to the most
difficult and valuable Problems, which were scarce to be resolved without the Differential Calculus, *AUT SIMILI*,
or another like it. What he meant by the Words *AUT
SIMILI* was impossible for the *Germans* to understand
without an Interpreter. He ought to have done Mr. *Newton*
justice in plain intelligible Language, and told the *Germans*
whose was the *Methodus SIMILIS*, and of what Extent
and Antiquity it was, according to the Notices he had received from *England* ; and to have acknowledged that his own
Method was not so ancient. This would have prevented
Disputes, and nothing less than this could fully deserve the
Name of Candor and Justice. But afterwards, in his Answer to Mr. *Fatio*, to tell the *Germans* that in the Year 1684,
when he first published the Elements of his Calculus, he
knew nothing of a *Methodus SIMILIS*, nothing of
any other Method than for drawing Tangents, was very
strange and wants an Explanation.

It lies upon him also to satisfy the World why, in his Answer to Dr. *Wallis* and Mr. *Fatio*, who had published that
Mr. *Newton* was the oldest Inventor of that Method by many
Years, he did not put in his Claim of being the oldest Inventor thereof, but staid till the old Mathematicians were dead,
and then complained of the new Mathematicians as Novices ;
attacked Mr. *Newton* himself, and declined to contend with
any Body else, notwithstanding that Mr. *Newton* in his Letter of *Octob.* 24. 1676 had told him, that for the sake of Quiet, he had Five Years before that time laid aside his Design
of publishing what he had then written on this Subject, and
has ever since industriously avoided all Disputes about Philosophical and Mathematical Subjects, and all Correspondence by Letters about those Matters, as tending to Disputes.;

putes; and for the fame Reafon has forborn to complain of Mr. *Leibnitz*, untill it was fhewed him that he ftood accufed of Plagiary in the *Acta Lipfiæ*, and that what Mr *Keill* had publifhed was only in hisDefence from theGuilt ofthatCrime.

It has been faid the Royal-Society gave judgment againft Mr. *Leibnitz* without hearing both Parties. But this is a Miftake. They have not yet given judgment in the Matter. Mr. *Leibnitz* indeed defired the Royal-Society to condemn Mr. *Keill* without hearing both Parties ; and by the fame fort of Juftice they might have condemned Mr. *Leibnitz* without hearing both Parties; for they have an equal Authority over them both. And when Mr. *Leibnitz* declined to make good his Charge againft Mr. *Keill*, the Royal-Society might in juftice have cenfured him for not making it good. But they only appointed a Committee to fearch out and examin fuch old Letters and Papers as were ftill extant about thefe Matters, and report their Opinion how the Matter ftood according to thofe Letters and Papers. They were not appointed to examin Mr. *Leibnitz* or Mr. *Keill*, but only to report what they found in the ancient Letters and Papers: and he that compares their Report therewith will find it juft. The Committee was numerous and fkilful and compofed of Gentlemen of feveral Nations, and the Society are fatisfied in their Fidelity in examining the Hands and other Circumftances, and in printing what they found in the ancient Letters and Papers fo examined, without adding, omitting or altering any thing in favour of either Party. And the Letters and Papers are by order of the Royal-Society preferved, that they may be confulted and compared with the *Commercium Epiftolicum*, whenever it fhall be defired by Perfons of Note. And in the mean time I take the Liberty to acquaint him, that by taxing the Royal-Society with Injuftice in giving Sentence againft him without hearing both Parties, he has tranfgreffed one of their Statutes which makes it Expulfion to defame them.   N n   The

(222)

The Philofophy which Mr. *Newton* in his *Principles* and *Optiques* has purfued is Experimental ; and it is not the Bufinefs of Experimental Philofophy to teach the Caufes of things any further than they can be proved by Experiments. We are not to fill this Philofophy with Opinions which cannot be proved by Phænomena. In this Philofophy Hypothefes have no place, unlefs as Conjectures or Queftions propofed to be examined by Experiments. For this Reafon Mr. *Newton* in his Optiques diftinguifhed thofe things which were made certain by Experiments from thofe things which remained uncertain, and which he therefore propofed in the End of his Optiques in the Form of Queries. For this Reafon, in the Preface to his *Principles*, when he had mention'd the Motions of the Planets, Comets, Moon and Sea as deduced in this Book from Gravity, he added : *Utinam cætera Naturæ Phænomena ex Principiis Mechanicis eodem argumentandi genere derivare liceret. Nam multa me movent ut nonnihil fufpicer ea omnia ex viribus quibufdam pendere poffe, quibus corporum particulæ per caufas nondum cognitas vel in fe mutuo impelluntur & fecundum figuras regulares cohærent, vel ab invicem fugantur & recedunt : quibus viribus ignotis Philofophi hactenus Naturam fruftra tentarunt.* And in the End of this Book in the fecond Edition, he faid that for want of a fufficient Number of Experiments, he forbore to defcribe the Laws of the Actions of the Spirit or Agent by which this Attraction is performed. And for the fame Reafon he is filent about the Caufe of Gravity, there occurring no Experiments or Phænomena by which he might prove what was the Caufe thereof. And this he hath abundantly declared in his *Principles*, near the Beginning thereof, in thefe Words ; *Virium caufas & fedes Phyficas jam non expendo.* And a little after : *Voces Attractionis, Impulfus, vel Propenfionis cujufcunque in centrum indifferenter & pro fe mutuo promifcue ufurpo, has Vires non Phyfice fed Mathematice tantum confiderando. Unde caveat Lector ne per hujufmodi voces cogitet me*

I   *fpeciem*

312

(223)

*speciemvel modum actionis, caufamve aut rationem phyficam alicubi defi-
nire, vel Centris (quæ funt puncta Mathematica) vires verè & phyficè
tribuere, fi forte aut Centra trahere aut vires Centrorum effe dixero.* And
in the End of his Opticks : *Qua caufa efficiente hæ attractiones* [fc.
gravitas, vifque magnetica & electrica] *peragantur, hic non inquiro.
Quam ego Attractionem appello, fieri fane poteft ut ea efficiatur impulfu
vel alio alique modo nobis incognito. Hanc vocem Attractionis ita hic ac-
cipi velim ut in univerfum folummodo vim aliquam fignificare intelliga-
tur qua corpora ad fe mutuo tendant, cuicunque demum caufæ attribuenda
fit illa vis. Nam ex Phænomenis Naturæ illud nos prius edoctos oportet
quænam corpora fe invicem attrahant, & quænam fint leges & proprietates
iftius attractionis, quam in id inquirere par fit quanam efficiente caufa pe-
ragatur attractio.* And a little after he mentions the fame Attractions as Forces which by Phænomena appear to have a Being in
Nature, tho' their Caufes be not yet known ; and diftinguifhes
them from occult Qualities which are fuppofed to flow from the
fpecifick Forms of things. And in the Scholium at the End of
his *Principles*, after he had mentioned the Properties of Gravity,
he added : *Rationem vero harum Gravitatis proprietatum ex Phænomenis nondum potui deducere, & Hypothefes non fingo. Quicquid enim ex
Phænomenis non deducitur Hypothefis vocanda eft ; & Hypothefes feu Me-
taphyficæ feu Phyficæ, feu Qualitatum occultarum, feu Mechanicæ, in Phi-
lofophia experimentali locum non habent. —— fatis eft quod Gravitas
revera exiftat & agat fecundum leges à nobis expofitas, & ad Corporum
cœleftium & Maris noftri motus omnes fufficiat.* And after all this, one
would wonder that Mr. *Newton* fhould be reflected upon for not
explaining the Caufes of Gravity and other Attractions by Hypothefes ; as if it were a Crime to content himfelf with Certainties and let Uncertainties alone. And yet the Editors of the
Acta Eruditorum, (*a*) have told the World that Mr. *Newton* denies
that the caufe of Gravity is Mechanical, and that if the Spirit or
Agent by which Electrical Attraction is performed, be not the
Ether or *fubtile Matter* of *Cartes,* it is lefs valuable than an Hypothefis, and perhaps may be the Hylarchic Principle of Dr. *Henry Moor* :
and Mr. *Leibnitz* (*b*) hath accufed him of making Gravity a
natural or effential Property of Bodies, and an occult Quality
and Miracle. And by this fort of Railery they are perfwa-
ding the *Germans* that Mr. *Newton* wants Judgment, and was not
able to invent the Infinitefimal Method.

(*a*) *Anno* 1714; *menfe Martio,* p. 141. 142. (*b*) *In tractatu de Bonitate Dei
& in Epiftolis ad D. Hartfoeker & alibi.* Ttt

(224)

It muft be allowed that thefe two Gentlemen differ very much in Philofophy. The one proceeds upon the Evidence arifing from Experiments and Phænomena, and ftops where fuch Evidence is wanting; the other is taken up with Hypothefes, and propounds them, not to be examined by Experiments, but to be believed without Examination. The one for want of Experiments to decide the Queftion, doth not affirm whether the Caufe of Gravity be Mechanical or not Mechanical : the other that it is a perpetual Miracle if it be not Mechanical. The one (by way of Enquiry) attributes it to the Power of the Creator that the leaft Particles of Matter are hard : the other attributes the Hardnefs of Matter to confpiring Motions, and calls it a perpetual Miracle if the Caufe of this Hardnefs be other than Mechanical. The one doth not affirm that animal Motion in Man is purely mechanical : the other teaches that it is purely mechanical, the Soul or Mind (according to the Hypothefis of an *Harmonia Præftabilita*) never acting upon the Body fo as to alter or influence its Motions. The one teaches that God (the God in whom we live and move and have our Being) is Omniprefent; but not as a Soul of the World : the other that he is not the Soul of the World, but *INTELLIGENTIA SUPRA MUNDANA,* an Intelligence above the Bounds of the World; whence it feems to follow that he cannot do any thing within the Bounds of the World, unlefs by an incredible Miracle. The one teaches that Philofophers are to argue from Phænomena and *Experiments* to the Caufes thereof, and thence to the Caufes of thofe Caufes, and fo on till we come to the firft Caufe : the other that all the Actions of the firft Caufe are Miracles, and all the Laws impreft on Nature by the Will of God are perpetual Miracles and occult Qualities, and therefore not to be confidered in Philofophy. But muft the conftant and univerfal Laws of Nature, if derived from the Power of God or the Action of a Caufe not yet known to us, be called Miracles and occult Qualities, that is to fay, *Wonders* and *Abfurdities?* Muft all the Arguments for a God taken from the Phænomena of Nature be exploded by *new hard Names?* And muft Experimental Philofophy be exploded as *miraculous* and *abfurd,* becaufe it afferts nothing more than can be proved by Experiments, and we cannot yet prove by Experiments that all the Phænomena in Nature can be folved by meer Mechanical Caufes? Certainly thefe things deferve to be better confidered.

ERRATA.   *Pag.* 199. *line.* 14. *put an* Afterisk (*) *after the Word* Letter.

NOTES

SHORT TITLES

Aiton, *Vortex Theory*

E. J. Aiton, *The Vortex Theory of Planetary Motions* (London and New York: Macdonald and American Elsevier, respectively, 1972).

Johann Bernoulli, *Briefwechsel*

O. Spiess, *Der Briefwechsel 'von Johann Bernoulli* (Basel: Birkhäuser Verlag, 1955).

Boyer, *History*

Carl B. Boyer, *The History of the Calculus and Its Conceptual Development* (New York: Dover, 1949).

Brewster, *Memoirs*

Sir David Brewster, *Memoirs of the Life, Writings and Discoveries of Isaac Newton* (Edinburgh and London: Thomas Constable and Hamilton, Adams, respectively, 1855).

Cohen, *Introduction*

I. Bernard Cohen, *Introduction to Newton's "Principia"* (Cambridge: Cambridge University Press, 1971).

Cohen, Papers and Letters

I. Bernard Cohen et al., *Isaac Newton's Papers and Letters on Natural Philosophy* (Cambridge, Mass.: Harvard University Press, 1958).

Commercium Epistolicum

[Isaac Newton] *Commercium Epistolicum D. Johannis Collins, et aliorum de Analysi promota* (London: Pearson, 1712; 2nd ed., London: J. Tonson and J. Watts, 1722).

Des Maizeaux, *Recueil*

Pierre Des Maizeaux, *Recueil de diverses Pièces sur la Philosophie*, 2 vols. (Amsterdam: For H. du Sauzet, 1720).

D.N.B.

The Dictionary of National Biography, ed. Sir Leslie Stephen and Sir Sidney Lee (1885–1901; reprint ed., Oxford: Oxford University Press, 1967–8).

D.S.B.

Dictionary of Scientific Biography, ed. C. C. Gillispie (New York: Charles Scribner's Sons, 1970–   ).

Edleston, *Correspondence of Newton and Cotes* — Joseph Edleston, *Correspondence of Sir Isaac Newton and Professor Cotes* (London and Cambridge: John W. Parker and John Deighton, respectively, 1850).

Gerhardt, *Briefwechsel* — C. I. Gerhardt, *Der Briefwechsel von G. W. Leibniz mit Mathematikern* (Berlin: Mayer & Müller, 1899, all published).

Gerhardt, *Math. Schriften* — C. I. Gerhardt, *G. W. Leibniz Mathematische Schriften*, 7 vols. (Halle, 1849–63; reprint ed., Hildesheim: Georg Olms, 1962).

Gerhardt, *Phil. Schriften* — C. I. Gerhardt, *Die Philosophischen Schriften von G. W. Leibniz*, 7 vols. (Berlin, 1875–90; reprint ed., Hildesheim: Georg Olms, 1960).

Hofmann, *Paris* — J. E. Hofmann, *Leibniz in Paris, 1672–76* (Munich: R. Oldenbourg, 1949; English translation, Cambridge: Cambridge University Press, 1974).

Huygens, *Oeuvres* — *Oeuvres Complètes de Christiaan Huygens publiées par la Société Hollandaise des Sciences*, 22 vols. (The Hague: Martinus Nijhoff, 1888–1950).

Leibniz, *Correspondence* — J. E. Hofmann, ed., *Gottfried Wilhelm Leibniz Sämtliche Schriften und Briefe*, III Reihe, I Band (Berlin: Akademie Verlag, 1976).

Manuel, *Portrait* — Frank Manuel, *A Portrait of Isaac Newton* (Cambridge, Mass.: Harvard University Press, 1968).

More, *Newton* — Louis Trenchard More, *Isaac Newton, a Biography* (New York and London: Charles Scribner's Sons, 1934).

Newton, *Correspondence* — H. W. Turnbull, J. F. Scott, A. Rupert Hall, and Laura Tilling, eds., *The Correspondence of Isaac Newton*, 7 vols. (Cambridge: Cambridge University Press, 1959–77).

Oldenburg, *Correspondence* — A. Rupert Hall and Marie Boas Hall, eds., *The Correspondence of Henry Oldenburg*, 11 vols. (Vols. I–IX, Madison: University of Wisconsin Press, 1965–73; X– (in progress), London: Mansell, 1975–).

Principia, 1687 — *Philosophiae Naturalis Principia Mathema-*

tica Autore Is. Newton. (London: Jussu Societatis Regiae ac Typis Josephi Streater, 1687.) I have used the facsimile issued by William Dawson, London [1953].

Principia, 1713   2nd ed., Roger Cotes, ed. (Cambridge: Cambridge University Press, 1713.

Principia, 1726   3rd ed., Henry Pemberton, ed. (London: William Bowyer, 1726).

Rigaud, *Correspondence*   S. P. and S. J. Rigaud, *Correspondence of Scientific Men of the Seventeenth Century*, 2 vols. (Oxford: Oxford University Press, 1841).

Turnbull, *Gregory*   H. W. Turnbull, *James Gregory Tercentenary Memorial Volume* (Edinburgh: G. Bell, 1939).

Wallis, *Works*   John Wallis, *Opera Mathematica*, 3 vols. (Oxford: Oxford University Press, 1695, 1693, 1699).

Whiteside, *Math. Papers*   D. T. Whiteside, *The Mathematical Papers of Isaac Newton*, 8 vols. (Cambridge: Cambridge University Press, 1967–　).

Whiteside, *Math. Works*   D. T. Whiteside, *The Mathematical Works of Isaac Newton*, 2 vols., (New York and London: Johnson Reprint, I: 1964; 2: 1967).

1. INTRODUCTION

1   Hofmann's writings on Leibniz as a mathematician and on the priority dispute are numerous; most useful for English readers is *Leibniz in Paris, 1672–76* (Cambridge, 1974). James Gregory's published writing influenced Leibniz rather than Newton, and his unpublished work splendidly demonstrates convergence with that of his two more celebrated contemporaries; see H. W. Turnbull, *James Gregory Tercentenary Memorial Volume* (London: George Bell, 1939). There is no complete modern edition of either the published or unpublished mathematical writings of Leibniz, hence C. I. Gerhardt's *Mathematische Schriften* (Halle, East Germany, 1849–63; reprinted, 1962) are still invaluable, as are his editions of Leibniz's correspondence save where these have been replaced in the *Sämtliche Schriften und Briefe*

of the Akademie der Wissenschaften der DDR. Newton's previously unpublished mathematical papers will all appear in D. T. Whiteside, *The Mathematical Papers of Isaac Newton* (Cambridge, 1967–), 8 vols.

2 See the lively J. M. Levine, *Dr Woodward's Shield* (Berkeley, Calif.: University of California Press, 1977).

3 Literally: "If I have seen further it is by standing on ye sholders of Giants". Newton to Hooke, 5 February 1676, *Correspondence* I: 416; Robert K. Merton, *On the Shoulders of Giants* (New York: Free Press, 1965).

4 See Newton's letters to Halley from 27 May 1686 onward (Newton, *Correspondence* II:433ff.).

5 Hofmann, *Paris*, pp. 74–5.

2. BEGINNINGS IN CAMBRIDGE

1 Pierre Des Maizeaux, *Recueil de diverses Pièces sur la Philosophie* (Amsterdam, 1720). Vol. I contains the correspondence between Leibniz and Newton's friend (and parish priest) Samuel Clarke on matters of Newtonian philosophy, which had been previously published in 1717, Vol. II, the letters relating to mathematics exchanged between Newton and Leibniz and various other correspondents. Voltaire, *Letters concerning the English Nation*, Letter XIV.

2 Newton, *Correspondence* VI:454–7; Cohen, *Introduction*, p. 291; Whiteside, *Math. Papers* I:7–8, 10.

3 Whiteside, *Math. Papers* I:112, 146, 369–70, 382, 383, 386.

4 Ibid., pp. 145, 152–3.

5 Newton, *Correspondence* I:15, 26, 32–3, 36–7 and *passim*; II:114, 212ff.; A. Rupert Hall, "Newton's First Book," *Arch. Int. d'Hist. des Sciences* 13 (1960): 39–61.

6 Whiteside, *Math. Papers* III:20–1, draft Preface to the *Commercium Epistolicum*. *On Analysis* appears in Vol. II of this edition; *On the Methods of Series and Fluxions* in Vol. III. I have relied heavily on Whiteside's rich commentary and annotation.

7 "The 'Account,'" *Phil. Trans.* 29, No. 342 (1715): 204–5; Whiteside, *Math. Papers* II: 165–7; Newton, *Correspondence* II: 111, 114. P. Kitcher, "Fluxions, Limits and Infinite Littlenesse," *Isis* 64 (1973): 33–49.

8 Whiteside, *Math. Papers* III:33, 73; Cohen, *Introduction*, p. 291.

9 Cohen, *Introduction*, p. 291; Newton, *Correspondence* II:114. He probably means that only haste persuaded him to agree readily to the publication of his famous optical letter of 1672, which entailed so much controversy.

3. NEWTON STATES HIS CLAIM:
1685

1 There is no sign of any interest on Newton's part in the dramatic events that followed soon after the accession of James II (February 1685), when he was deep in the first draft of the *Principia*; but it is curious that Newton was involved in the resistance to the Catholicization of Cambridge by James II in the spring of 1687 (Macaulay, *History of England*, Chap. VIII; Brewster, *Memoirs* II:104–10) at about the time Book III of the *Principia* was sent to London. Still later, as is well known, Edmond Halley used the *Principia* in an attempt to ingratiate himself with the already deeply unpopular monarch (Cohen, *Papers and Letters*, pp. 403–4).

2 I follow the familiar story recorded by Conduitt; see Brewster, *Memoirs* I:296–7; Cohen, *Introduction*, pp. 47–52 and Newton, *Correspondence* II: Letter 289. On 20 June 1686 Newton told Halley that Book II had been "finished last summer being short"; Book I was already being printed in London. However, Newton again refers to Book II in February 1687 as being "made ready for you [Halley] last Autumn [1686], having wrote to you in Summer that it should come out with ye first & be ready against ye time you might need it . . ." This indicates further work on Book II in summer/autumn 1686. A. Rupert Hall, "Newton and his editors," *Proc. R. Soc. London A*, 338 (1974): 397–417.

3 Newton, *Correspondence* II: Letter 278. Newton's early printed papers reproduced in Cohen, *Papers and Letters*; unpublished mathematical writings in Whiteside, *Math. Papers*, I–IV.

4 Brewster, *Memoirs* II:91ff. Cf. Manuel, *Portrait*.

5 D. T. Whiteside, "The Mathematical Principles underlying Newton's *Principia Mathematica*," *Jour. Hist. Ast.* 1 (1970): 116–38; quotation p. 133, note 17.

6 Ibid. and Whiteside, *Math. Papers* VI; Cohen, *Introduction*, pp. 236–8; Hall, "Newton and his editors," 411.

7 Whiteside, *Math. Papers* IV:522, note 1; VI:25.

8 Whiteside, *Jour. Hist. Ast.* 1 (1970): 119; *Math. Papers* IV:410–13, 521–5; Boyer, *History*, pp. 198–201.

9 *Principia*, 1687, pp. 253–4.

10 G. W. Leibniz, "Nova methodus pro maximis et minimis," *Acta Eruditorum*, October 1684, 467–73; Whiteside, *Math. Papers* IV:411–12.

11 Whiteside, *Math. Papers* IV: 412; VII: 3–13.

12 Newton, *Correspondence* II: 396; Whiteside, *Math. Papers* IV: 413–16.

13 Whiteside, *Math. Papers* IV:526ff.

14 Ibid., VII:3–20.

15 Ibid., IV:673 (applying Whiteside's words more generally).

4. LEIBNIZ ENCOUNTERS
NEWTON: 1672–1676

1 W. H. Barber, *Leibniz in France: From Arnauld to Voltaire* (Oxford: Clarendon Press, 1955), p. ix.
2 Arthur Lovejoy, *The Great Chain of Being* (Cambridge: Harvard University Press, 1936); Manuel, *Portrait*; Lord Keynes, "Newton: The Man," *Royal Society Tercentenary Celebrations* (Cambridge: Cambridge University Press, 1947); B. J. T. Dobbs, *The Foundations of Newton's Alchemy* (Cambridge: Cambridge University Press, 1976).
3 Barber, loc. cit., pp. 27–8.
4 R. S. Westfall, *Force in Newton's Physics*, London and New York: Macdonald and American Elsevier, respectively, 1971; Aiton, *Vortex Theory*.
5 Oldenburg, *Correspondence* VII:112.
6 Newton, *Correspondence* IV: Letter 544. To be exact, Newton denied in the letter that he was to be Comptroller of the Mint; five days afterward the office of Warden of the Mint was offered to him and speedily accepted. The warden's chief responsibilities for catching and prosecuting currency offenders were executed with his customary efficiency.
7 A. E. Burtt, *Metaphysical Foundations of Modern Physical Science*, (reprint ed., London: Routledge & Kegan Paul, 1949); I. Bernard Cohen, *Franklin and Newton*, Philadelphia: American Philosophical Society, 1956.
8 Oldenburg, *Correspondence* VII:66–7 (13 July 1670). Leibniz was sent to Paris in 1672 to pursue a scheme devised by himself whereby Louis XIV was to be persuaded to seek, like St. Louis, glory in defeating the Turks, thereby removing his forces from the Rhineland and the Turks from Austro-Hungary.
9 Ibid., IX: Letters 2196, 2208.
10 Ibid., VIII:xxiii–xxiv, Letter 1799.
11 See my paper "Huygens and Newton" in *The Anglo Dutch Contribution to the Civilization of Early Modern Society* (London: British Academy, 1976). A much more elaborate study of this theme should be made.
12 The test series is formed from the reciprocals of the triangular numbers, which are (in turn) the numbers formed by adding to each preceding number (starting from 1) the next in the series of ordinal numbers, starting with 2, thus 1, $(1 + 2) = 3$, $(3 + 3) = 6$, $(6 + 4) = 10$, etc. See Hofmann, *Paris*, Chap. 2.
13 Oldenburg, *Correspondence* XI: Letter 2511 (15 July 1674); Huygens, *Oeuvres*, VII: Letter 1999 (= Leibniz, *Correspondence*: Letters 30, 40).
14 Oldenburg, *Correspondence* XI: Letter 2550.

15 Oldenburg, *Correspondence* IX: Letter 2140.
16 Oldenburg, *Correspondence* XI: Letters 2511, 2550.
17 Ibid., Letter 2576.
18 Ibid., Letters 2633, 2641, 2642.
19 Ibid., Letter 2660; Newton, *Correspondence*, VI: Letters 1005, 1009.
20 Oldenburg, *Correspondence* XI: Letters 2697, 2754; Hofmann, *Paris*, pp. 184–6.
21 Leibniz, *Correspondence* (18/28 December 1675), pp. 327–34; Oldenburg, *Correspondence* (XII, in press), Letter 2804. Compare Leibniz to Oldenburg, 17 August 1676, Leibniz, *Correspondence*, p. 583 Newton, *Correspondence* II: 63.
22 Oldenburg, *Correspondence* XI: Letter 2642 (2 May 1676); Newton, *Correspondence* II; Letter 158.
23 Rigaud, *Correspondence* II:21–2, 280.
24 Hofmann, *Paris*, pp. 214–5, 225–6; Rigaud, *Correspondence* II:453–4; Leibniz, *Correspondence*, pp. 434–84 (The "Historiola"), pp. 504–16 (the "Abridgement"), pp. 518–54 (Oldenburg to Leibniz, 26 July 1676, containing the content of the "Abridgement" and Newton's *First Letter*).
25 Newton, *Correspondence* II:20; Hofmann, *Paris*, p. 230; Whiteside, *Math. Papers* VII:15–16.
26 Newton, *Correspondence* II: Letter 188; Hofmann, *Paris*, p. 260; Whiteside, *Math. Papers* IV:637.
27 Hofmann, *Paris*, Chap. 19; J. O. Fleckenstein, *Der Prioritätsstreit zwischen Leibniz und Newton* (Elemente der Mathematik, Beiheft Nr. 12, 1956), p. 23.

5. THE EMERGENCE OF
THE CALCULUS: 1677–1699

1 Gerhardt, *Briefwechsel*, Letter 49; Newton, *Correspondence*, Letter 209 (21 June 1677 – old style).
2 Newton, *Correspondence* II:227, note 10; 234, note 2; *Commercium Epistolicum* (1712), pp. 89, 96.
3 Hofmann, *Paris*, pp. 279–87; Whiteside, *Math. Papers* II:170, 248–59; Leibniz, *Correspondence*, pp. 485–503, 170, 664–81.
4 Gerhardt, *Briefwechsel*, pp. 249–52, 253–5; Newton, *Correspondence*, II:237. A letter inviting continued correspondence with the Royal Society was sent to Leibniz in the spring of 1678.
5 Newton, *Correspondence* II:241–4.
6 "Extrait d'une Lettre de M. Leibniz . . . ," 1678, in Gerhardt, *Math. Schriften* V:116–17; "De vera proportione circuli . . . ," 1682, ibid., pp. 118–22; Boyer, *History*, pp. 207–8; "Nova methodus pro maximis et minimis . . . ," 1684, Gerhardt, *Math. Schriften* V:220–6;

"De geometria recondita et analysi indivisibilium et infinitorum," 1686, ibid., pp. 226–33. The 1682 paper was published in English translation in Robert Hooke's *Philosophical Collections*, no. 7 (April 1682): 204–10.

7 J. Craig, *Tractatus mathematicus de figurarum curvilinearum quadraturis* (London, 1693), p. 1; *Methodus figurarum . . . quadraturas determinandi* (London, 1685), p. 27. Writing of the mathematical preparation requisite for studying the *Principia*, Craige told William Wotton to study for the method of tangents Descartes, Sluse, Barrow "and Mr Leibnitz his Method, which is the best of all." For indivisibles, he cites Cavalieri and Barrow's *Geometrical Lectures*; for quadratures, Wallis and David Gregory. Nothing of Newton's was yet available (24 June 1691), and Craige obviously does not think of the differential calculus as appropriate under indivisibles or appropriate to mention at all (except under tangents).

8 *Phil. Trans.* 16, no. 183: 185, 189; J. Craig, *De calculo fluentium libri duo* (London, 1718), Preface; Newton, *Correspondence* II:417, 452, 501. Newton took a different view from Craige and emphasized (about this same time in an abortive draft) that he had in 1685 explained Leibniz's 1684 paper to Craige, and "told him that the method was mine in a new dress . . . Mr Craige is still alive & remembers this" (Whiteside, *Math. Papers* VIII:2, Introduction). It may have been so, but such was not Craige's report.

9 Gerhardt, *Phil. Schriften* III:179; Johann Bernoulli, *Briefwechsel*, 97; Huygens, *Oeuvres* IX:224–9.

10 Gerhardt, *Math. Schriften* II:216–8, 250. André Robinet in *Rev. Hist. Sci.* 12 (1959): 1–18; Malebranche, *Oeuvres*, XVII–2, P. Costabel, ed. (Paris: J. Vrin, 1968); Suzanne Delorme in *D.S.B*, s.v. "Fontenelle" (V, 60, col. 2).

11 Gerhardt, *Math. Schriften* IV:91, 100; Boyer, *History*, p. 209.

12 Huygens, *Oeuvres* IX:448–52, 470–3, 496–9, 517–19, 536–40; X: 129–34, 221–5.

13 Huygens, *Oeuvres* X:609–15.

14 Ibid., pp. 383–5.

15 Wallis, *Algebra* (1685), pp. 346–7; *Opera* II: Preface.

16 Wallis, *Opera* II:390–6; Whiteside, *Math. Papers* VII:170–82; compare the English version in Newton, *Correspondence* III:222–8.

17 Gerhardt, *Phil. Schriften* III:171 note (22 November 1695); *Math. Schriften* IV:6–7.

18 Newton, *Correspondence* III: Letter 392; Wallis, *Opera* II:396.

19 Whiteside, *Math. Papers* VII:xxii–iii.

20 Gerhardt, *Math. Schriften* V:231–2.

21 Gerhardt, *Phil. Schriften* III:186, 189–90, 199, 222, 316.

22 Gerhardt, *Math. Schriften* IV:70–2.

23 Huygens, *Oeuvres* XXII:125–51; correspondence in Vol. IX; work in common in Vol. XX.
24 Newton, *Correspondence* III: 45, 231, 245, 263, 279–80, 284, 308; Manuel, *Portrait*, pp. 191–205.
25 Newton, *Correspondence* IV: Letter 561. In his account of the affair Johann Bernoulli related that the solution of his brachistochrone problem in the *Philosophical Transactions* (no. 224 (1697), p. 384) readily identified itself as coming from Newton's pen: "ce serait assez de le connoître par cet échantillon, comme *ex ungue Leonem*." This learned man, he wrote, thoroughly deserved all the praise due to those who should solve the problem, small though this was compared with the praise due to Newton's incomparable work "in which he displays so much intellectual depth and penetration." (H. Basnage de Beauval, *Histoire des Ouvrages des Savans*, June–August 1697, pp. 454–5.)
26 Gerhardt, *Math Schriften* V:331–6.
27 More, *Newton*, p. 570.
28 *Lineae brevissimi descensus investigatio geometrica duplex*, London 1699, p. 18; Newton, *Correspondence* V:98; More, *Newton*, p. 575.
29 Huygens, *Oeuvres* X:213–5, 257–9.
30 Newton, *Correspondence* III: 257–8.

6. THE OUTBREAK: 1693–1700

1 Newton, *Correspondence* III: Letter 464; Huygens, *Oeuvres* X:257–9; ibid., Letter 383.
2 Newton, *Correspondence* III: Letter 376; Whiteside, *Math. Papers* VII:8–9, 21–3; Newton, *Correspondence* III: Letter 427 (16 October 1693).
3 Newton, *Correspondence* IV: Letter 498 (10 April 1695); Wallis, *Opera* I (Oxford 1695): Preface.
4 Newton, *Correspondence* IV:24, note 5; Letter 567 (1 July 1697, quoting Leibniz to Wallis, 28 May 1697).
5 Gerhardt, *Math. Schriften* III/2:529, 549 and passim.
6 Whiteside, *Math. Papers* VI:305–7, note 126.
7 Gerhardt, *Briefwechsel*, 730 (19 May 1694), 745 (14 August 1694), 746 (14 September 1694); Huygens, *Oeuvres* X: 610, 669, 675; Whiteside, *Math. Papers* VII:180–1, note 26.
8 Gerhardt, *Math. Schriften* III/1:301, 312, 316–7.
9 Ibid., pp. 320, 332ff.
10 Ibid., III/2:596–621.
11 Bernoulli to Leibniz, 7 August 1699, Gerhardt, *Math. Schriften* III/

2:602–9, printed first in *Acta Eruditorum*, August and November 1699: 354–9, 510–16.

12 *Acta Eruditorum*, May 1700, 198–202.
13 Ibid., pp. 203, 206.
14 "The 'Account,'" *Phil. Trans.* 29, no. 342 (January–February 1715): 201.

7. OPEN WARFARE: 1700–1710

1 "The 'Account,'" *Phil. Trans.* 29, no. 342 (January–February 1715): 200.
2 *Fluxionum methodus inversa, sive quantitatum fluentium leges generaliores* (London, 1703); Gerhardt, *Math. Schriften* III/2:723–4 (29 September 1703).
3 Gerhardt, *Math. Schriften* III/2:724, 766.
4 Ibid., pp. 727, 744.
5 W. G. Hiscock, ed., *David Gregory, Isaac Newton and their Circle. Extracts from David Gregory's Memoranda 1677–1708.* (Oxford: Oxford University Press, 1937), pp. 43–5.
6 Ibid., pp. 13–17.
7 Whiteside, *Math. Papers* VII:20.
8 Ibid., pp. 566, 578.
9 Gerhardt, *Math. Schriften* III/2:760, 762, 771.
10 *Acta Eruditorum*, January 1705, 30–6.
11 "The 'Account,'" *Phil. Trans.* 29, no. 342 (January–February 1715): 202; *Commercium Epistolicum*, 1712, p. 108, note.
12 *D.S.B.* IV:506 (col. 1): E. A. Fellman, "Honoré Fabri."
13 Gerhardt, *Math. Schriften* III/2:766.
14 Ibid., pp. 825, 822 (15 March 1708), Whiteside, *Math. Papers* V:xiv, 16–17, 24, 30.
15 John Keill, *Epistola . . . de Legibus Virium Centripetarum, Phil. Trans.* 26 (1708): 185.

8. THE PHILOSOPHICAL DEBATE

1 Huygens, *Oeuvres* X:316–8 (26 September 1692).
2 Leibniz, *Hypothesis physica nova* (Mainz, 1671); Gerhardt, *Math. Schriften* VI:23–6.
3 Newton, *Correspondence* III:3–5, from *Acta Eruditorum*, 1689; "Tentamen de motuum coelestium causis," *Acta Eruditorum*, February 1689; Gerhardt, *Math. Schriften* VI: 144–61; Leibniz read the *Princi-*

pia in Rome in the summer of 1689 (see E. A. Fellmann, ed., *G. W. Leibniz: Marginalia in Newtoni Principia Mathematica* [Paris, 1973]).

4 Huygens, *Oeuvres* X:297, 316, 383; Aiton, *Vortex Theory*, Chap. VI; for a more critical view of Leibniz's treatment of orbital motion, see R. S. Westfall, *Force in Newton's Physics*, Chap. 4, note 4; pp. 308–10.

5 Huygens, *Oeuvres* IX: 523, 524, 538; X:260.

6 L. Couturat, *Opuscules et fragments inédits de Leibniz* (Paris 1903, Hildesheim: Georg Olms, 1961), pp. 11–12.

7 *Phil. Trans.* 133 (1843): 108. I owe this quotation to Mr. V. M. D. Hall. Mary B. Hesse, *Forces and Fields* (London: Nelson, 1961), pp. 157ff.

8 Pierre Costabel, *Leibniz et la Dynamique* (Paris: Hermann, 1960), p. 55; Gerhardt, *Phil. Schriften* VII:338–9.

9 Richard Bentley, *A Confutation of Atheism* (London 1693), pp. 28–9, 31–2. W. G. Hiscock, *David Gregory, Isaac Newton and Their Circle* (Oxford: Oxford University Press, 1937), p. 30 (21 December 1705). Gerhardt, *Phil. Schriften* III:228, 260. Hesse, op. cit., pp. 166–7. Gerhardt, *Phil. Schriften* VI:61–2.

10 Gerhardt, *Phil. Schriften* III:261–2.

11 *Acta Eruditorum* 22 (November 1703): 504–10. The English version of Keill's book was published under the prosaic title: *An Introduction to Natural Philosophy* (1720).

12 D. Gregory, *Astronomiae physicae, et geometricae elementa* (Oxford 1702). Gerhardt, *Math. Schriften* IV:98 (14 April 1702); III/2:713, 715, 739, 760 (25 January 1705), 762. D. Gregory, *The Elements of Astronomy, physical and geometrical* (London 1715), pp. 172–82. *Acta Eruditorum* 22 (October 1703): 452–62. In Leibniz's "layer" vortex the linear velocities of the layers or deferents of ether bearing a single planet are inversely proportional to the distance of the layer from the sun; but the linear velocities of different layers bearing two different planets are inversely proportional to the square roots of their distances from the sun, so that the Leibnizian vortex is doubly discontinuous.

13 John Freind, *Chymical Lectures, in which almost all the operations of Chymistry are reduced to their true Principles, and the Laws of Nature* (London 1712), pp. 82–4; the first edition (1709) was in Latin. J. Keill, "The Laws of Attraction," in *Phil. Trans.* 26, no. 315 (1708). A. Rupert Hall, *From Galileo to Newton* (London: Collins, 1963), pp. 324–6.

14 *Acta Eruditorum* 29 (September 1710): 412–13; (October 1710): 455–9; 30 (May 1711): 221.

15 Emil Ravier, *Bibliographie des Oeuvres de Leibniz* (Paris, 1937; Hildesheim: Georg Olms, 1966), p. 106, no. 244; pp. 90–1. Gerhardt, *Math. Schriften* IV:398, 403, 409; *Phil. Schriften* III:328.

9. THRUST AND PARRY:
1710–1713

1 Newton, *Correspondence* V:96–7.
2 Ibid., pp. 115–17.
3 Ibid., pp. 132– 49. Sloane to Leibniz and Keill to Sloane, May 1711. Note that on p. 134 Keill improperly makes $\dot{x}$ (rather than $o\dot{x}$), the equivalent of dx.
4 Ibid., II:20 –41; 110 –49.
5 Ibid., V:207–8.
6 Ibid., V:xxiv–v, 212–14.
7 Ibid., V:xxv–vii; Whiteside, *Math. Papers* III:20; Newton, *Correspondence* VI: 455 (draft for Des Maizeaux) quoting correctly *Commercium Epistolicum* (1712), p. 122.
8 Newton, *Correspondence* V:144; Hofmann, *Paris*, pp. 280, 284.
9 *Acta Eruditorum* (February 1712), 76 –7; Newton, *Correspondence* V:383.
10 Newton, *Correspondence* V:298–300; A. De Morgan, *Essays on the Life and Work of Newton* (London, 1914), pp. 70 –5; *Commercium Epistolicum* (1712), pp. 47, 121; Leibniz, *Correspondence*, 434, 467–9, 495, 507, 523; Newton, *Correspondence* I:247–8; II:18, 53; Newton, *Principia* (1713), Preface, *ci verso*.
11 Gerhardt, *Math. Schriften* IV:194, 195 (28 June and 9 August 1713).
12 *Commercium Epistolicum* (1712), pp. 1–2; Newton, *Correspondence* I:13–14; P. Kitcher, "Fluxions, Limits and Infinite Littlenesses," *Isis*, 64 (1973): 38, note 17; Newton, *Correspondence* II:3–4 (2 May 1676). C. J. Scriba in *Archive for History of Exact Sciences*, 2 (1964): 118.
13 *Commercium Epistolicum* (1712), pp. 1, 27, 42–3, 88, 96, 98, 100; Hofmann, *Paris*, pp. 59, 73. Many refutations of the *Commercium Epistolicum* are contained passim in this book. Leibniz, *Correspondence*, 235; Newton, *Correspondence* I:340; the arc-tangent series of Gregory (1671) includes as a special case Leibniz's arithmetic quadrature sent by him to London in August 1676.
14 Turnbull, *Gregory*, p. 170; Newton, *Correspondence* VI:7. The series $\frac{\pi}{4} = 1 - \frac{1}{3} + \frac{1}{5} - \frac{1}{7}\ldots$ has been found in the work of Nilakantha, a fifteenth-century Hindu mathematician. *Commercium Epistolicum* (1712), pp. 62, 65; Newton, *Correspondence* II; 64. The points noted were all sharpened and elaborated in the revised edition of 1722.
15 Newton, *Correspondence* VI: 217 (24 April 1715); 222.
16 Whiteside, *Math. Papers* VI passim; Gerhardt, *Math. Schriften* III/2:854–5; *Mémoires de l'Académie Royale des Sciences*, 1711 (Paris, 1714), pp. 47–54. Newton, *Correspondence* V:349 (18 October 1712).
17 D. T. Whiteside, "The Mathematical Principles underlying Newton's *Principia Mathematica*," *Jour. Hist. Ast.* 1 (1970): 128–9. Newton, *Correspondence* V:xxxiv–vii, 347–50.

18 Whiteside, *Math. Papers* VI:548–9, note 25; 147–9, note 124; 348–51, note 209; 358–9, note 198. An equivalent theorem was first printed by Varignon in 1704.
19 *Mémoires de l'Académie Royale des Sciences*, 1711 (Paris, 1714), pp. 54–6. *De quadratura curvarum* (ed. 1706), p. 39; in *Opuscula* (Lausanne and Geneva, 1744), I:241–2; Whiteside, *Math. Works* I:158. The scholium is not to be found in the very different *On Quadrature* of 1691 (Whiteside, *Math. Papers VII*).
20 The series of the Bernoullis' critical articles begins with Johann's "Letter to Hermann" of 7 October 1710, *Mémoires de l'Académie Royale des Sciences*, 1710 (Paris, 1714), pp. 519–33, and continues with the articles cited above (notes 16 and 19) and those in *Acta Eruditorum* for February and March 1713, pp. 77–97, 115–32. J. Bernoulli to Leibniz, 27 May 1713, Newton, *Correspondence* VI:1–5; Whiteside, "Mathematical Principles," op. cit., p. 130.
21 Newton, *Correspondence* VI:7–8, 15–17.
22 Hofmann, *Paris*, Chap. 2.

10. THE DOGS OF WAR:
1713–1715

1 To Varignon, January 1721 (draft), *Correspondence* VII:123; VI:62–3, 71, 79–80.
2 *Correspondence* VI:79–80. *Journal Literaire de la Haye*, November–December 1713, pp. 444–53.
3 *Correspondence* VI:30–32.
4 Ibid., pp. 89, 81–82.
5 Whiteside, *Math. Papers* II:264.
6 Ibid., p. 85. It is perhaps just worth noting here that Newton primed Keill with a very brief summary of this abortive paper in a letter of 20 April 1714.
7 "Réponse de M. Keill . . . aux Auteurs des Remarques sur le Différent entre M. de Leibnitz et M. Newton," *Journal Literaire de la Haye*, July–August 1714, pp. 319–58. *Correspondence* VI:113–4, 129.
8 D. T. Whiteside, private communication; Edleston, *Correspondence of Newton and Cotes*, pp. 307–14; *Commercium Epistolicum* (1712), p. 97.
9 Keill, loc. cit., pp. 351–2; Aiton, *Vortex Theory*, p. 138, writes that Newton's criticism "that Leibniz made errors involving second-order small quantities in almost every proposition" is in one case based on a misconception, "and in all the others is demonstrably false"; see also pp. 144–5; *Acta Eruditorum*, January 1689, pp. 36–47

("De lineis opticis" and "Schediasma de Resistentia Medii"); February 1689, pp. 82–96 ("Tentamen de motuum coelestium causis"). Newton, *Correspondence* II:64.

10 Newton, *Correspondence* VI:151–2, 180.

11 Ibid., pp. 37; 38, note 2; 153–4 (24 July 1714). Whiteside, *Math. Papers* VI:146–8 provides some justification of Keill's indignant outburst that Bernoulli had really added nothing to the *Principia*.

12 Newton, *Correspondence* VI:131–2 (12 May 1714); 173 (14 August 1714).

13 Ibid., pp. 159, 181–2. Keill's "Observations" were not read by Bernoulli before the summer of 1715.

14 Ibid., pp. 145; VII:48.

15 Newton, *Correspondence* VI:86; 92–3, note 24; 146, note 1.

16 I. Bernard Cohen, "Isaac Newton, Hans Sloane and the Académie Royale des Sciences" in *Mélanges Alexandre Koyré*, I, *L'Aventure de la Science* (Paris: Hermann, 1964), pp. 61–116.

17 Newton, *Correspondence* VI:187–9 (7 and 12 November 1714); 201–3 (26 January 1715); 171 (6 August 1714); 215, note 1. On 30 November 1715 the Duke of Bolton, Lord Chamberlain, ordered the release to Flamsteed of the remaining bulk of the copies of *Historia Coelestis*, which he at once destroyed (ibid., pp. 255–6). Halley, as editor, is often made responsible for this edition of Flamsteed's observations, brought about by Newton's *force majeure*, and it is true that the two astronomers were enemies; but Halley acted throughout under Newton's direction as president of the Royal Society.

18 Ibid., pp. 201, 202.

19 See A. R. Forsyth, "Newton's problem of the Solid of Least Resistance," in W. J. Greenstreet, ed., *Isaac Newton, 1642–1727*, (London: George Bell, 1927), pp. 76–86, and Whiteside, *Math. Papers* VI:456–80.

20 Newton, *Correspondence* VI:206, 211, 216–7, 222–3, 228–9, 234, 239–40, 243–4, 253.

21 Paolo Casini, "Les Debuts du Newtonianisme en Italie, 1700–40," *Dix-huitième Siècle* 10 (1978); Gerhardt, *Phil. Schriften* III: 653–6; Newton, *Correspondence* VI:215–6, 295–6.

22 H. G. Alexander, *Leibniz-Clarke Correspondence* (Manchester: Manchester University Press, 1956); Alexandre Koyré and I. Bernard Cohen, "Newton and the Leibniz-Clarke Correspondence," *Archives Int. d'Histoire des Sciences*, 15 (1962): 64–126; Brewster, *Memoirs* I:248; *D.N.B.* The story that Clarke was selected to succeed Newton at the Mint must surely be a mistake.

23 Koyré and Cohen, op. cit., pp. 65, note 5; 66. I was myself formerly of the opinion that Newton had drafted arguments for Clarke's benefit, i.e., taken "part in the fight."

24 Koyré and Cohen, op cit., p. 112; Newton, *Correspondence* VI:285; 460, note 8.

25 Newton, *Correspondence* VI:458, note 1; 285–6; Koyré and Cohen, op cit., p. 73.

26 Newton, *Correspondence* VI:212, 251–3, 261, 354. [Samuel Clarke], *A Collection of Papers which passed between the late learned Mr Leibniz and Dr Clarke* (London, 1717), pp. 3, 11, 41.

27 Leibniz found the first passage in *Optice* (1706) at p. 315 and (if he read on) the second at p. 346. In this edition these two Queries are numbered 20 and 23. In later editions of *Opticks*, to meet Leibniz's criticism, Newton added the words in Query 31 (formerly 23): "God has no need of such Organs, he being everywhere present to the Things themselves." When Leibniz read *Optice* (1706), he found no such explicit denial that whatever Newton understood by God's *sensorium*, it was not a divine organ of sensation. It has been suggested that Leibniz's copy (like some others of the 1706 *Optice*) lacked in the former passage the Latin word *tanquam* ("as it were"). But no *tanquam* appears in the second passage where God is given an infinite *sensorium* without qualification; see A. Koyré and I. B. Cohen, "The case of the missing *tanquam*," *Isis* 52 (1961): 555–66).

28 *Phil. Trans.* 29, no. 342: 173–225 (see in Appendix). All copies of Raphson's book have the same 1715 title page, but as will be seen later, some are later reissues (1718) with additional material inserted by Newton himself. It was issued in both English and Latin.

29 Newton, *Correspondence* V:95; Joseph Raphson, *History of Fluxions* (London, 1715), Preface, pp. 1–3, 14.

30 Raphson, *Fluxions*, pp. 19, 92.

31 D. T. Whiteside, private communication of Cambridge University Library MS. Add 3968.36, fo. 505r and Trinity College MS. NQ. 16. 173; Newton, *Correspondence* VII:80.

32 James Wilson, *Math. Tracts of B. Robins* (London 1761), II, 368 note; Newton, *Correspondence* II: 57–64, 199–200, 213; Hofmann, *Paris*, p. 74, note 55.

33 "The 'Account,'" *Phil. Trans.* 29, No. 342 (1715): 196. Newton quotes Barrow's *Lectiones Geometricae*, Lectio X (*Mathematical Works*, Cambridge 1860, p. 247).

34 Newton, *Correspondence* VI:242, 256–7, 295.

11. WAR BEYOND DEATH:
1715–1722

1 Newton, *Correspondence* VI:250–3, 285–8 (to Conti, 26 February 1716), 304–12; VII:138.

2 See Whiteside, *Math. Papers* VIII:2, Introduction, notes 132 to 134 for Newton's involvement with Raphson's book; Newton, *Correspondence* VI:341–9, 458.

3 Newton, *Correspondence* VI:253–4, 301–3, 322–3.

4 Ibid. pp. 257, 285, 319, 327–9.

5 Ibid. pp. 383, 447; VII:12–13 (from *Acta Eruditorum* [May 1719], pp. 216–26).

6 Newton, *Correspondence* VII:16, 43–4, 48–9, 69–70, 75–85, 160–2, 196–8, 218–20; Cambridge University Library, MS. Add. 3968.29, fos. 434–5 (quoted in Whiteside, *Math Papers* VIII:2, Introduction, note 167; and *Correspondence* VII:459, note 1).

7 *Correspondence* VI:454–7; VII:70, 73, 84, 124, and notes; Whiteside, loc. cit., note 159. Des Maizeaux, *Recueil*, 3rd. ed., (Amsterdam, 1759), pp. 208–11 (from Leibniz, 11 August 1716).

8 *Correspondence* VI:460–3 (in the last quotations I have transposed the original order of Newton's draft); Whiteside *Math. Papers* VIII:2, note 166.

9 Whiteside, *Math. Papers* VIII:2, *Introduction ad fin.*; Clifford Truesdell, "Rational Fluid Mechanics, 1687–1765," Editor's Introduction to Leonhard Euler, *Opera Omnia*, series 2, vol. 12, XII–XIII.

10 A. Rupert Hall, "Newton in France: A New View," *History of Science* 13 (1975): 233–50; Newton, *Correspondence* VII:116–8. The first edition of Algarotti's book has the subtitle *Ovvero dialoghi sopra la luce e i colori*, to which from the third revised edition (1739) onward was added *e l'attrazione*.

11 Newton, *Correspondence* VI:3, 145–6, 157–8.

12 Pierre Brunet, *L'Introduction des Théories de Newton en France au XVIIIᵉ Siècle* (Paris, 1931); P. L. M. de Maupertuis, *Discours sur les differentes figures des Astres* (Paris, 1732); Voltaire, *Lettres sur les Anglais* (Amsterdam, Rouen, 1734), *Elemens de la Philosophie de Neuton* (Amsterdam, 1738).

13 [G. L. Le Clerc de Buffon], *La Methode des Fluxions et des Suites infinies* (Paris, 1740), Preface.

14 Robert K. Merton, "Priorities in Scientific Discovery" in *The Sociology of Science*, ed. B. Barber and W. Hirsch (New York, 1962), p. 475; in this address of 1957 Merton elegantly emphasizes the significance of priority in discovery as the sole occasion for a scientist's personal fame.

15 Oldenburg, *Correspondence* IV:361, 368 note 4; T. S. Patterson, "John Mayow in contemporary setting," *Isis* 15 (1931): 48–51; C. Darwin and A. R. Wallace, *Evolution by Natural Selection*, with a foreword by Sir G. de Beer (Cambridge: Cambridge University Press, 1958), pp. 257–8.

16 Marcus Marci, *Thaumantias Liber de arcu coelestis* (Prague 1648), Theorems 18–20. See J. A. Lohne in *Notes and Records of the Royal*

Society 23 (1968): 174–6; J. Marek in *Organon* 8 (1971): especially 188–9; and *D.S.B.*

17 N. R. Hanson, *Patterns of Discovery* (Cambridge: Cambridge University Press, 1961), pp. 5–8.

18 *Principia* (1687), p. 10, only trivially modified in later editions, my translation; Newton distinguishes absolute, unmeasurable time, etc., from "time," etc., quantified relatively by clocks, measuring sticks, etc., which we ordinarily mean by the word "time," etc. There is no doubt that in Newton's mind the fluxion is intimately related to the true flow of things, that is, absolute time. I am grateful to Barry Clive for at least partially opening my eyes to the role of time in the philosophies of Newton and Sir William Hamilton.

INDEX

"Abridgment," *see* Collins, J.
Acta Eruditorum, 35, 91, 97, 141, 169, 176, 240
  calculus in, 3, 34, 41, 76–8, 82, 133, 213
  Principia and, 149, 197, 198
  reviews in, 116, 122, 137, 138, 142, 159, 160–5, 177, 183
Aiton, E., 152
Algarotti, F.
  Newtonianismo per il Dame (1737), 248–9
Angeli, Stefano degl', 8
Apollonius, 28, 31
attraction, 103, 115, 143, 148, 149, 151–8, 161–4, 166, 169
  as a divine act, 155–8, 162, 184

Barrow, Isaac, 63, 72, 81
  and calculus, 7, 8, 79, 173, 214
  his *Geometrical Lectures* (1670), 55–6, 75, 96–7, 116
  and Leibniz, 117, 172, 228
  and Newton, 15, 16–20, 56, 187, 234
Barton, Catherine, 214–15, 250, 252
Bayle, Pierre, 157, 242
Bentley, Richard, 2, 155–7
  Confutation of Atheism (1693), 156, 163
Berkeley, George, 32
Berlin Academy, 99, 123
Bernoulli, Jakob, 30, 105–6, 195, 216
  and Leibniz, 80–2, 117
Bernoulli, Johann, 30, 96, 225
  accuses Newton, 116–19, 198–9
  on attraction, 115, 161
  on Cheyne, 131, 134
  and *Commercium Epistolicum*, 187, 199, 206
  on Fatio, 121–2, 141
  on D. Gregory, 160
  influence of, 119
  and integral calculus, 73, 81, 83–4, 192, 236
  and inverse problem of forces, 235–6
  and Keill, 197, 212, 235–6, 238–40
  and Leibniz, 59, 80–2, 105–6, 116–19, 123, 193–4, 199–201, 215–16, 217, 222
  and L'Hospital, 83

and Newton, 211, 215–16, 228, 232, 240
and *Principia*, 115, 194–9
and Varignon, 239–41
as "very eminent mathematician," 200, 236–7, 243, 244
on Wallis's *Works*, 116–17
see also challenge problems
Bernoulli, Nikolaus, 195–8, 238
Bignon, J.-P., 249
Boyer, Carl B., 77
Boyle, Charles, 156
Boyle, Robert, 6, 54
Brouncker, William, Viscount, 53
Buffon, Comte de
  La Methode des Fluxions (1740), 253–4
Bürgi, Jost, 5
Burnet, Thomas, 95–99–100, 157, 158, 166, 169
Byzance, Fr. L. R. L., 83

calculus:
  Bernoulli's accusations, 116–17, 198–9
  binomial expansion, 7, 13, 175
  circle quadrature, 7, 53–9, 61, 76, 79, 92, 190–1, 212, 227
  concepts of, 13, 31–2, 61–2, 70–1, 78–82, 84–6, 89–90, 94, 130
  Craige and, 77–81
  differentials spoiled copy of fluxions, 108, 111, 174
  diffusion of, 80–4, 91–2
  dispute begun by Fatio, 101, 107–8, 110–11, 118, 119–20
  dispute continued by Keill, 145, 170
  exponential, 225
  integration, 78–9, 130, 191, 236
  methods of tangents, 7, 11, 79, 130, 169, 179–82, 190, 228
  origins of, 7–8, 11–21, 90–1, 200
  and *Principia*, 27–34, 89, 108
  second differentials, 30, 108, 170, 197–8, 207–8, 209–11, 228
  series, methods of, 19–20, 41, 52–4, 63, 66, 72, 75–6, 79–80, 91, 92, 132–3, 188, 191, 226–7
  solid of least resistance, 121, 122, 123

332

on calculus (1686), 77
Charta Volans, 59, 200–7
on circle quadrature (1682), 76, 92, 96
Combinatory Art, 48
Essais de Theodicée (1710), 157–8, 161, 184, 219
Essay on the causes of the heavenly motions (1689), 150–2, 160, 208–10
History and Origin of the Calculus (1714), 193, 213
Hypothesis Physica Nova, 48, 148
"Remarks on the Dispute" (1713), 204
reply to Fatio (1700), 123–6
reviews *Arithmetica Universalis*, 142
reviews *On the Quadrature of Curves* (1705), 138–41, 164, 169, 234
reviews Wallis (1695), 116
letters quoted (*to avoid ambiguity, dates are here given in both Old and New Styles, the year beginning on 1 January*)
Barrow to Collins: 20/30 July 1669, 187; 31 July/10 August 1669, 18
Johann Bernoulli to Leibniz: 5/15 August 1696, 117; 8/18 July 1699, 121; 28 July/7 August 1699, 121, 122; 19/29 September 1703, 131–2; 27 May/7 June 1713, 198–200, 237
Burnet to Leibniz: 30 November/10 December 1696, 100; 25 April/4 May 1697, 100
Chamberlayne to Leibniz: 27 February/10 March 1714, 202
Cheyne to Gregory: 10/20 January 1702, 133, 134
Collins to Strode: 24 October 1676, 63–4
Collins to Wallis: ? 1677/8, 75–6
Cotes to Jones: 15/26 February 1711, 223–4
Daguesseau to Newton: 17/28 September 1721, 251
De Moivre to Bernoulli: 16/27 July 1705, 196
Fatio to Huygens: 14/24 June 1687, 103; 18/28 December 1691, 107–8; 5/15 February 1692, 108, 111
Fatio to Newton: 11/21 April 1693, 110
Fontenelle to Newton: 29 May/9 June 1714, 250
D. Gregory to Newton: 9 June 1684, 36; 27 August 1691, 36, 39; 11 October 1691, 36, 39
Hermann to Leibniz: 11/22 November 1715, 165
Hooke to Newton: 17/27 January 1680, 109
Huygens to Fatio: 26 January/5 February 1692, 108
Huygens to Leibniz: 27 October/6 November 1674, 53; 14/24 August 1690, 87; 29 September/9 October

1690, 87; 22 August/1 September 1691, 88; 19/29 May 1694, 89, 116; 14/24 August 1694, 116
Keill to Newton: 3/13 April 1711, 164
Keill to Sloane for Leibniz: [May 1711], 169–75, 199
Leibniz to Johann Bernoulli: 13/23 August 1696, 118; 15/25 July 1699, 121; 21 September/2 October 1703, 132; 11/22 November 1703, 132; 14/25 January 1705, 138; 17/28 July 1705, 138; 4/15 March 1708, 142; 17/28 June 1713, 59; 29 March/9 April 1715, 222; 1/12 April 1716, 237
Leibniz to Bridges: October 1694, 113
Leibniz to Burnet: 12/22 November 1695, 95; ? 1698, 100; ? 1699, 158; 6/16 March 1708, 100; 13/23 August 1713, 166, 167
Leibniz to Conti: 25 November/6 December 1715, 216, 220, 235; 29 March/9 April 1716, 233
Leibniz to Hermann: 30 April/11 May 1708, 142; 6/17 September 1715, 165; 22 November/3 December 1715, 166
Leibniz to Huygens: 11/21 July 1690, 86–7; c. 3/13 October 1690, 152; 4/14 September 1694, 116
Leibniz to Baroness von Kilmansegge: 7/18 April 1716, 237, 242
Leibniz to Mencke: [1688] 149
Leibniz to Newton: 7/17 March 1693, 108–9, 112
Leibniz to Oldenburg: 15/25 October 1671, 51; 3/13 February 1673, 54; 16/26 April 1673, 50; 5/15 July 1674, 57; 6/16 October 1674, 54, 57; 20/30 March 1675, 58; 10/20 May 1675, 58; 2/12 July 1675, 59; 18/28 December 1675, 61–2; 2/12 May 1676, 63; (for Newton), 17/27 August 1676, 34, 66–7, 98, 189, 191, 227; (for Newton) 11/21 June 1677, 17, 70–2, 74, 98, 172, 183, 226; 12/22 July 1677, 74
Leibniz to Sloane: 21 February/3 March 1711, 168, 224; 8/11 December 1711, 176, 177
Leibniz to Tournamine: 17/28 October 1714, 213
Leibniz to Wallis: 28 May/17 June 1697, 113; 25 July/4 August 1699, 101
L'Hospital to Leibniz: 4/14 December 1692, 20/30 November 1694, 82
Monmort to Newton: 14/25 February 1717, 250
Newton to Bernoulli: 29 September/10 October 1719, 240
Newton to Collins: 20/30 August 1672, 74; 10/20 December 1672, 169, 179–82, 190, 204
Newton to Conti: 26 February/7 March